The Chemical Bond

John N. Murrell
Professor of Chemistry, University of Sussex

Sydney F. A. Kettle
Professor of Chemistry, University of East Anglia

John M. Tedder
Professor of Chemistry, University of St. Andrews

JOHN WILEY & SONS
Chichester . New York . Brisbane . Toronto

Library of Congress Cataloging in Publication Data:

Murrell, John Norman.
The chemical bond.

Includes index.
1. Chemical bonds. I. Kettle, Sidney Francis Alan, joint author. II. Tedder, John Michael, joint author. III. Title.
QD461.M84 541'.224 77-21728
ISBN 0 471 99577 0 (cloth)
ISBN 0 471 99578 9 (paper)

Typeset in IBM Press Roman by Preface Ltd., Salisbury, Wilts.
Printed in Great Britain by Unwin Brothers Ltd.,
The Gresham Press, Old Woking, Surrey.

Contents

Preface

The advance and extension of that organized body of knowledge which is physical science involves an interplay between experiment and observation on the one hand and theory with its deductions and predictions on the other. In the development of some branches of science there are periods when theory lags behind experiment: this is characterized by the accumulation of facts which have seemingly little relationship to one another. The measurement of the frequencies of atomic spectral lines in the period just before the work of Balmer in 1885, and later before the development of the Bohr quantum theory in 1913, is a good example. There are other periods when experiment has lagged behind theory: some aspects of astronomy before the age of radio telescopes and space probes would typify this.

Whilst not wishing to underplay the importance in experimental science of lucky accidents which often initiate new areas of study, it would be generally agreed that experiments should be undertaken with specified objectives and with a reasonable expectation of achieving those objectives. In the broadest sense it is theory which provides the basis for these expectations. Theory can be anything from a highly developed and mathematically based model capable of giving a quantitative prediction of the result of the experiment, to a qualitative expectation based upon past experience with similar experiments.

Chemistry, or perhaps one should take a wider brief and say molecular science, is not for the most part susceptible to accurate quantitative predictions. For many scientists that is one of its attractions. It is nevertheless clear that chemists are extraordinarily successful in the development of their subject and ingenious in the new directions which they take. The two growth industries of the past thirty years, of polymers and pharmaceuticals, are evidence of that. In the field of organic chemistry over a million distinguishable compounds have been prepared and characterized and their number is increasing by several hundred each year, yet this development has occurred in a period when we cannot predict with accuracy the rate of a simple chemical reaction and we are in many cases uncertain about its course.

The majority of advances in chemistry have therefore been made with the support of qualitative theories. In many cases there would appear to be no theory at all in the usual sense in which we use that term, but if we have some 'explanation' or 'understanding' of an experimental result then that is a theory in its simplest form. We can say that in this respect a theory is part of the language of scientific communication. The concepts or ideas on which we rationalize our observations become our jargon. In the development of science this jargon changes partly by fashion but mainly by whether particular ideas are widely accepted as useful for communication. Concepts are continually being tested against new experiments or the results of more fundamental and reliable theories.

The language of modern chemistry stems largely from the beginning of this century and the formulation of atomic theory. Our understanding of molecular structure and reactivity is based upon the distribution of electrons in molecules, their movement in chemical reactions, and the energies associated with these distributions. In the 1920s modern quantum theory provided for the first time the basis for a quantitative description of molecular properties, and yet it was not until the 1960s that one could say that for molecular electron energy levels the results of this theory were fully tested against accurate experimental results. This was achieved through the agreement, within experimental accuracy, of the experimental and calculated dissociation energy of the hydrogen molecule.

The period 1930–1960 was one in which the basic equations that described molecular phenomena and the methods of solving them were known but their solution was beyond the ability of computational technology. However, it was not a period of inaction on the theoretical front. In retrospect it appears that the failure to solve the equations was a stimulus to the development of approximate solutions or models for which exact solutions could be found.

It is in this period that many of our concepts associated with the chemical bond and molecular structure were introduced. In contrast, during the past 15 years, in which the emphasis in theoretical chemistry has been to obtain computer based solutions of the quantum mechanical equations of increasing accuracy, few new concepts have been introduced. Some of the earlier ideas have failed the test of this recent period, but perhaps surprisingly a large number have maintained their usefulness to chemistry. The concept of hybridization is one example of an established and widely used concept of chemistry which has, however, little role in modern computer based calculations.

Most books on the chemical bond adopt a historical development of the subject in which the two basic theories of valence, molecular orbital and valence bond, are introduced and applied in the first place to simple systems like H_2^+ and H_2 for which a mathematically rigorous treatment was possible even in the 1930s. Such texts usually proceed with the discussion of empirical theories such as Hückel theory and ligand field theory, and, depending on the level at which the books are written, some of the more advanced empirical and non-empirical theories developed in recent years may be described. Our book *Valence Theory* published in 1965 was written to this pattern.

In this book, which is intended as an introduction to the subject at the undergraduate level, we adopt a different approach. We shall concentrate on those concepts of the subject which can be considered important within chemistry as a whole, but we will examine the validity of the concepts in the light of the most recent quantitative calculations. Although the computational techniques by which these calculations are made are not relevant at this level of the subject the principles underlying such calculations are no more difficult to understand than those on which the simple empirical theories are founded. The mathematical requirements for a reading of this book are not as high as those needed for *Valence Theory*, as the only computational technique which will be carried through in detail is the solution of a set of linear simultaneous equations.

Constants and Units

Physical constants

Constant	Symbol	Value	Units (SI)	Units (c.g.s.)
Speed of light in vacuum	c	2.998 x	10^{8} m s^{-1}	10^{10} cm s^{-1}
Electronic charge	$-e$	−1.602	10^{-19} C	10^{-20} e.m.u.
		−4.803		10^{-10} e.s.u.
Electron rest mass	m_e	9.110	10^{-31} kg	10^{-28} g
Proton rest mass	m_p	1.673	10^{-27} kg	10^{-24} g
Planck's constant	h	6.626	10^{-34} J s	10^{-27} erg s
	$\hbar = h/2\pi$	1.055	10^{-34} J s	10^{-27} erg s
Bohr radius	a_0	5.292	10^{-11} m	10^{-9} cm
Boltzmann's constant	k	1.381	10^{-23} J K^{-1}	10^{-16} erg K^{-1}
Avogadro's number	N_A	6.022	10^{23} mol^{-1}	10^{23} mol^{-1}
Permittivity of free space	ϵ_0	8.854	10^{-12} F m^{-1}	

Other common units:
Length: Angstrőm, Å = 10^{-10} m
Energy: Electron volt, 1 eV = 1.602 x 10^{-19} J
Calorie, 1 cal = 4.184 J
Energy per mole: 1 eV per molecule is equivalent to 96.49 kJ mol^{-1}
kT at 300 K is equivalent to 2.494 kJ mol^{-1}
Wavenumber (reciprocal wavelength): 1 eV is equivalent to 8066 cm^{-1} based upon
$$E = hc/\lambda$$
Dipole moment: debye = 10^{-18} e.s.u. cm = 3.334 x 10^{-30} C m

Atomic units

This system of units is chosen to avoid cluttering the quantum-mechanical equations with fundamental constants. It is based upon the choice $\hbar = m_e = e = 1$. The principal quantities of interest in this system are:

Unit of length: Bohr radius, a_0 = 5.292 x 10^{-11} m = 0.529 2 Å
Unit of energy; the hartree, E_H = 27.21 eV, which is equivalent to 2 626 kJ mol^{-1}

Chapter 1
The Chemical Bond – Early Concepts

'When the formulae of inorganic compounds are considered even a superficial observer is struck with the general symmetry of their construction; the compounds of nitrogen, phosphorous, antimony and arsenic especially exhibit the tendency of these elements to form compounds containing three or five equivalents of other elements, and it is in these proportions that their affinities are best satisfied.'

This comment by Frankland in 1852† illustrates the body of knowledge which was available to chemists in the middle of the 19th century and which led to rapid advances in the concept of valence. The rapid development of organic chemistry at that time led Kekulé in 1857 to deduce that carbon was tetravalent, and he also introduced the important idea that carbon atoms could form bonds with one another. The tetravalence of carbon was postulated independently by Couper in 1858 and he made use of structural formulae for molecules with lines between atoms linked together.

The existence of multiple links between carbon atoms was postulated by Kekulé in 1859, and in 1865 he gave a structural formula for benzene consisting of a flat hexagonal ring of carbon atoms with alternating single and double bonds.

One of the successes of Kekulé's formula for benzene was in explaining isomerism. Disubstituted benzenes, for example, have three isomeric forms. The next important development in molecular structure also came from a study of isomerism, in this case the optical isomerism of tartaric acid and similar compounds which had been investigated by Pasteur. The interpretation of Pasteur's results was given independently by van't Hoff and le Bel in 1874 with their model of the tetrahedral orientation of the valences of carbon. According to van't Hoff: 'In the case where four affinities of an atom of carbon are saturated by four different univalent groups, two and only two different tetrahedra can be obtained of which one is the mirror image of the other'. This development marked the beginning of our picture of a molecule of atoms joined by bonds: what one might call a ball and stick model. A satisfactory theory of valence must explain the number of bonds, their length and the angles between them.

The early attempts at an electronic theory of valence, following Thomson's discovery of the electron in 1897, suffered from the limitation that the electrons

†Frankland, *Phil. Trans.*, **67**, 417 (1852).

were considered to be at rest. Electron sharing within such a model conflicts with the electrostatic result that particles carrying like charges repel one another.

It was not until 1913 that Bohr introduced a dynamic model for the electrons in an atom which gave a satisfactory explanation of many features of atomic spectral lines. This model was based upon the laws of classical (Newtonian) dynamics, but the new principle introduced by Bohr was that only certain orbits of the electrons around the nucleus were allowed. Although a recipe was given for identifying these stable orbits, Bohr theory must be considered as fundamentally unsatisfactory because no explanation for the stability was forthcoming from within classical dynamics. Nevertheless, the work of Bohr indicated that an explanation for the chemical bond could be found in a dynamical model of the electron although no quantitatively satisfactory results were ever obtained. These had to await the development of the new principles of wave mechanics as we shall see in the next chapter.

The early electronic theories of bonding supported by the Bohr model of electron dynamics were successfully developed by Lewis into a broad rationalization of chemical bonding types. This work can be said to culminate in the publication of his book *Valence and the Structure of Atoms and Molecules* in 1923. In this book Lewis developed a symbolism for the electronic bonding in which electrons are represented as dots. Dots between atoms represented shared electrons thus

$$\mathrm{H:\overset{..}{\underset{..}{Cl}}:} \qquad\qquad \begin{array}{c}\mathrm{H}\\ \mathrm{H:\overset{..}{\underset{..}{C}}:H}\\ \mathrm{H}\end{array}$$

This symbolism was universally adopted by chemists for many years and is even now used in elementary texts.

In the Lewis theory of valence† there are two main types of chemical bond: ionic and covalent. The driving force for bond formation is identified as the pairing of electrons between atoms so as to obtain stable octets: the inert gas electronic structure. This idea of electron pairing had an important influence on the first quantitatively successful theories of the chemical bond which were a description of electron sharing in wave mechanical terms. As we shall see in later chapters, electron pairing is closely identified with a property of the electron which was unrecognized in 1923, namely its spin. Before we discuss the modern view of the concepts used by Lewis we must examine the development of new ideas in physics in the period 1900–1930.

†We should not ignore the contributions of others like Langmuir, Kossel, and Sidgwick to what is generally called Lewis theory.

Chapter 2
Matter Waves

2.1. Wave mechanics

In this chapter we introduce the concept of the wave nature of atomic particles. This is the foundation of the mathematical discipline of wave mechanics from which we can understand and predict the properties of molecules as individual entities (the so-called microscopic state). The properties of molecules in bulk (the macroscopic state) can be obtained by applying statistical techniques to these microscopic results.

Wave mechanics plays for atomic particles the role that classical mechanics plays for material objects. We interpret the motion of celestial bodies and we can predict the trajectories of space probes from the equations of classical mechanics developed by Newton, Lagrange, and Hamilton. We can understand and predict the properties of the hydrogen molecule from the equations of wave mechanics developed by de Broglie, Schrödinger, and Dirac. For both classical and atomic systems success in these endeavours depends to a large extent on the computational technology available at the time: the ability to land a man on the moon depended as much on the development of the digital computer as on the development of the rocket motor. Present computational resources are sufficient to enable us to understand many aspects of molecular behaviour for quite complicated molecules and to make accurate predictions for the smallest molecules, but they are insufficient for us to make accurate predictions for most larger molecules of interest to chemists. It is, however, important to appreciate that the scientific limitations we face at present do not suggest that the fundamental concepts of wave mechanics are inadequate or that the equations of wave mechanics are wrong.

We can identify the birth of wave mechanics either with the year 1924, when de Broglie postulated that material particles would show wave-like characteristics, or with the year 1926, when Schrödinger introduced an equation to define these characteristics. The conception, however, occurred much earlier, and is probably identified with the work of Planck in 1900. In attempting to explain the distribution of energy, as a function of frequency, of the radiation emitted by a so-called 'black-body' he made the hypothesis that atomic oscillators in equilibrium with electromagnetic radiation could only take up or give out energy in discrete amounts or 'quanta'. Following the hypothesis, which explained the experimental results, physicists developed first the 'old' quantum mechanics which we associate with the Bohr model of the atom, and then the 'new' quantum mechanics which we associate mainly with the work of Heisenberg. We shall have little need in this book to make specific reference to the ideas inherent in the Heisenberg approach to quantum mechanics but it is perhaps worth noting that in 1930 Dirac showed that the mathematical approaches of wave and quantum mechanics were complementary: which is not to say they are equally easy to apply.

The revolutionary postulate of de Broglie received direct experimental verification in 1927 by Davisson and Germer. They showed that mono-energetic electrons scattered from crystalline nickel foil gave a diffraction pattern analogous to that shown by X-rays. Similar experiments were carried out independently by G. P. Thomson, and later Stern showed that beams of heavier particles (H_2,He, etc.) showed diffraction patterns when reflected from the surfaces of crystals. De Broglie's expression for the wavelength of these matter waves, which we shall meet in the next section, was confirmed with high accuracy.

In retrospect, de Broglie's postulate was not such a bold step, as it was strongly suggested by the position that had been reached at that time regarding the nature of light or electromagnetic radiation. For this reason a brief interlude on the nature of light is appropriate.

2.2. The wave–particle duality of light

A satisfactory scientific description of light has presented a challenge to physicists over several centuries. In the 17th century there was a great controversy between the schools of Newton and Huygens over whether light was a stream of particles or a wave.

The fact that light travels in straight lines, is reflected and refracted and has the ability to impart momentum to anything it strikes, suggests a particulate (corpuscular) model. In contrast, the phenomena of diffraction and interference are most readily explained by a wave model. At the time when quantum theory was proposed the wave model was dominant because what was known at that time about the particulate behaviour of light could be largely understood from the wave model, although the carrier of the wave, the 'ether', was proving rather elusive.

Visible light is one part of a family of electromagnetic radiation whose members include X-rays, infrared, and ultraviolet radiation. The speed of light in a vacuum is a constant (3×10^8 m s^{-1}) independent of its frequency (ν) or wavelength (λ). Figure 2.1 shows the part of the electromagnetic spectrum of interest to scientists today, and we show the common names associated with different wavelength regions. Note that visible light is a very small band of the whole spectrum.

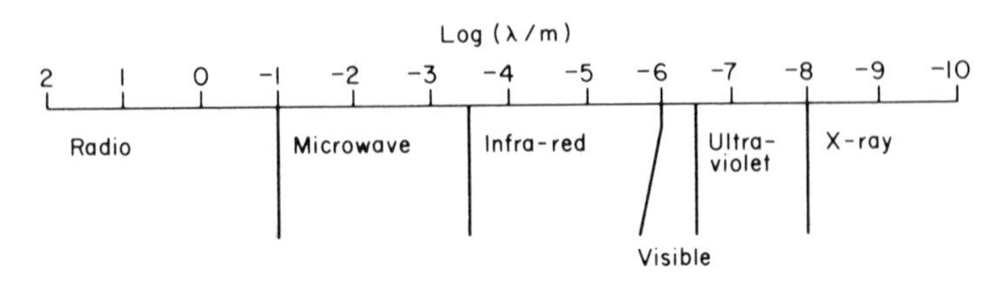

Figure 2.1 The electromagnetic spectrum.

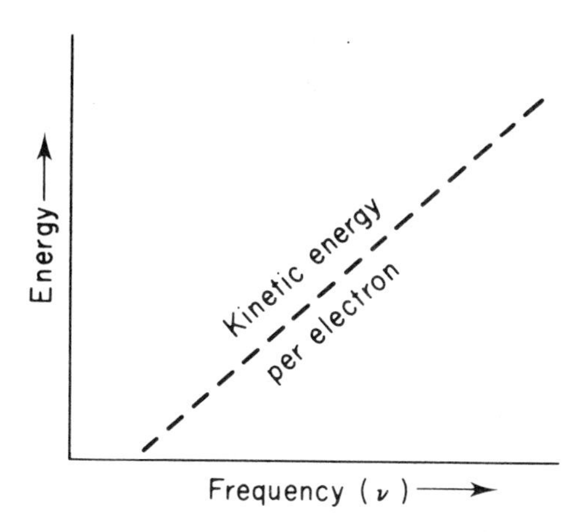

Figure 2.2 The relationship between the kinetic energy of the electrons emitted from a metal surface and the frequency of the light incident upon it.

The word electromagnetic is used to describe this radiation because its wave description is represented by electric and magnetic fields which fluctuate with the frequency of the radiation. Light is absorbed or scattered by matter either through the interaction of its electric field with the electric charges of atomic particles or through the interaction of its magnetic field with the magnetic moments of atomic particles. The former is by far the stronger effect and when in this book we consider the absorption of light by matter it is only the electric interaction that is important.

In the last few years of the 19th century the wave model of light was to experience a challenge to its superior position by experiments related to the observation that light is able to cause electrons to be ejected from the surface of metals. This so-called photoelectric effect was not only to have a big impact on the development of quantum theory, but as we shall see in Chapter 5 it has been developed in recent years as an important tool for probing the electron energies of molecules.

The importance of the photoelectric effect became apparent when Lenard in 1902 published his investigation of the relationship between the frequency and the intensity of light on the one hand and the number and kinetic energy of the ejected electrons on the other. Figure 2.2 shows the relationship between the frequency and the kinetic energy per electron. No electrons are emitted until the frequency of the light exceeds a value characteristic of the metal, and above this the number increases rapidly but then levels off to a constant value. In the latter region the kinetic energy of the electrons is increasing linearly with ν, but not their number; the number of electrons released depends on the intensity of the light but not its frequency.

These results are not explicable in terms of the wave description of light, for which an increase in intensity should lead to an increase in the magnitude of the electric field and hence to an increase in the energy of the ejected electrons. Further, a lower cut-off frequency for the electron ejection would not be expected; a lower cut-off intensity would be more likely. In contrast, the corpuscular model of light gives a ready explanation of the results as we now show.

In 1905 Einstein extended Planck's hypothesis (Section 2.1) that atomic oscillators could only take up or give out energy in discrete quanta, by regarding the radiation itself as consisting of indivisible quanta or photons. The energy associated with individual photons has to be proportional to the frequency of the light because the energy associated with a beam of light of constant intensity is proportional to its frequency. The proportionality constant relating the energy of a photon to its energy turned out to be the one introduced by Planck in his theory. This constant is given the symbol h. It is one of the fundamental constants of nature, taking its place alongside the electron charge and the velocity of light. It has the units energy x time and the value 6.6×10^{-34} J s. The relationship

$$E = h\nu, \tag{2.1}$$

is usually called the Planck–Einstein relationship.

The particulate interpretation of the photoelectric effect is straightforward. Each photon absorbed by a metal can lead to emission of one electron providing that the energy of the photon, when transferred to the electron, is sufficient to enable the electron to escape from the surface of the metal. Increasing the intensity of the light increases the number of photons but not their energy and so will lead to an increase in the number of electrons escaping but not to an increase in their energy.

Remembering Einstein's result, that the energy of the incident photon is given by $h\nu$, where ν is the frequency of the incident light, and that the kinetic energy of the ejected electron will be $\frac{1}{2}mv^2$, where m is the mass of the electron and v its velocity, we may write an equation for the energy balance in the experiment:

$$h\nu = A + \tfrac{1}{2}mv^2, \tag{2.2}$$

where A is an energy characteristic of the metal surface. For several years this expression provided the best method of evaluating the magnitude of Planck's constant.

This interpretation of the photoelectric effect restored the balance between the wave and particle models for light, and the position adopted today is that light has a wave–particle duality and that for any experiment one uses whichever model leads to the simpler interpretation. For example, the so-called Compton scattering of X-rays by the electrons in solids is best treated as a collision of two particles, the photon and the electron. There is no conflict here: light is light and it is only for convenience that we use familiar terms like wave and particle.

An important aspect of the treatment of the Compton effect is the conservation of the momentum of the colliding particles. But how can a photon, which has no mass, have momentum? Similarly, in writing the energy of the photon in equation (2.2) as $h\nu$ we avoided any discussion of the form of this energy. If a photon has momentum can it not also have kinetic energy? The fact that the photon has momentum but no mass can be understood within the framework of relativity. In general, for a particle of rest mass, m_0 the momentum is

$$p = c^{-1}\sqrt{E^2 - m_0^2 c^4}. \tag{2.3}$$

Therefore, for the photon with zero rest mass

$$p = \frac{E}{c} = \frac{h\nu}{c} = \frac{h}{\lambda}. \tag{2.4}$$

When de Broglie postulated that material particles would show wave-like characteristics, expression (2.4), which relates the momentum to the wavelength, was taken to apply not only to photons but to matter waves also. However, although the de Broglie postulate appears to give a close parallelism between light and matter, one must not lose sight of the essential difference that matter has mass and the photon does not. With this proviso however we can now examine the wave equation that applies to matter waves.

2.3. The Schrödinger equation

In discussions of elementary mechanics, one commonly starts by making definitions – such as defining force as the rate of change of linear momentum. This approach tends to conceal the fact that what is really involved is a postulate; a postulate which is accepted because it leads to results which agree with the measurements we make on large bodies. Similarly there is no way of deducing by strict logic the form of the equation that describes matter waves. Any relationship which we may obtain has to be tested, like any other fundamental equation of physics, by the fact that the results obtained from it are in accord with experiment. In the present case one can either derive the wave equation starting from certain postulates or, alternatively, argue by analogy with other established principles of physics. We shall adopt the latter approach. Our derivation is not the most general or elegant one to the fundamental equations of quantum mechanics but it has the advantage of relative simplicity and is sufficient for the objectives of this book.

First let us rehearse the characteristics of the equation which we are seeking. Most important of all is the fact that it is a wave equation and so may be expected to have some of the characteristics of more familiar wave equations, such as those describing the vibrations of a violin string. However, for the most part we shall not be interested in time-dependent quantities associated with the systems to which we shall apply our equation. So, we shall be interested in the allowed energy levels of an atom or molecule but will not consider whether a level only persists for a short time because the atom or molecule emits radiation or some other energy-changing process occurs. It follows, then, that the wave equation will not be time-dependent. In particular, it will not contain quantities differentiated with respect to time (in contrast to the more common mathematical description of the wave motion of systems such as a vibrating violin string). Another feature which we expect to find is that all of those quantities which we would have included in a classical treatment – the kinetic energies of the particles, the repulsions between particles of like electrical charge and the attraction between those of opposite charge – will have their counterparts in the wave treatment. Finally, and in the light of our earlier discussion, we would expect that the de Broglie relationship (equation 2.4) will be in some way involved.

Let us write down a general expression for the sort of wave that we are considering but, for simplicity, consider a wave motion along only one coordinate axis. Note that the wave will be a stationary one, that is, for a given wave motion the nodes will not move with time. The reader may find it easiest to visualize this equation as representing, for example, the relative amplitude of excursion of points along a vibrating violin string at any instant in time (note that these relative amplitudes are themselves time independent); but the equation is really much more general than this. A suitable expression is

$$y(x) = A \sin \frac{2\pi x}{\lambda} . \tag{2.5}$$

In this expression x is the coordinate axis along which the wave motion occurs with wavelength λ. The relative amplitude of the wave at any point along x is represented by the function $y(x)$, and as the maximum value of $\sin(2\pi x/\lambda)$ is unity, A represents the maximum amplitude of vibration. We shall refer to a function like $y(x)$ as a *wavefunction*. In equation (2.5) we have chosen the origin from which we measure x to lie at a node of the wave (for instance, one end of a violin string). A more general point could have been chosen as origin but there would have been an additional, constant, term on the right hand side of equation (2.5).

We next derive a differential equation for y by twice differentiating each side of (2.5) with respect to x, obtaining

$$\frac{d^2 y}{dx^2} = -\frac{4\pi^2 A}{\lambda^2} \sin \frac{2\pi x}{\lambda} , \tag{2.6}$$

or

$$\frac{d^2 y}{dx^2} = -\frac{4\pi^2}{\lambda^2} y. \tag{2.7}$$

This equation has an infinite number of solutions like (2.5). Firstly, equation (2.7) in no way places any limitations on A so that an infinite number of acceptable values exist. Secondly, the equation does not limit λ in any way. Thirdly, we note that a solution of equation (2.7) more general than (2.5) is

$$y(x) = A \sin \frac{2\pi}{\lambda} (x + \varphi), \tag{2.8}$$

where φ is a phase angle which can have any value. However, we can remove this freedom by the specification that $y(0) = 0$ so that

$$A \sin \frac{2\pi}{\lambda}(0 + \varphi) = 0. \tag{2.9}$$

It follows that we may set φ equal to 0 as in the function (2.5).

We can restrict the infinite number of values of λ which are acceptable as solutions in a similar manner by specifying the relative amplitude at some other point along the x axis. The simplest way of doing this is to require there to be a node at another specified value of x, say at l, so that $y(l) = 0$.

A general solution of

$$y(l) = A \sin \frac{2\pi l}{\lambda} = 0 \tag{2.10}$$

is

$$\frac{2\pi l}{\lambda} = n\pi, \tag{2.11}$$

where n is an integer. It follows that

$$\lambda = \frac{2l}{n}. \tag{2.12}$$

Thus in fixing the ends of a violin string and forcing these points to be nodes we restrict the value of the wavelengths which can be excited when the string is played. We refer to conditions which constrain the form of the wave function as *boundary conditions*. Boundary conditions are important in wave mechanics because as we shall see they are the origin of the quantization of energy: quantization is analogous to the requirement that n be an integer in equations (2.11) and (2.12).

We have gone into some detail over the form of equation (2.7) and its solutions because the pattern is followed closely for matter waves to which we now turn. The derivation is a simple one: we merely replace the wavelength, λ, by the momentum, p, in equation (2.7) using the de Broglie relationship (2.4). For a single particle moving in a one-dimensional space (x) with momentum p_x, we therefore have a wave function [which is traditionally represented by the Greek letter psi (ψ)] which satisfies the equation

$$\frac{d^2\psi}{dx^2} = -\frac{4\pi^2}{h^2} p_x^2 \psi. \tag{2.13}$$

Equation (2.13) is not yet of such a form that its solutions will give directly the energy levels of the system because it does not yet contain any description of the forces acting on the particle. Force is defined by the derivative of the potential energy and hence we need to introduce into (2.13) some function of the potential energy. The momentum of the particle is related to its kinetic energy (T) by the expression

$$T = \frac{1}{2} mv^2 = \frac{p^2}{2m}, \tag{2.14}$$

where m is the mass of the particle. The kinetic energy is the difference between the total energy (E) and the potential energy (V)

$$T = E - V, \tag{2.15}$$

hence the momentum in (2.13) can be replaced according to the expression

$$p_x^2 = 2m(E - V), \tag{2.16}$$

to give the equation

$$\frac{d^2\psi}{dx^2} = -\frac{8\pi^2 m}{h^2}(E - V)\psi, \tag{2.17}$$

which on rearrangement gives

$$\frac{-h^2}{8\pi^2 m}\frac{d^2\psi}{dx^2} + V\psi = E\psi. \tag{2.18}$$

Equation (2.18) is the equation proposed by Schrödinger for a particle moving in one dimension in a potential V. To be more useful we need to extend the equation to three-dimensional space and we also require an equation which is applicable to more than one particle. The extension to three dimensions is made by letting ψ be a function of the three Cartesian coordinates and writing the appropriate partial derivatives instead of the complete derivatives in (2.17) as follows

$$\frac{-h^2}{8\pi^2 m}\left(\frac{\partial^2\psi}{\partial x^2} + \frac{\partial^2\psi}{\partial y^2} + \frac{\partial^2\psi}{\partial z^2}\right) + V\psi = E\psi. \tag{2.19}$$

Equation (2.19) can be written in a more compact form as

$$\mathscr{H}\psi = E\psi, \tag{2.20}$$

where $\mathscr{H}$ is defined by

$$\mathscr{H} = \frac{-h^2}{8\pi^2 m}\left(\frac{\partial^2}{\partial x^2} + \frac{\partial^2}{\partial y^2} + \frac{\partial^2}{\partial z^2}\right) + V, \tag{2.21}$$

and is called the Hamiltonian of the system for reasons which will soon be clear. Note that $\mathscr{H}$ contains differential operators $\partial^2/\partial x^2$ etc. and for this reason it makes no sense to cancel the ψ from both sides of equation (2.20).

Having established the general structure of Schrödinger's equation for atomic particles as (2.20) it is not difficult to give its precise form for any number of particles, by comparison with the corresponding equations that would describe the motion of the particles in classical mechanisms.

2.4. The Hamiltonian

A link must exist between the equations which describe the motion of atomic (quantum) particles and those which describe the motion of heavy (classical) particles because the results obtained by applying the quantum equations to heavy particles must be the same as those obtained from the classical equations. This idea was first proposed by Bohr in what is referred to as the correspondence principle.

The function which represents the total energy of a system in classical mechanics expressed in terms of the coordinates and momenta of all the particles is called Hamilton's function.† For a particle having mass m and moving under the influence of a potential V which is a function of the position of the particle, Hamilton's function is

$$\mathscr{H}_{\mathrm{f}} = \frac{1}{2m}(p_x^2 + p_y^2 + p_z^2) + V(x, y, z), \tag{2.22}$$

the first term being the kinetic energy according to expression (2.1). We can set up a correspondence between Hamilton's function (2.22) and the operator, $\mathscr{H}$, defined as the Hamiltonian in (2.21), if we make the substitution

$$p_x \rightarrow \frac{-ih}{2\pi}\frac{\partial}{\partial x}, \tag{2.23}$$

or

$$p_x^2 \rightarrow \frac{-h^2}{4\pi^2}\frac{\partial^2}{\partial x^2}. \tag{2.24}$$

It has been found that the connection (2.23) provides a general recipe for constructing the quantum mechanical Hamiltonian for any number of particles from the appropriate Hamilton function. It is because of this relationship that the Schrödinger equation is formally written as (2.20) and the operator $\mathscr{H}$ is called the Hamiltonian. For example, for a set of i particles of mass m_i interacting with a potential energy V which is a function of the relative position of the particles, Hamilton's function will be

$$\mathscr{H}_{\mathrm{f}} = \sum_i \frac{1}{2m_i}(p_{x_i}^2 + p_{y_i}^2 + p_{z_i}^2) + V(x_i, y_i, z_i), \tag{2.25}$$

the total kinetic energy being a sum of the kinetic energies of each particle. The appropriate many-particle Hamiltonian is obtained by making the substitution in accord with (2.24)

$$\mathscr{H}_{\mathrm{f}} = -\sum_i \frac{h^2}{8\pi^2 m_i}\left(\frac{\partial^2}{\partial x_i^2} + \frac{\partial^2}{\partial y_i^2} + \frac{\partial^2}{\partial z_i^2}\right) + V(x_i, y_i, z_i). \tag{2.26}$$

†Hamilton was born in Dublin in 1805. According to the Dictionary of National Biography he had by the age of twelve studied Hebrew, Latin, Greek, and the four leading continental languages and could profess a knowledge of Syriac, Persian, Arabic, Sanskrit, Hindustani, and Malay. The choice of languages owed much to his father's intention to obtain for him a clerkship in the East India Company. In mathematics he appears to have been mainly self-taught. In 1823 he become a student of Trinity College, Dublin, and was first in all subjects and at all examinations. In 1824 when only a second year student Hamilton read a paper on Caustics (the shapes produced by focused light rays) before the Royal Irish Academy. In 1827 he was appointed Andrews Professor of Astronomy and superintendent of the observatory and soon after Astronomer Royal for Ireland.

The Schrödinger equation for such a system is then simply

$$\mathcal{H}\psi = E\psi. \tag{2.27}$$

The solutions, ψ, of such an equation are called the *eigenfunctions* of $\mathcal{H}$ and the associated energies, E, are the *eigenvalues*.

2.5. The physical significance of the wavefunction

The wavefunction of a particle, ψ, is clearly an important quantity which is the amplitude of the particle wave in three-dimensional space. But just what is its connection to measurable properties of the particle? The step from a classical, particulate, to a wave picture (the step from equation (2.16) to (2.18)) makes the representation unclear. We gain some insight by looking at the wave nature of light. Suppose a beam of monochromatic light is incident on two narrow, closely spaced, slits. If one of the slits is covered, part of the beam of light will pass through the open slit, which itself acts as a secondary source, and illuminates a screen placed behind the slit. If we now cover this slit but open up the adjacent one a similar illumination of the screen will result. How will the illumination of the screen be altered when both slits are open? It is well known that an interference pattern will be seen on the screen. To describe this effect the wave amplitudes originating in each slit must be summed, leading to constructive interference when their phases are identical and destructive interference when they are of opposite phase. The light intensity and therefore the distribution of energy density on the screen is given by the *square* of the function which results when the wave amplitude from the two slits are summed: this energy density must always be positive.

There are similar interpretations of other phenomena related to wave motion and so we are led to look for a connection between the *squares* of the particle-wavefunctions ψ which are solutions of the Schrödinger equation and some physical quantity. In particular, we have to give a physical interpretation of the connection between ψ^2, a continuous (cloud like) function† distributed in space, and the position of the particle. This interpretation was given by Born in 1926 and is now generally accepted. Born suggested that ψ^2 is interpreted as a *probability distribution* for the particle such that the probability of finding the particle is a small element of space $\mathrm{d}v$ is proportion to $\psi^2\ \mathrm{d}v$. The proportionality constant can be evaluated when we recognize that there is unit probability of finding the particle somewhere in space. That is, the integral of the probability density over all space must be unity

$$\int \psi^2\, \mathrm{d}v = 1. \tag{2.28}$$

Expression (2.28) is called the *normalization* condition for the wavefunction. It can be seen from equation (2.27) that the energy of the particle is unaffected by this

†Or $\psi^*\psi$ if ψ is a complex quantity as in some situations it may be. We shall assume in this book that wave functions are real unless we specify otherwise.

normalization. If ψ is a solution of (2.27) so too is $k\psi$ where k is any constant. If we have any solution of (2.27) which is not normalized, it is easy to normalize it by multiplying by the number N defined by

$$N^{-2} = \int \psi^2 \, \mathrm{d}v. \tag{2.29}$$

We must recognize that the interpretation of $\psi^2 \, dv$ as a probability distribution means that we cannot say exactly where the particle is at any specific time. Although the point is not clear from the derivation we have used in this chapter it is a fundamental aspect of quantum mechanics that there are limitations to the accuracy with which some quantities can be measured which have nothing to do with the accuracy of the instruments we use in their measurement. The position of an electron (or any other atomic particle) is one such quantity. The physical rationale is simple enough. In order to measure the position of an electron (assumed stationary) we would have to bounce something off it (like a photon) and then observe the photon. Unfortunately, the collision would impart some momentum to the electron and so observation of the photon would only tell us where the electron was when it collided (some time ago) with the photon, not where it is now (when we observe the photon). Even this discussion simplifies the problem because one such measurement would not be sufficient to locate uniquely a point in space. What is clear, however, is that in this example uncertainty in position and changes (and, therefore, uncertainty) in momentum are linked.

A mutual and simultaneous uncertainty in the position and momentum of atomic particles is an integral part of quantum mechanics. This was first stated formally in 1927 by Heisenberg with his famous *Uncertainty Principle*. He proved that the product of the limits to our knowledge of the position of a particle and the limits to our knowledge of its momentum is of the order of Planck's constant h.

There has been much philosophical debate over whether the Uncertainty Principle is inherent in nature or only a consequence of our attempts to make measurements on nature (Einstein held the latter view). Such debates can never be more than philosophical; in this book we are concerned with our understanding of atomic and molecular structures based on measurements made on these systems. For us, then, it is good to know that this uncertainty is covered by our theory; in particular we note that the wavefunction which is a continuous function in three dimensional space provides us with all the information that can be determined about the particle by experimental observation.

Chapter 3
Atomic Orbitals

3.1. The Schrödinger equation for the hydrogen atom

The wavefunctions which are the solutions of the Schrödinger equation for atoms and molecules are in general functions of extreme complexity. We note from expression (2.26) that the Schrödinger equation involves a summation over all the particles of the system and hence the solution of this equation leads to wavefunctions ψ which are functions of the positions of all the particles. Even if we had knowledge of such functions this would not be of any great interest to us. From the square of the wavefunction $\psi^2(x_i, y_i, z_i)$ we determine the probability of finding particle 1 at the point with coordinates x_1, y_1, z_1, particle 2 at x_2, x_2, z_2, etc.† However, there is no physical measurement that could possibly test whether such a probability distribution would be correct or not. The type of distribution we *can* test from experiment is the distribution of the net charge of a molecule in space. In other words, we are interested in probability distributions in real 3-dimensional space rather than in abstract $3n$-dimensional space for n particles.

There is no exact relationship between the total n-particle wavefunction and individual wavefunctions for each particle, but, as we shall see later, an approximate relationship does exist. The theory of the electron distribution and energies of atoms and molecules is based upon such an approximate relationship and we shall have no cause in this book to look beyond such an approximation. In other words, our concern is primarily with one-particle wavefunctions and the way in which we build up approximate many-particle wavefunctions from these.

For a full description of molecular properties we must be concerned with the wavefunctions of both nuclei and electrons. In Chapter 5 we shall show that because nuclei are several thousand times heavier than electrons their motion in a molecule is slow compared with the electronic motion, and we can obtain electronic wavefunctions by assuming that the nuclei are at rest. For atoms we have only a single nucleus and we are therefore only interested in the position of the electrons relative to the nucleus. The one-electron wavefunctions for atoms which define these positions are called *atomic orbitals*.

We shall, in later chapters, examine the one-electron wavefunctions for molecules, which are called *molecular orbitals*, and shall find that there are important similarities between atomic and molecular orbitals.

†More precisely the probability that particle 1 is between x_1 and $x_1 + dx_1$, y_1 and $y_1 + dy_1$, etc.

Atoms or molecules (or their ions) which have only a single electron are clearly in a rather special category, in so far as the solution of the Schrödinger equation is concerned, because the orbital wavefunctions are also the total electronic wavefunctions. For such systems the Schrödinger equation can be solved exactly. Although for chemists such one-electron systems are not very important for their own sake, they have general importance because the orbitals of many-electron systems have strong similarities to the orbitals of one-electron systems. It is therefore appropriate that we begin our examination of atomic orbitals by looking at an exactly solvable problem, namely, the wavefunctions of the electron of a hydrogen atom.

The task of solving the Schrödinger equation for the electrons of an atom or molecule can be made simpler by a judicious choice of the coordinate system which defines the position of the electrons relative to the nuclei. For an isolated atom, without external fields, all directions in space are equivalent. At a fixed distance r from the nucleus (that is over the surface of a sphere of radius r) we expect to find a uniform electron density. Changing r, however, would lead to a change in electron density. We conclude that it may well be profitable to work in a coordinate system in which r is one coordinate, rather than in the familiar x, y, z (Cartesian) system. Such a coordinate system is the set (r, ϑ, φ) known as spherical polar coordinates. The angles ϑ and φ serve to locate a point on the surface of a sphere defined by a given value of r.

Spherical polar coordinates are defined in Figure 3.1 and are related to Cartesian coordinates by the following expressions

$$x = r \sin \vartheta \cos \varphi, \tag{3.1}$$

$$y = r \sin \vartheta \cos \varphi, \tag{3.2}$$

$$z = r \cos \vartheta. \tag{3.3}$$

It is clear by summing the squares that

$$r^2 = x^2 + y^2 + z^2. \tag{3.4}$$

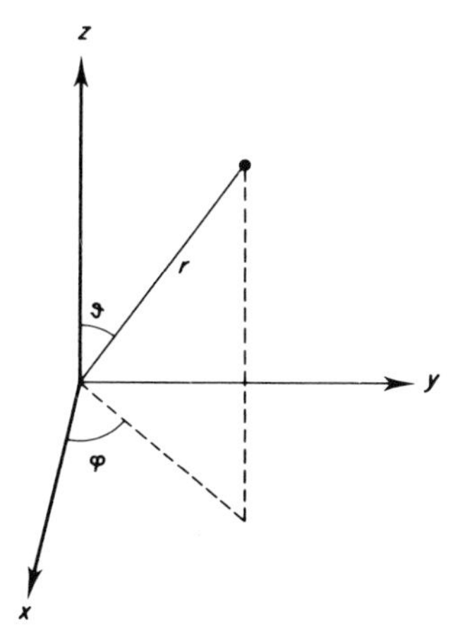

Figure 3.1 Spherical polar coordinates and their relationship to Cartesian coordinates.

It can be seen from Figure 3.1 that the polar angles ϑ and φ are defined with origins along the z axis and in the xz plane respectively. However, for an atom in free space, for which there is no unique choice of axes, we must find from the solutions of the Schrödinger equation that the properties of the electrons, if not their individual wavefunctions, are independent of the choice of axes.

The advantage of using spherical polar coordinates rather than Cartesian coordinates can be seen if we write down the explicit form of the Schrödinger equation for the hydrogen atom in both cases. The potential energy of the electron moving in the field of the nucleus is, in Cartesian coordinates,

$$V = -\left(\frac{e^2}{4\pi\epsilon_0}\right)\frac{1}{(x^2+y^2+z^2)^{1/2}}. \tag{3.5}$$

We have here defined the potential in SI units where $-e$ is the electron charge (e the proton charge) and ϵ_0 the permittivity of free space. Introducing this potential into the Schrödinger equation (2.18) we obtain

$$\frac{-h^2}{8\pi^2 m}\left(\frac{\partial^2\psi}{\partial x^2}+\frac{\partial^2\psi}{\partial y^2}+\frac{\partial^2\psi}{\partial z^2}\right)-\frac{e^2}{4\pi\epsilon_0}\left(\frac{1}{(x^2+y^2+z^2)^{1/2}}\right)\psi = E\psi, \tag{3.6}$$

where m is the mass of the electron.†

Although the kinetic energy part of the Hamiltonian is quite simple, being the sum of separate terms involving x, y, and z, the potential energy is a rather complicated function of the three coordinates.

In spherical polar coordinates the potential energy is much simpler

$$V = \frac{-e^2}{4\pi\epsilon_0}\,\frac{1}{r}. \tag{3.7}$$

The kinetic energy part of the Hamiltonian can be transformed from Cartesian to spherical polar coordinates by the standard rules for changing the variables in partial differentiation. It is not difficult, but rather tedious to obtain the result and we therefore content ourselves with quoting it:

$$\frac{\partial^2}{\partial x^2}+\frac{\partial^2}{\partial y^2}+\frac{\partial^2}{\partial z^2}=\frac{1}{r^2}\left[\frac{\partial}{\partial r}\left(r^2\frac{\partial}{\partial r}\right)+\frac{1}{\sin^2\vartheta}\frac{\partial^2}{\partial\varphi^2}+\frac{1}{\sin\vartheta}\frac{\partial}{\partial\vartheta}\left(\sin\vartheta\frac{\partial}{\partial\vartheta}\right)\right]. \tag{3.8}$$

The Schrödinger equation for the hydrogen atom in spherical polar coordinates is therefore

$$\frac{-h^2}{8\pi^2 mr^2}\left[\frac{\partial}{\partial r}\left(r^2\frac{\partial\psi}{\partial r}\right)+\frac{1}{\sin^2\vartheta}\frac{\partial^2\psi}{\partial\varphi^2}+\frac{1}{\sin\vartheta}\frac{\partial}{\partial\vartheta}\left(\sin\vartheta\frac{\partial\psi}{\partial\vartheta}\right)\right]-\frac{e^2}{4\pi\epsilon_0 r}\psi = E\psi, \tag{3.9}$$

†More precisely m should be the reduced mass of the electron and the proton which is $m_e m_p/(m_e + m_p)$.

which on multiplying throughout by $[-8\pi^2 mr^2/h^2]$ becomes

$$\frac{\partial}{\partial r}\left(r^2\frac{\partial\psi}{\partial r}\right)+\frac{1}{\sin^2\vartheta}\frac{\partial^2\psi}{\partial\varphi^2}+\frac{1}{\sin\vartheta}\frac{\partial}{\partial\vartheta}\left(\sin\vartheta\frac{\partial\psi}{\partial\vartheta}\right)+\frac{2\pi me^2 r}{h^2\epsilon_0}\psi=-\frac{8\pi^2 mr^2}{h^2}E\psi. \tag{3.10}$$

Despite the fact that the differential part of this equation is more complicated than is that of (3.6) it is more amenable to solution because each term is either a function of r or a function of the angles ϑ and φ. It can be shown, as a result, that the solutions ψ may be written as a product of a function of r only, and a function of ϑ and φ as follows:

$$\psi(r,\vartheta,\varphi)=R(r)Y(\vartheta,\varphi), \tag{3.11}$$

and if this expression is substituted back into (3.10) we obtain instead of one differential equation in three variables, two separate differential equations, one in r only and one in ϑ and φ. It can further be shown that the functions Y may themselves be written as a product of separate functions of ϑ and φ

$$Y(\theta,\varphi)=\Theta(\vartheta)\,\Phi(\varphi), \tag{3.12}$$

and that Θ and Φ are solutions of separate one-variable equations. These three one-variable differential equations, for $R(r)$, $\Theta(\vartheta)$ and $\Phi(\varphi)$ are readily solved. We will not give in this book the detailed form of the differential equations nor of their solutions, both of which can be found in more mathematically oriented texts. We shall, however, elaborate on the form of the functions R and Y which define the solutions (orbitals) of the hydrogen atom.

3.2. The shapes of atomic orbitals

Because of the separation of variables that has been made in (3.11) we can discuss separately the dependence of the atomic orbitals on the distance of the electron from the nucleus (r) and the angular variation over a spherical shell, which is described by $Y(\vartheta,\varphi)$. The functions Y can be said to define the 'shape' of the orbitals.

The functions Y are called spherical harmonics. They occur frequently in problems in physics involving central fields, that is those in which the potential depends only on the distance from a centre. For example, they describe the vibrations of a spherical shell. Spherical harmonics are all simple products of sines and cosines of the two angles ϑ and φ. This is perhaps obvious from the fact that functions which describe the vibrations of a sphere must satisfy continuity conditions that they are unchanged on making a complete rotation about the sphere ($\vartheta\to\vartheta+2\pi$ for example). Such continuity conditions will be most simply satisfied by trigonometric functions. If we also rule out any infinities in the wavefunctions ($\tan\vartheta$ is thus unacceptable) we are left with acceptable functions that can be expressed as polynomials in sines and cosines of ϑ and φ.

The continuity conditions on a sphere have the same role for the wavefunctions of central field problems as the boundary condition of fixed ends has for a vibrating string. Thus there are restrictions on the actual polynomials that make up the spherical harmonics analogous to the condition (2.12) imposed on the wavelengths of a stretched string.

The functions Y can be grouped into sets each of which is characterized by an integer l. This integer is the number of zeros of the function $\Theta(\vartheta)$ in the range $0 \leqslant \vartheta < \pi$. These zeros lie on nodal planes which pass through the centre of the system (the nucleus of the atom).

The function which has no nodal planes ($l = 0$) is spherically symmetrical. That is, $Y(l = 0)$ is a constant. Atomic orbitals with such an angular function are called s orbitals for historical reasons.

There are three spherical harmonics which have one nodal plane ($l = 1$), and they have the following form:

$$Y(l=1) = \begin{cases} \sin\vartheta\cos\varphi, & (3.13) \\ \sin\vartheta\sin\varphi, & (3.14) \\ \cos\vartheta & (3.15) \end{cases}$$

Atomic orbitals having such angular behaviour are, again for historical reasons, called p orbitals. As can be seen from equations (3.1) to (3.3) these functions represent the projections of a unit radius on the x, y, and z axes respectively, we therefore label the corresponding atomic orbitals p_x, p_y, and p_z. These have as nodal planes yz, zx, and xy respectively.

At this point we should ask why there must be three functions having $l = 1$. Individually the three functions are not spherically symmetric. An electron having any one of the wavefunctions does not therefore have a spherical distribution in space. However, as we have already noted, there are no unique axes for a free atom so that any wavefunctions of the atom which are not spherically symmetric must occur in sets so that no axis is unique. We can see that the three functions (3.13)–(3.15) satisfy this criterion. An electron with a wavefunction p_x has a probability distribution over a sphere of $\sin^2\vartheta\cos^2\varphi$. The corresponding distributions for p_y and p_z are $\sin^2\vartheta\sin^2\varphi$ and $\cos^2\vartheta$ respectively. In a spherically symmetric situation there is equal chance of having the electron in any one of the three and hence the resultant probability distribution is spherically symmetric because

$$\sin^2\vartheta\cos^2\varphi + \sin^2\vartheta\sin^2\varphi + \cos^2\vartheta = 1. \tag{3.16}$$

How can we represent these p orbitals by diagrams? The vibrating sphere analogy is not too much help here because the functions we are considering do not correspond to displacements from some equilibrium position. The simplest graphical description is obtained as follows. Consider the p_z orbital and draw a z axis through the nucleus in some purely arbitrary direction (were we to apply an electric or magnetic field this would conveniently define the direction of z). The purpose of this construction is to define an axis from which we can measure the

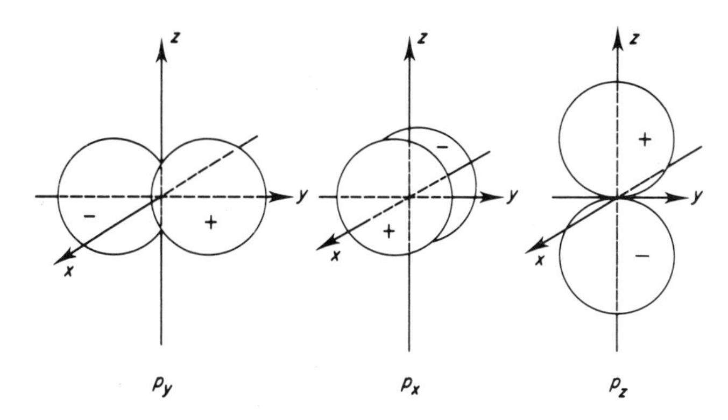

Figure 3.2 Polar diagrams for p orbitals.

angle ϑ (see Figure 3.1). Note that the angle φ is not involved in the function p_z. Now draw lines radially from the nucleus, like the spines of a defensive porcupine. Along each radial line mark off a length proportional to $\cos\vartheta$ but ignore the sign (so that at $\vartheta = 0$ and π the length marked off is unity). It will be found that the points so marked off define the surfaces of two spheres which touch at the nucleus. Finally, label each sphere with the sign (+ or −) of p_z (i.e. $\cos\vartheta$) for that sphere. The result of this procedure is shown in Figure 3.2.

In a similar fashion the functions p_x and p_y can be constructed (placing the x axis at some arbitrary position perpendicular to the z axis but through the nucleus). They look just like p_z but oriented along the x and y axes respectively (and hence their suffixes).

The set of spherical harmonics with $l = 2$ are five in number and atomic orbitals having such angular variation are called d orbitals. They are inevitably more complicated than the s and p orbitals. It is customary to give the d orbitals suffixes according to their Cartesian equivalences as follows:

$$d_{z^2} \propto 3\cos^2\vartheta - 1, \tag{3.17}$$

$$d_{zx} \propto \sin\vartheta\cos\vartheta\cos\varphi, \tag{3.18}$$

$$d_{yz} \propto \sin\vartheta\cos\vartheta\sin\varphi, \tag{3.19}$$

$$d_{x^2-y^2} \propto \sin^2\vartheta\cos 2\varphi \tag{3.20}$$

$$d_{xy} \propto \sin^2\vartheta\sin 2\varphi. \tag{3.21}$$

It is easy to confirm from (3.1)–(3.3) that, for example, $zx \propto \sin\vartheta\cos\varphi\cos\varphi$. The suffix on z^2 is not very precise; better is $3z^2 - r^2$, but writing d_{z^2} is a conventional abbreviation.

Pictures of d orbitals, drawn in a similar way to the p orbitals described earlier, are shown in Figure 3.3. The orbitals d_{xy}, d_{yz}, d_{zx} and $d_{x^2-y^2}$ are all of the same 'shape' differing only in their orientation in space but d_{z^2} is different. It may seem odd that the shape of d_{z^2} is different from the others. This is due to the fact that

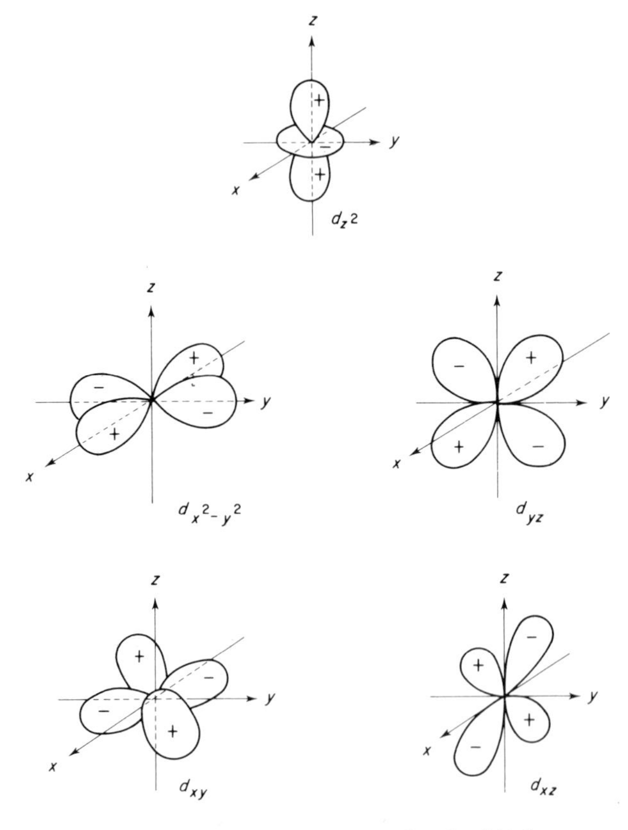

Figure 3.3 Polar diagrams for d orbitals.

one can only choose five different combinations of the quadratic forms ($x^2, y^2, z^2, xy, yz, zx$), which are independent and non-spherical, a sixth combination $x^2 + y^2 + z^2$ is spherical and therefore not a d orbital. It would therefore be wrong to take $x^2 - y^2$, $y^2 - z^2$ and $z^2 - x^2$ as three independent d orbitals. Any one of them can clearly be written in terms of the other two; thus (using the Cartesian functions directly)

$$x^2 - y^2 = (x^2 - z^2) + (z^2 - y^2), \tag{3.22}$$

whilst

$$(z^2 - y^2) - (x^2 - z^2) = 3z^2 - x^2 - y^2 - z^2 = 3z^2 - r^2 = \text{“}z^2\text{”}. \tag{3.23}$$

In our discussion of the physical significance of ψ we made brief reference to the fact that ψ may be complex, and to the fact that in such a case the probability density is the product of ψ and its complex conjugate, ψ^*, which is always real.

For an atom it is sometimes convenient to use an alternative complex form of the spherical harmonics in which Φ has the general form

$$\Phi = e^{\pm im\varphi}. \tag{3.24}$$

This is particularly true if we are interested in the interaction between atoms and external magnetic fields. The existence of these alternative complex forms which are also solutions of the Schrödinger equation is not difficult to understand. In the first place we note the mathematical relationship

$$e^{\pm im\varphi} = \cos m\varphi \pm i \sin m\varphi, \tag{3.25}$$

so that the complex functions can be obtained from suitable combinations of the real functions. Secondly, the standing waves of any vibrating system can always be represented by combinations of travelling waves, or vice versa. The real forms of the spherical harmonics which we have described are standing waves and they can readily be visualized. The complex forms are travelling waves and their representation is more difficult.

We note from expression (3.24) that the complex functions are classified by a parameter m. In view of expression (3.25), and the fact that the real form of the wavefunctions only involve sines or cosines of integer multiples of φ, m must, like l, be an integer. These integers are called *quantum numbers* or to be more specific they have been called the azimuthel (l) and magnetic (m) quantum numbers through their relationship to similar numbers that arise in the Bohr theory of the atom. These integers are used to label the angular part of atomic wavefunctions as follows:

$$\Theta_{lm}(\vartheta)\Phi_m(\varphi), \tag{3.26}$$

which means that there is a set of functions Φ_m which are identified by the number m and a set Θ_{lm} which are different for each pair of numbers l and m. The m value depends on the l: for a given l value m can take the $2l + 1$ values

$$l, l-1, l-2, \ldots -l+1, -l. \tag{3.27}$$

These $2l + 1$ orbitals have the same energy unless the atom is put into an external electric or magnetic field which destroys the spherical symmetry of the problem. A set of solutions of the Schrödinger equation which have the same energy is called a *degenerate* set.

If $l = 0$ then m can only have the value 0. If $l = 1$ then m can have the values 1, 0, -1. This is the set of three p orbitals in complex form. If $l = 2$ then $m = 2, 1, 0, -1, -2$ and we have the set of five d orbitals in complex form. For a free atom s orbitals are not degenerate, p orbitals are 3-fold degenerate, d orbitals are 5-fold degenerate, etc.

3.3. The radial wavefunctions of the hydrogen atom

Whereas the angular properties of atomic orbitals that we have discussed in the last section are common to the orbitals of all atoms, the radial wavefunctions $R(r)$

(see equation 3.11) are unique for each atom. Only in the case of hydrogen or other one-electron atoms (He^+ etc.) can the functions $R(r)$ be expressed exactly in terms of simple analytical functions. In other cases they must be defined numerically or by series expansions. However, the radial functions for many-electron atoms have features in common with those of the hydrogen atom and we shall therefore look at these first.

We are only interested in those functions $R(r)$ that can represent the state of an electron bound to the proton. We must therefore impose the restriction that $R(r)$ becomes zero as r approaches infinity; if this were not so then there would be greatest probability of the electron being at infinity, that is, of it not being bound to the nucleus. This is a further boundary condition on the solutions of the Schrödinger equation. The radial functions which satisfy this boundary condition are the product of a *finite* polynomial (P) and an exponential and are of the form

$$R_{nl}(r) = P(r) \exp(-kr). \tag{3.28}$$

Note that the exponential decays to zero at infinity faster than the increase of any finite polynomial. Note also that these solutions are labelled by l and by another quantum number n, which is related to the principal quantum number of Bohr theory and which is associated with the boundary condition at infinity. The number of nodes in the radial wavefunction between $r = 0$ and $r = \infty$ is $n - l - 1$ and as this must be zero or a positive integer the l values for a specified integer n are restricted by

$$l \leqslant n - 1.$$

In other words, the allowed values are:

$$\begin{aligned} n &= 1; \quad l = 0, \\ n &= 2; \quad l = 0, 1, \\ n &= 3; \quad l = 0, 1, 2, \text{etc.} \end{aligned} \tag{3.29}$$

Each complete orbital (radial and angular function multiplied together) is given a label which is a combination of its n value and the symbol for the l value, thus

$$1s, 2s, 2p, 3s, 3p, 3d, \text{etc.} \tag{3.30}$$

These labels are frequently extended to include suffixes, thus $3p_x$, $3p_y$ and $3p_z$ are the three (real) $3p$ orbitals whilst $3p_1$, $3p_0$ and $3p_{-1}$ are the corresponding three complex p orbitals (the subscripts in the latter case simply being the m values).

Although we shall have no need for the precise form of the hydrogen atom radial wavefunctions in this book they are given in Table 3.1 for the $1s$, $2s$, and $2p$ orbitals for the reader's interest. These functions are shown diagramatically in Figure 3.4. We also show the probability density obtained by squaring the wavefunction (R_{nl}^2) and this quantity weighted by the factor r^2. This latter function measures the probability of finding the electron at a distance r from the nucleus; the factor r^2 accounts for the fact that the surface area of a spherical shell

Table 3.1. The radial wavefunctions $R(r)$ of hydrogen atomic orbitals. The functions are not normalized and the parameter ρ is related to r by $\rho = r/a_0$ where $a_0 = h^2\epsilon_0/\pi me^2$ has the dimensions of length and is called the Bohr Radius

1*s*	$\exp(-\rho)$
2*s*	$(2 - \rho)\exp(-\rho/2)$
2*p*	$\rho\exp(-\rho/2)$

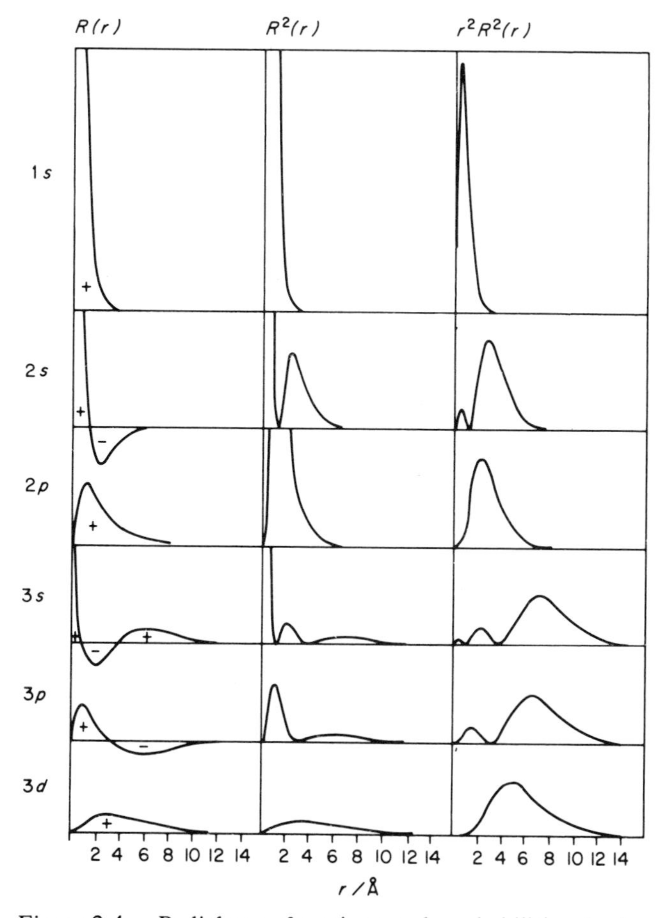

Figure 3.4 Radial wavefunctions and probabilities for 1*s*, 2*s*, 2*p*, 3*s*, 3*p*, and 3*d* hydrogen atomic orbitals. A common vertical scale is used for each orbital but different scales are used for $R(r)$, $R^2(r)$, and $r^2R^2(r)$.

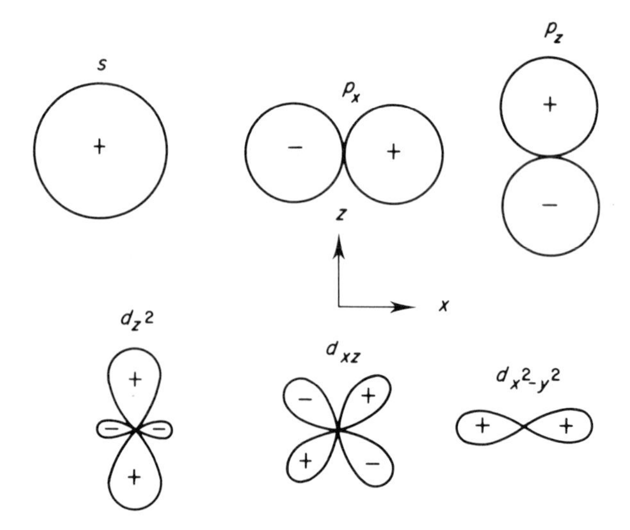

Figure 3.5 Sections through the xz plane ($y = 0$) of the s, p, and d orbital polar diagrams. p_y, d_{xy}, and d_{yz} are zero in this plane.

of radius r is proportional to r^2, so that the increasing size of this shell as we move away from the nucleus increases the chance of finding the electron at a distance r.

Figures 3.2 and 3.3 describe angular functions; Figure 3.4 radial functions. Can we combine them in some way to obtain a picture of a complete orbital? Unfortunately, not within three dimensional space. It is therefore the practice to use diagrams such as those of Figures 3.2 and 3.3 on their own to represent an orbital. More commonly still sections through these orbitals, taken so as to contain the nucleus, are used. Such sections are shown in Figure 3.5. An alternative does exist, however, and that is to sketch contour surfaces of the wavefunction. Insert

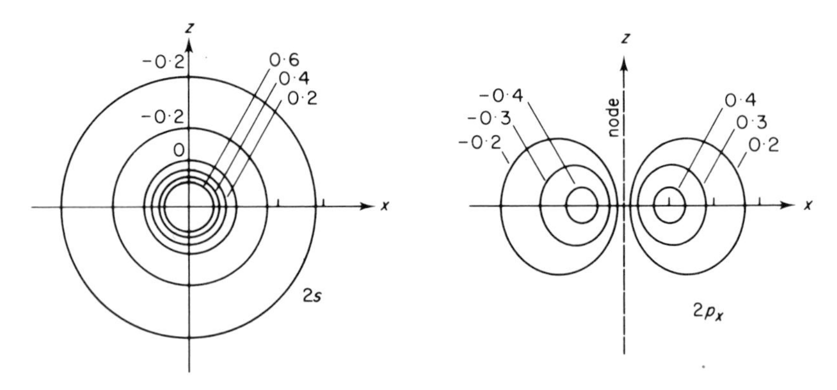

Figure 3.6 Radial wavefunction contours for hygrogen $2s$ and $2p_x$ orbitals in the xz plane.

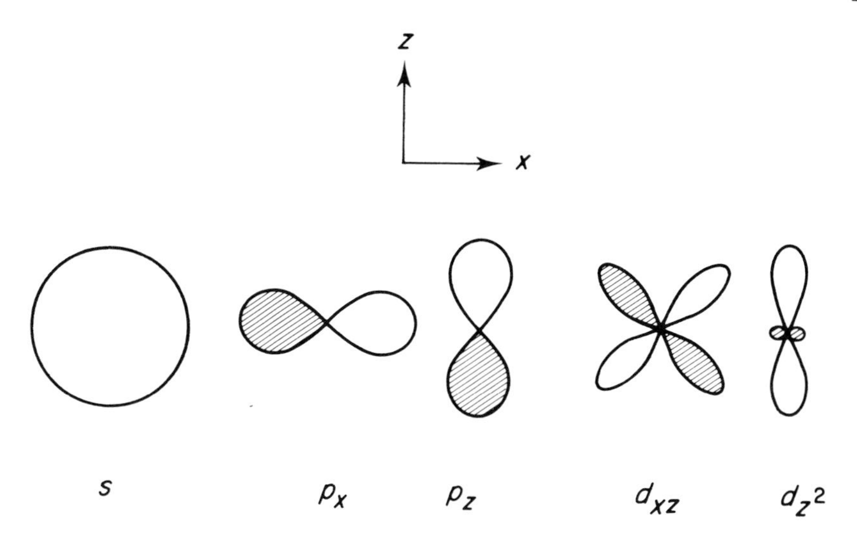

Figure 3.7 Schematic polar diagrams of *s*, *p*, and *d* atomic orbitals. They approximate to polar diagrams of the square of the wavefunction but the regions associated with negative parts of the wavefunction are hatched.

values of r, ϑ, and φ into expression (3.11) and a number results. If we evaluate $\psi(r, \vartheta, \varphi)$, for a large number of values of the variables we can imagine contour surfaces being drawn in three-dimensional space, each surface being the locus of points with the same value of $\psi(r, \vartheta, \varphi)$. Again, it is convenient to take a suitable section through the shells of contour surfaces. Such sections are shown in Figure 3.6. In contrast, Figure 3.7 shows a representation of orbitals which is often encountered in books or used in discussion between chemists. Their theoretical basis is very uncertain but, at least, they have the twin virtues of being easy to draw and also showing the general nodal pattern of an orbital.

Finally, we have seen that the wavefunctions of the atomic orbitals have positive and negative regions and it might be asked what is the significance of the sign. The observable properties of the electron are all related to the probability density ψ^2 and this is everywhere positive. However, the signs are important for our understanding of the chemical bond because, as we shall see, this involves the overlap or interference of two atomic orbitals. The signs then determine whether this interference is constructive or destructive, in a similar manner to the interference of light waves.

3.4. The orbitals of many-electron atoms

It is perhaps worth repeating the definition of an orbital which we gave at the beginning of this chapter; it is the wavefunction of an electron which moves under the influence of the nuclear attraction and the *average* repulsion of all other electrons. We shall see in the next chapter and again in Chapter 8, how to construct the wavefunctions of a whole set of electrons from the one-electron wavefunction (those of its orbitals), but at this point we will examine the orbitals themselves and discuss qualitatively the method by which they are derived.

An electron (let us label it with the number 1) in a many-electron atom interacts with the nucleus (of charge Ze) and with all of the other i electrons. The actual magnitude of the potential of electron 1 at any instant depends on its separation from the nucleus, r_1, and on its distance from each of the other electrons, i.e. on the various r_{1i}. Mathematically, the potential, V, is given by

$$V(r_1, r_i) = \frac{e^2}{4\pi\epsilon_0}\left[-\frac{Z}{r_1} + \sum_{i\neq 1}\frac{1}{r_{1i}}\right]. \tag{3.31}$$

We now introduce two symbolic steps. Although we shall not concern ourselves with just how to carry them out mathematically it is to be emphasized that the two steps lead to enormous simplifications in the mathematics and also allow us to define an orbital in a many-electron context. We formally take an average of the electron repulsion over the position of all electrons i and obtain (the bar indicating an average)

$$V(\mathbf{r}_1) = \frac{e^2}{4\pi\epsilon_0}\left[\frac{-Z}{r_1} + \overline{\sum_{i\neq 1}\frac{1}{r_{1i}}}\right]. \tag{3.32}$$

The simplification which this step represents is shown more on the left-hand side of the equation than the right. V is now only a function of the position of electron 1.

The potential given by (3.32) will not generally have spherical symmetry because the electrons i are not necessarily distributed spherically around the nucleus. We therefore make a second average which is over all angles to give a spherically symmetric potential

$$V(r_1) = \overline{V(\mathbf{r}_1)}. \tag{3.33}$$

Because (3.33) is spherically symmetric, the wavefunctions of an electron moving in this potential may be separated, according to expression (3.11), into a product of a radial function and the spherical harmonic functions Y which we have already examined for the hydrogen atom. We have now simplified – and therefore approximated – the problem so that we now have to solve the Schrödinger equation in the form

$$\frac{-h^2}{8\pi^2 m}\left(\frac{\partial^2\psi}{\partial x_1^2} + \frac{\partial^2\psi}{\partial y_1^2} + \frac{\partial^2\psi}{\partial z_1^2}\right) + V(r_1)\psi = E\psi. \tag{3.34}$$

Even with the simplification, this equation does not admit of algebraic solution because $V(r_1)$ is generally a rather complicated function of r_1. However, it may be solved approximately and we can make the approximate solutions as close as we wish to the accurate ones. The method for doing this is called the self-consistent field (SCF) method and hinges upon the fact that we have to perform the double average in (3.33). It turns out that this can be done fairly easily but before it can be done we need to know the orbitals for all of the electrons other than that of electron 1. But we cannot know these until we know the orbital of electron 1 [because this knowledge will be needed to evaluate all the other $V(r_i)$ values, where $i \neq 1$]. The way out of this circular dilemma is an iterative procedure.

1. We make an intelligent guess of the form of the orbital wavefunctions of each electron.
2. We use these orbitals to determine the average potential that occurs in expression (3.33) and in the analogous expression for other electrons.
3. We now solve equation (3.34) numerically to obtain a set of orbital wavefunctions ψ_k. The precise method of doing this is a technical problem that need not concern us, but it is something that can be handled easily on a digital computer. There is a large number (in general infinite) of solutions to (3.34) but in practice we usually try to obtain the ones which have low energies. We shall discuss the energies of orbitals in more detail shortly. For the present it is sufficient to point out that, as is evident from (3.34), associated with each orbital (each ψ_k) there is an energy E_k. The way that a computer tackles the problem of solving (3.34) is first to determine these energies and second to determine the associated wavefunctions.
4. We have to decide which of the orbitals ψ_k shall contain electrons. There are certain rules to determine this as we shall see. If we are interested in the lowest energy (ground) state of the atom then we generally allocate electrons to the lowest energy orbitals consistent with these rules.
5. We can now calculate a new average electron repulsion which we hope is nearer to the truth than that obtained in 2 from our intelligent guess of 1.

3a. Carry out the same procedure as in 3 to obtain new orbitals ψ_k.

4a. Carry out the same procedure as in 4.

5a. Carry out the same procedure as in 5, etc.

It can be seen that we are carrying out a series of operations 3, 4, 5 in a cycle and the results of one cycle are used to start the next. We hope the procedure converges, that is the changes in the solutions get smaller as the procedure progresses. Perhaps after 50 cycles no further change can be detected and we have reached convergence; sometimes it does not and we usually then have to go back to step (1) and make a *more* intelligent guess.

The self-consistent aspect of the calculation should be obvious, it is that the field or potential determined by a step like 5 leads through step 3 to a set of orbitals which are consistent with this field.

The SCF orbitals of many atoms were determined in the period 1930–1950 mainly according to procedures developed by Hartree and Fock. Exact solutions of equations like (3.34) are therefore usually known as Hartree–Fock orbitals. From 1950 onwards the computational technology improved to the point at which SCF orbitals could also be obtained for molecules, which is a more difficult problem because there is no overall spherical symmetry when many nuclei are present. We will examine this in more detail later.

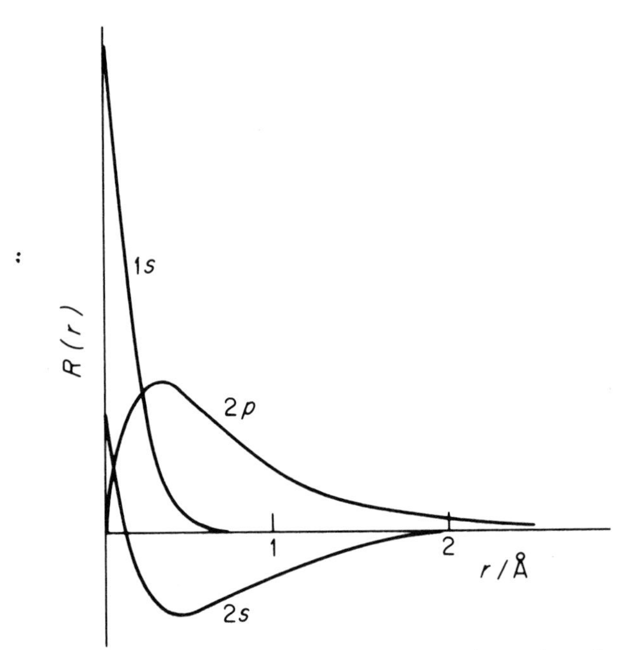

Figure 3.8 The radial wavefunctions for carbon 1s, 2s, and 2p atomic orbitals.

Figure 3.8 shows the radial parts of the 1s, 2s, and 2p atomic orbitals of the carbon atom for its ground state. It is seen that there is a strong similarity with the hydrogen orbitals shown in Figure 3.4. In particular, the number of nodes (zeros) in the wavefunction is determined by n and l for all atoms although their positions vary from one atom to another. We shall see in the next section that for the hydrogen atom the quantum number n can be defined by a simple expression for the energy of the orbital. For other atoms there is no such simple relationship between n and E so that n can only be defined by the nodal properties of the radial wavefunction; the number of such nodes between 0 and ∞ is $n - l - 1$.

A considerable amount of work in valence theory has been based upon approximate solutions of equation (3.34) which have a simple analytical form. The most widely used are known as Slater orbitals or Slater functions. The radial parts of such functions have the form

$$R_n = r^{n-1} \exp(-\zeta r) \tag{3.35}$$

where the exponent ζ depends on the atom and on the quantum number l. Although rules have been given for choosing the best value of ζ, these rules are not widely used today. We show instead in Table 3.2 values that have been deduced by accurate calculations for the atoms H–Ne.

From expression (3.35) we see that the Slater 1s and 2s orbitals have the form:

$$R_{1s} = \exp(-\zeta_1 r), \tag{3.36}$$

$$R_{2s} = r \exp(-\zeta_2 r). \tag{3.37}$$

Table 3.2. The best atomic orbital exponents (ζ) for the free atoms as defined by expression (3.35) [Clementi and Raimondi, *J. Chem. Phys.*, **38**, 2686 (1963)].

	1*s*	2*s*	2*p*
H	1.0	–	–
He	1.687 5	–	–
Li	2.690 6	0.639 6	–
Be	3.684 8	0.956 0	–
B	4.679 5	1.288 1	1.210 7
C	5.672 7	1.608 3	1.567 9
N	6.665 1	1.923 7	1.917 0
O	7.657 9	2.245 8	2.226 6
F	8.650 1	2.563 8	2.550 0
Ne	9.642 1	2.879 2	2.879 2

When (3.36) is plotted as a function of r it looks very similar to the real 1s orbital for any atom but R_{2s}, as given by (3.37), is different from real 2s wavefunctions because it is zero at $r = 0$, whereas real 2s radial functions are not. However, it is very easy to form a function that is qualitatively correct by combining Slater orbitals. For example, for positive k,

$$kR_{1s} - R_{2s} = k \exp(-\zeta_1 r) - r \exp(-\zeta_2 r) \tag{3.38}$$

is positive at $r = 0$, has a node at some finite value of r and decays exponentially as $r \to \infty$; it thus behaves as the 2s orbitals of Figures 3.4 and 3.8.

More accurate representations for atomic orbitals can be obtained by combining even more functions of the type (3.35) and standard sets of such functions have also been published.

3.5. The energies of atomic orbitals

The end of the 19th century was a period of great activity in the study of the discrete wavelengths of light emitted by hot atoms. These data were eventually to be rationalized by the Bohr theory of the quantized energy states of the atom, but an important first step was made by Balmer, Rydberg, Ritz, and others.

Balmer in 1855 made the inspired deduction that the spectral lines which were emitted by hydrogen atoms, and which could be detected in solar and stellar emission, had wavelengths which could be fitted to the general formula

$$\frac{1}{\lambda} = R\left[\frac{1}{n_1^2} - \frac{1}{n_2^2}\right], \tag{3.39}$$

where n_1 and n_2 were integers. The wavelengths of all lines known at that time were reproduced with high accuracy by taking $n_1 = 2$ and $n_2 = 3,4,5 \ldots$ for the first, second, third . . . members of the series. The value of R, later to be known as the Rydberg constant, was 10 967 800 m^{-1}.

Expression (3.39) was later generalized by Rydberg and Ritz to fit the spectral lines of other atoms and, in particular, Ritz suggested that other series of hydrogen lines would be expected by letting the integer n_1 take running values 1, 2, 3 etc. Spectral lines were later observed in the far ultra-violet and in the infra-red regions of the spectrum which confirmed this expectation.

The prediction of new spectral lines from expression (3.39) is based upon the Ritz combination principle. As expression (3.39) is the difference between two quantities R/n_1^2 and R/n_2^2 it is clearly possible from one series R/n^2 to construct, by taking differences, many spectral series. Thus from k values of n one can obtain $k(k-1)/2$ combination differences. The Ritz principle is therefore an important rationalization of atomic spectral data, and it has in fact a general application to all spectroscopy.

Expression (3.39) can be put in energy units by applying the Planck–Einstein relationship (2.1)

$$\Delta E = h\nu = \frac{hc}{\lambda} = hcR\left[\frac{1}{n_1^2} - \frac{1}{n_2^2}\right]. \tag{3.40}$$

We can now apply the condition of energy conservation to the emission of the photon so that if the energy of the emitted photon is $h\nu$, the atom must lose this same energy in the process. From the solutions of the Schrödinger equation we determine the energy levels of the atom and, applying the Ritz combination principle, we can take differences of these energy levels to obtain the required frequencies according to expression (3.40).

The solution of the Schrödinger equation for the hydrogen atom leads to energy levels which are given by the expression (in SI units)

$$E_{nl} = -\left(\frac{me^4}{8h^2\epsilon_0^2}\right)\left(\frac{1}{n^2}\right). \tag{3.41}$$

Note that this expression is a function of the principal quantum number n but is independent of l. The degeneracy of all m levels for a given (n, l) has already been noted and is a property of all atoms, associated with the fact that there are no unique axes in space. The degeneracy of l values for a given n is surprising and is in fact a property only of the hydrogen atom, or any other one-electron atomic system like He^+, Li^{2+}, etc.

Expression (3.41) is easily confirmed for a 1s orbital by substituting the wavefunction [$\psi = \exp(-r/a_0)$, $a_0 = h^2\epsilon_0/\pi me^2$] into equation (3.10) and carrying out the differentiation; as ψ is not a function of ϑ and φ the differentials with respect to these angles are zero.

By applying the Ritz combination principle to the energies (3.41), we deduce that if the hydrogen atom makes a transition from a state with $n = n_2$ to one having $n = n_1$ the change in energy will be

$$\Delta E = \frac{me^4}{8h^2\epsilon_0^2}\left[\frac{1}{n_1^2} - \frac{1}{n_2^2}\right], \tag{3.42}$$

and a comparison with expression (3.40) shows that this is in agreement with experiment if the Rydberg constant is given by

$$R = \frac{me^4}{8h^3c\epsilon_0^2}. \tag{3.43}$$

On substituting the values of the appropriate constants in SI units, the value of the Rydberg constant deduced from experiment, which has already been quoted, is reproduced exactly.

For atoms with more than one electron the orbital energies depend on l as well as on n, although there is no simple relationship between the energy and these two quantum numbers. For the same value of n the energy increases as l increases. The reason is that as l increases the orbital penetrates less into the region close to the nucleus so that the effective screened nuclear charge experienced by the electron decreases as l increases. This is illustrated in Figure 3.9 where the potential of K^+ calculated by the SCF method is plotted as a function of r. The radial electron densities, r^2R^2, for hydrogen-like orbitals $3s$, $3p$, and $3d$ are also shown and it is clear that an electron in the $3s$ orbital would experience the greatest attraction to the nucleus, on average, and one in the $3d$ orbital, on average, the least.

We shall see in the next chapter that the difference in energy between atomic orbitals of the same n but different l values has an important influence on the periodicity of the elements. It is also important for our understanding of the energies of molecular orbitals. For this reason we give in Table 3.3 the difference between the $2s$ and $2p$ orbital energies of the atoms Li–F. These are average

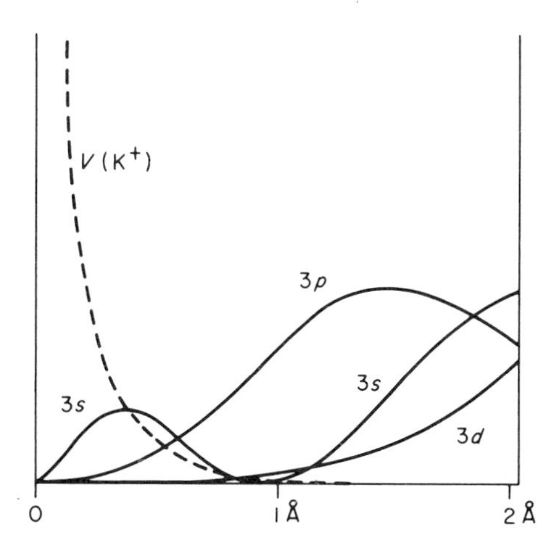

Figure 3.9 The attractive potential of K^+ and its relationship to the hydrogen atomic orbital radial densities $r^2R^2(r)$.

Table 3.3. The difference in energy between $2s$ and $2p$ atomic orbitals

	Li	Be	B	C	N	O	F
$E(2p) - E(2s)$	1.85	3.36	5.75	8.77	12.39	16.53	21.54 eV
	178	324	554	846	1 195	1 595	2 078 kJ mol^{-1}

differences because the actual energies depend on other factors of electron occupation of these orbitals which we have not yet discussed.

3.6. Atomic units

The appearance of so many fundamental constants in the orbital energy expression (3.41) makes such expressions rather cumbersome. Further, it is clear that the mathematical manipulation of Schrödinger's equation does not rely on any preknowledge of these quantities; for instance, if a new value were to be produced for the charge of the electron there would be no need to repeat all our calculations of atomic orbitals. For this reason it is the custom in molecular quantum mechanics to use units which simplify the equations. These units are called atomic units (a.u.). They are defined in such a way that the electron mass and charge are unity, as also is $h/2\pi$ and $4\pi\epsilon_0$. Expression (3.41) for the energy of a hydrogen orbital then becomes

$$E_{nlm} = -\frac{1}{2n^2}, \tag{3.44}$$

and the unit of energy is called the hartree (E_{H}). It has the value 27.21 eV, or 2 626 kJ mol^{-1}. The unit of length in this scheme is called the bohr (a_0) and is 0.529 7 Å or 0.529 7 x 10^{-10} m. In terms of fundamental constants it is given in SI units by the expression

$$a_0 = \frac{h^2\epsilon_0}{\pi m e^2}. \tag{3.45}$$

a_0 happens to be equal to the radius of the $1s$ orbit in the Bohr theory of the hydrogen atom, and it is also the average distance of the electron from the nucleus in the hydrogen $1s$ orbital. In terms of the bohr unit the radial wavefunctions of the hydrogen atom (3.28) have a simple form in that the constant k that appears in the exponential takes the value n^{-1}, where n is the principal quantum number.

In atomic units the Schrödinger equation for the hydrogen atom becomes

$$\left(-\frac{1}{2}\nabla^2 - \frac{1}{r}\right)\psi = E\psi, \tag{3.46}$$

where ∇^2 is the standard symbol for the partial derivatives

$$\nabla^2 \equiv \frac{\partial^2}{\partial x^2} + \frac{\partial^2}{\partial y^2} + \frac{\partial^2}{\partial z^2}. \tag{3.47}$$

Because we have set $h/2\pi$ (usually written as $\hbar$ equal to unity, and h has the dimensions energy x time, then the unit of time is not the second but 2.419×10^{-17} s. This is the time taken for an electron in the first Bohr orbit of hydrogen to travel one bohr radius.

Chapter 4
Periodicity of the Elements

4.1. The Periodic Table

The Periodic Table represents one of the outstanding organizational developments of the physical sciences. Amongst the 100 or so elements, there are groups which have very similar chemical and physical properties; the halogens (F, Cl, Br, I) and the alkali metals (Li, Na, K, Rb, Cs) are typical of such groups. There were many attempts in the middle of the 19th century to find some property of the elements which could be the basis for an ordering in which these groups would naturally appear. The solution is generally credited to Mendeleef who in 1869 pointed out that the physical and chemical properties of the elements and their compounds are a periodic function of their atomic weights.

With this classification he was able to postulate the existence of some elements unknown at that time (e.g. germanium) and to predict some of their properties with what proved to be remarkable precision. Apart from the replacement of atomic weight by atomic number (charge on the nucleus) as the basis of the ordering, and the insertion of new elements, the modern periodic table is basically that of Mendeleef. A modern representation of it is given in Table 4.1. The elements increase by one in atomic number with each entry as the table is read from left to right down the page.

Bohr, in 1921, was the first to give a satisfactory explanation of the periodic table in terms of the electronic theory of the atom. He introduced what is now known as the *aufbau* or *building-up* principle.

This states that the electronic structure of an element is built up from that of the preceding element by adding one electron to the lowest energy orbital (orbit in Bohr theory) which can accept an electron. However, to obtain periodicity Bohr found it necessary to postulate that there is a maximum to the number of electrons that could occupy an orbital. The first orbital ($1s$) can take only two electrons (these make up the so-called K-shell of the atom). The next group of orbitals ($2s, 2p$) may take eight electrons, which make up the L-shell, and so on. A justification of this postulate was offered later by Pauli in his famous exclusion principle.

In 1925 Uhlenbeck and Goudsmit proposed that the electron behaves like a spinning particle and has an intrinsic angular momentum and associated magnetic moment. This hypothesis enabled them to explain some of the small splittings that had been observed in atomic spectral lines. They found it necessary to postulate

Table 4.1. A periodic classification of the elements

	The main group elements								The transition metals										The rare earths													
	1a	2a	3b	4b	5b	6b	7b	0	3a	4a	5a	6a	7a	8a			1b	2b														
1s	H							He																								
2s	Li	Be																														
2p			B	C	N	O	F	Ne																								
3s	Na	Mg																														
3p			Al	Si	P	S	Cl	Ar																								
4s	K	Ca																														
3d									Sc	Ti	V	Cr	Mn	Fe	Co	Ni	Cu	Zn														
4p			Ga	Ge	As	Se	Br	Kr																								
5s	Rb	Sr																														
4d									Y	Zr	Nb	Mo	Tc	Ru	Rh	Pd	Ag	Cd														
5p			In	Sn	Sb	Te	I	Xe																								
6s	Cs	Ba																														
4f																			La	Ce	Pr	Nd	Pm	Sm	Eu	Gd	Td	Dy	Ho	Er	Tm	Yb
5d									Lu	Hf	Ta	W	Re	Os	Ir	Pt	Au	Hg														
6p			Tl	Pb	Bi	Po	At	Rn																								
7s	Fr	Ra																	The actinides													
5f																			Ac	Th	Pa	U	Np	Pu	Am	Cm	Bk	Cf	Es	Fm	Md	No

that the quantum number of this spin angular momentum (s) was ½, in contrast with the values $l = 0$, 1, 2 etc. for the quantum number of orbital angular momentum of the electron. We have seen in the last chapter that orbitals with a given l value are $2l + 1$ fold degenerate, each of the different $2l + 1$ values corresponding to a different m value. By analogy, if the electron has $s = ½$ we expect $2s + 1 = 2$ possible components of this spin: that is, m_s either has the value ½ or –½. This was the hypothesis of Uhlenbeck and Goudsmit. It was later realized that experiments performed about three years earlier by Stern and Gerlach supported this deduction. They had passed a beam of silver atoms through an inhomogeneous magnetic field and found that it split into two beams as though the silver atoms had just two possible directions for their magnetic moments relative to the direction of the magnetic field. Because silver atoms have a single electron in an s orbital outside a closed (and therefore spherical) shell of electrons, the behaviour of the silver atoms in a magnetic field is determined by the properties of this electron. The splitting observed by Stern and Gerlach is therefore seen to arise from the two possible m_s values for the electron.

At the time when the phenomenon of electron spin was proposed Pauli was also working on the theory of atomic spectra and in particular on the question of why certain lines which might have been expected were not observed. He explained these absences by an *exclusion principle*, that in an atom, no two electrons could have the same values for the four quantum numbers n, l, m, and m_s. He then realized that this exclusion principle accounted for the periodic electron structures deduced by Bohr. Thus the 1s orbital can take one electron (hydrogen) or two with opposite spins (helium) but the next electron must then go into the 2s orbital (lithium), which can also take a second electron (beryllium). We then start to fill the 2p orbitals, and the electronic structures of the next six elements (B, C, O, N, F, Ne) follow as the three 2s orbitals are filled. Table 4.2 gives as an example the values of the quantum numbers of the ten electrons of neon.

Table 4.2. The values of the four quantum numbers for the ten electrons of neon.

n	l	m	m_s
1	0	0	½
1	0	0	–½
2	0	0	½
2	0	0	–½
2	1	1	½
2	1	1	–½
2	1	0	½
2	1	0	–½
2	1	–1	½
2	1	–1	–½

When we come to the heavier elements in the table we start to occupy d or f orbitals. The point at which these start to be filled is important for an understanding of the position of the transition metals and the rare earth elements. The rule is as follows: orbitals are filled in order, those with the smallest $(n + l)$ values being filled first. When more than one orbital has the same value of $(n + l)$ they are filled in order of their l values, those with largest l values being filled first. This rule leads to the following order for filling orbitals, the values given in brackets being $(n + l/l)$.

> 1*s*(1/0), 2*s*(2/0), 2*p*(3/1), 3*s*(3/0), 3*p*(4/1), 4*s*(4/0),3*d*(5/2), 4*p*(5/1), 5*s*(5/0), 4*d*(6/2), 5*p*(6/1), 6*s*(6/0), 4*f*(7/3), 5*d*(7/2), 6*p*(7/1), 7*s*(7/0), 5*f*(8/3), etc.

In addition, the following general points should be noted:

1. nd orbitals have about the same energy as $(n + 1)s$;
2. nf orbitals have about the same energy as $(n + 1)d$.

Thus, when the 3*s* and 3*p* sub-shells have been filled (Ar) the next electron goes into 4*s* rather than 3*d* and this is the ground state of potassium (K). For this element it can be shown, from spectroscopy, that its 3*d* orbital is about 2.7 eV higher in energy than its 4*s*. After the 4*s* orbital has been filled (Ca) the 3*d* shell is filled in the first transition metal series (Sc – Zn) and following this the 4*p* shell is completed. The separation of the 3*d* and 4*s* levels in the transition metal series is very small and there are some cases where the ground state of the atom does not have a filled 4*s* orbital. For example, Cr has the ground state configuration $4s3d^5$ and not $4s^2 3d^4$; Cu is $4s3d^{10}$ not $4s^2 3d^9$. Completely filled or half-filled *d* sub-shells seem to have an extra stability. However, these subtleties are not important in the chemistry of the transition metals and their ions; it is usually a good starting point to suppose that all the outermost electrons (4*s* and 3*d*) are in the 3*d* orbitals. Thus, V^{2+} is considered as having a $3d^3$ configuration.

The pattern of the 4*s*, 3*d*, and 4*p* electrons is repeated through the 5*s*, 4*d*, and 5*p* sub-shells to give the second transition metal series. Then, after the 6*s* orbital has been filled electrons start to fill the 4*f* orbitals and the rare earth series is built up. Likewise at the end of the periodic table the actinides are constructed by filling the 5*f* orbitals. Beyond uranium (element 92) the elements are only produced by neutron bombardment of the heavy elements and are mostly short lived.

4.2. Ionization potentials, electron affinities, and electronegativities

It is not within the scope of this book to discuss all the periodic properties of the elements which are rationalized in the Periodic Table. There are, however, some that are particularly relevant to chemical bonding and which we shall find useful in later chapters. In this section we discuss the ionization potential, the electron affinity, and the electronegativity. Of these the first two are rigorously defined by experimental quantities but the third is only a useful concept and any value which may be assigned to electronegativity depends on the method of calculation or the experiment by which we choose to define it.

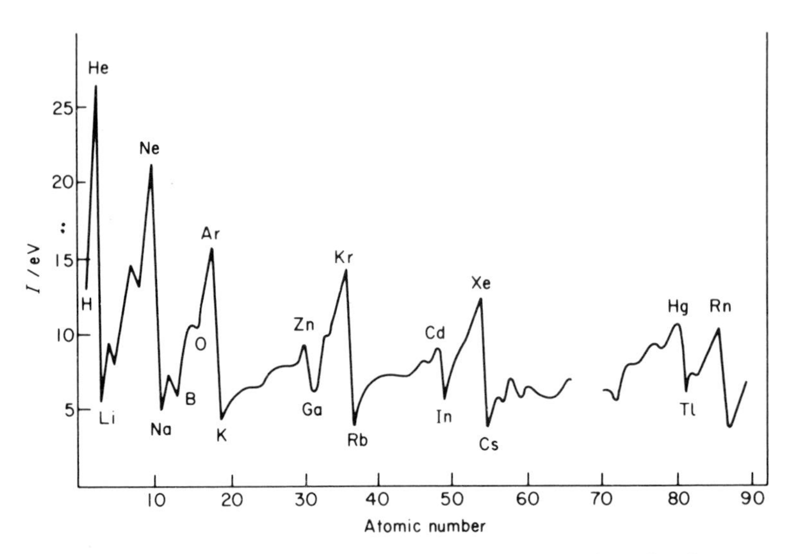

Figure 4.1 Variation of the first ionization potential of the elements with atomic number.

The ionization potential is the energy required to remove an electron from an atom in its lowest energy state and leave both electron and ion with zero kinetic energy. In other words, it is the minimum energy required to remove the electron. Figure 4.1 shows the variation of the ionization potential with atomic number through the Periodic Table. The dominant peaks are at the inert gas atoms and the troughs are at the alkali metals. The lower the ionization potential the more readily will the element form ionic compounds in which it is positively charged.

The reason for the steep rise in ionization potential from elements of group 1 to group 0 (the inert or rare gases) is that we are adding progressively one more nuclear charge which binds the outer electrons in the atom and the nucleus is only partly shielded by the extra electron that is added at each time. The concept of an effective nuclear charge was widely adopted in the early years of quantum chemistry for calculating approximate atomic wavefunctions, and recipes for obtaining effective charges were given by Slater. Table 4.3 gives the effective nuclear charges experienced by the $2s$ and $2p$ electrons of the elements Li – Ne, as

Table 4.3. Effective nuclear charges experienced by valence electrons as given by Slater's rules

Element	Li	Be	B	C	N	O	F	Ne
Effective nuclear charge	1.30	1.95	2.60	3.25	3.90	4.55	5.20	5.85

given by Slater rules. The larger this charge the greater will be the ionization potential.

The ionization potentials of the transition elements vary more slowly with atomic number than those of the main group elements and this is due to the greater shielding of the nuclear potential experienced by one *d* electron by the other *d* electrons.

Figure 4.1 shows small fluctuations within a period; such features usually find a ready explanation in the details of the filling of the orbitals by electrons. For example, the fact that boron has a lower ionization potential than beryllium is because beryllium has a filled 2*s* orbital whereas the most easily ionized electron in boron is that in the 2*p* orbital.

Just as it is possible to remove an electron from an atom so too, is it possible to remove a further electron from the resulting positive ion (except, of course, for H^+).

The removal of more than one electron from an element requires progressively higher energies due to the fact that the electron for the *n*th ionization has to leave an ion with a net charge of $n-1$. As an illustration, the values of the first three ionization potentials of the elements of the first transition series are shown in Table 4.4. It might be thought surprising that these elements commonly occur in compounds in a highly charged state (e.g. Fe^{3+}) in view of the large energy required to form such ions. However, these values refer to the energies to form isolated ions in the gas phase, and in solution there is a large compensating energy term which is the solvation energy of the ion. This solvation energy consists for the most part of the electrostatic interaction between the ion and the dipolar solvent (e.g. water) molecules.

The electron affinity (A) of an atom is the energy gained when an electron with no kinetic energy is added to the isolated atom. The electron affinity of an atom M is identical in magnitude and sign to the ionization potential of the negative ion M^-. Electron affinities are not known with such high precision as ionization potentials

Table 4.4. Ionization potentials of the neutral atoms, positive ions, and doubly positive ions of the transition metals

Atomic number	Element	I/eV	I^+/eV	I^{2+}/eV
21	Sc	6.56	12.89	24.75
22	Ti	6.83	13.57	28.14
23	V	6.74	14.2	29.7
24	Cr	6.76	16.49	31
25	Mn	7.43	15.64	33.69
26	Fe	7.90	16.18	30.64
27	Co	7.86	17.05	33.49
28	Ni	7.63	18.15	36.16
29	Cu	7.72	20.29	36.83
30	Zn	9.39	17.96	39.70

Table 4.5. Electron affinities for some of the elements

Element	Electron affinity/eV
H	0.754 2
C	1.25 ± 0.03
O	1.465 ± 0.005
F	3.448 ± 0.005
S	2.07 ± 0.7
Cl	3.613 ± 0.003
Br	3.363 ± 0.003
I	3.063 ± 0.003

as they are usually not determined directly by spectroscopic techniques but are inferred from electron attachment experiments.

Some atoms do not have positive electron affinities in that the negative ion is not stable. That the inert gases behave in this way is clearly due to the fact that the extra electron would have to go into an orbital outside the filled shell. In some such cases it is possible to infer values for *negative* electron affinities by an empirical extrapolation; values derived in this way are useful in determining the energetics of formation of ionic species. For example, O^{2-} is not a stable species in isolation and so it is not possible to measure an affinity for it, yet compounds such as CaO exist as ionic solids. From the measured heats of formation of such solids and calculated electrostatic energies of the lattice one can infer an electron affinity for the process

$$O^-(g) + e^- \rightarrow O^{2-}(g)$$

of about −9 eV. Notice the negative sign: electron affinities are defined as positive quantities if the negative ion is stable towards loss of an electron .

Table 4.5 shows the electron affinities of some of the more common elements.†

A concept of considerable importance for a qualitative understanding of the chemical bond is the *electronegativity*, χ, of an atom. This is a measure of the tendency of the atom to form ionic compounds, or, to be more precise, the ionic character of an A–B bond will depend on the difference in the electronegativities of the two atoms.

Although the concept is qualitatively valuable the attempts to derive a comprehensive quantitative scale of electronegativity have been disappointing because of the lack of correlation between experimental quantities and the scale over a wide front. The most popular scale at present in use, and the one whose derivation is most straightforward, is that suggested by Mulliken. This is

$$\chi = \tfrac{1}{2}(I + A). \qquad (4.1)$$

It is argued that the tendency of an atom to acquire electrons is a balance of the

† L. G. Christophorou, *Atomic and Molecular Radiation Physics,* New York, 1976.

Table 4.6. Mulliken electronegativities (χ/eV) for some elements. Values for the elements not listed in Table 4.5 are uncertain due to the lack of accurate values for their electron affinities

H	Li	Be	B	C	N	O	F
7.17	3.0	4.5	4.3	6.26	7.4	7.54	10.43
	Na	Mg	Al	Si	P	S	Cl
	2.8	4.1	3.5	4.8	5.9	6.21	8.31

readiness with which it gains electrons (measured by A) and the difficulty with which it loses electrons (measured by I). Table 4.6 gives the Mulliken electronegativities for some elements.

It is argued that the ionic character of the AB bond depends on the difference between the electronegativities of the two atoms. Qualitatively this is true, but there is a lack of quantitative correlation even, with properties such as molecular dipole moments which one might expect to show this correlation clearly. Figure 4.2 contains data for the dipole moments of the hydrogen halides and shows that there is some basis for the Mulliken scale because of the correlation

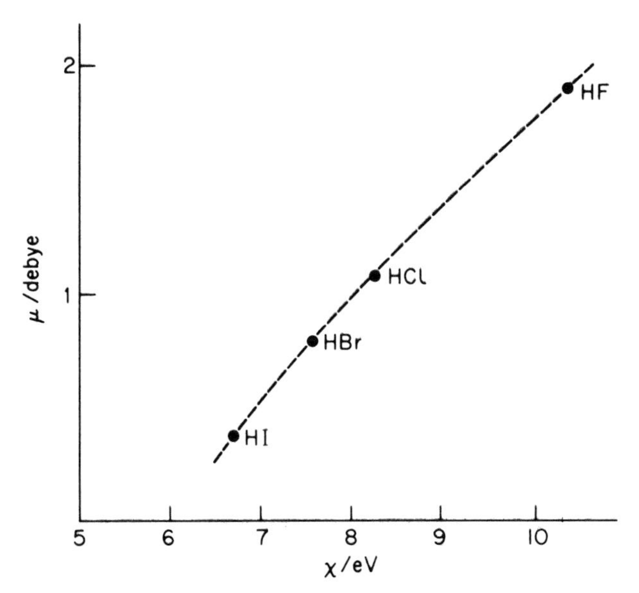

Figure 4.2 Correlation between dipole moment and electronegativity (Mulliken) of the halogen atom for the hydrogen halides. Dipole moments are given in debye units 1D = 10^{-18} e.s.u. cm = 3.334×10^{-30} Cm.

of these dipole moments with it. Such relationships are not, however, widespread and, indeed, this is not too surprising because it is clear that the electronegativity of an atom should depend on the type of bond involved. For example, acetylene is a much stronger acid than methane so we must deduce that the carbon atom in acetylene is more electronegative than the carbon atom in methane.

4.3. Atomic radii

The concept of an atomic radius is not well defined. Its value depends either on the method used to calculate it from atomic wavefunctions or on the experimental quantity from which it is deduced. However, it is a qualitatively useful concept as it gives a measure of how close atoms can normally approach in a bonding or non-bonding situation.

Three different radii have been found useful. Firstly, the so-called Van der Waals' radius, which measures the effective radius of an atom in a non-bonding situation. In other words, if two atoms in a molecule which are not bonded in a traditional sense are closer than the sum of their Van der Waals' radii then we

Table 4.7. Atomic radii (Å) according to different definitions

	Radius of[a] Maximum density	Van der[b] Waals' radius (Pauling)	Covalent[c] radius (single bond)	Ionic[b] Radius
H	0.53	1.2	0.28	1.54 (H^-)
He	0.30			
Li	1.50			0.68 (Li^+)
Be	1.19			0.35 (Be^{2+})
B	0.85			0.23 (B^{3+})
C	0.66		0.77	
N	0.53	1.5	0.70	
O	0.45	1.40	0.66	1.32 (O^{2-})
F	0.38	1.35	0.64	1.33 (F^-)
Ne	0.32			
Na	1.55			0.97 (Na^+)
Mg	1.32			0.66 (Mg^{2+})
Al	1.21			0.51 (Al^{3+})
Si	1.06		1.17	
P	0.92	1.9	1.10	
S	0.82	1.85	1.04	1.84 (S^{2-})
Cl	0.75	1.80	0.99	1.81 (Cl^-)
Ar	0.67			

[a] J. C. Slater, *Quantum Theory of Atomic Structure*, vol *1*, McGraw Hill, New York, 1960.
[b] *Handbook of Chemistry and Physics*, Chemical Rubber Co., 56th Edition.
[c] F. A. Cotton and G. Wilkinson, *Advanced Inorganic Chemistry*, Interscience, Cleveland, Ohio, 1975; New York, 1962.

would expect steric repulsion to force them apart until they reach this distance. The other two, covalent and ionic radii, measure the effective radii for covalent or ionic bonding situations. The fact that these two differ means that the predictive value of these radii for partly covalent–partly ionic compounds will always be limited.

Table 4.7 lists values which have been deduced for these atomic radii together with a calculated quantity which is the radius of the maximum electron density of the most diffuse SCF atomic orbital: this is the most probable distance from the nucleus at which to find the electron. The best correlation amongst these values is between the radius of maximum density and the Van der Waals' radius. For both the covalent and ionic radii there is the problem of finding a satisfactory recipe for taking observed internuclear distances and dividing them into two atomic radii, and different scales can be deduced from the same bond length. However, the concept of atomic size is not sufficiently rigorous for one to worry too much about the method for doing this. Thus, it has recently become clear that the values of ionic radii which are widely quoted (and given in Table 4.7) are inconsistent with values measured by X-ray studies on ionic crystals.† Such measurements also reveal that, contrary to the general assumption, the radius of an ion is not a constant quantity.

† D. F. C. Morris, *Structure and Bonding,* **4**, 63 (19).

Chapter 5

Molecular Electronic Wavefunctions

5.1. Separation of electron and nuclear motion

The simplest molecular species H_2^+ has one electron and two protons. A wavefunction which would define the position of each of these would be a function of nine variables, three position coordinates for each particle. However, in the absence of external fields it is sufficient to define the positions of the particles relative to each other and this can be done by choosing as origin for the coordinates the centre of mass of the system. Thus the wave function to describe the relative positions will be a function of only six variables. Nevertheless, this is still a large number and we are discussing only the simplest molecule.

It becomes increasingly difficult to obtain accurate solutions of the Schrödinger equation as the number of variables increases and any method which allows us to reduce the number that have to be considered at any one time is of considerable value. Luckily, for molecules there is an approximation which gives excellent results in almost all situations and this is to consider separately the motion of the electrons and that of the nuclei. The reason this can be done is that the nuclei are much heavier than the electrons ($m_H/m_e = 1836$), and to a good approximation the electrons immediately adjust their positions to follow the nuclei as they move. In other words, the wave function of the electrons depends on the positions of the nuclei, but not on the momenta of the nuclei.

This separation of the electron and nuclear motions was first formulated in a precise manner by Born and Oppenheimer in 1927. Although the mathematical details of the approximation are not of interest to most chemists, the results have considerable importance because it is only within this approximation that one can introduce the concept of a molecular potential energy surface and potential energy surfaces are central to our understanding of molecular structure, spectroscopy and chemical kinetics. The vast evidence of molecular spectroscopy shows that it is a good approximation to talk about rotational, vibrational and electronic energies as separate entities and this shows that the Born–Oppenheimer approximation is a good starting point for molecular wavefunctions. The importance of the Born–Oppenheimer approximation warrants at least a cursory glance at its mathematical foundations.

The total Hamiltonian for a molecule is a sum of kinetic energy (T) and potential energy (V) terms which we can write as follows:

$$\mathscr{H} = T_n + T_e + V_{ee} + V_{en} + V_{nn} \tag{5.1}$$

where e and n are suffixes to distinguish electrons and nuclei. We are here ignoring some small terms which depend on the spin of the electrons and nuclei. The kinetic energy operators are differential operators as defined in equation (2.21), and the potential energy terms are the same as in classical mechanics; for example, the repulsion between all pairs of electrons, V_{ee}, is, in atomic units,

$$V_{ee} = \sum_{i,j>i} (r_{ij})^{-1}, \tag{5.2}$$

where r_{ij} is the distance between the electrons i and j.

If the term representing the kinetic energy of the nuclei is removed from expression (5.1) the remainder is a Hamiltonian for stationary nucei which is called the electronic Hamiltonian, $\mathscr{H}_e$

$$\mathscr{H}_e = \mathscr{H} - T_n. \tag{5.3}$$

$\mathscr{H}_e$ is a function that depends on both the positions of the electrons and nuclei, because V_{en} depends on both, but for any specified configuration of the nuclei, $\mathscr{H}_e$ will contain as variables only the electronic coordinates. The solutions of the Schrödinger equation

$$\mathscr{H}_e \psi_i^e = E_i^e \psi_i^e, \tag{5.4}$$

define the electronic wavefunctions ψ_i^e, and the electronic energies E_i^e which are characteristic of the particular nuclear configuration being considered.

If equation (5.4) is solved for H_2^+ we obtain an electronic wavefunction for the electron, which is a function of the position vector of the electron, **r** (defined in Figure 5.2), for each internuclear separation (R). The lowest energy solution of (5.4) in this case leads to an energy $E_0^e(R)$ of the form shown in Figure 5.1, and this is a typical potential energy curve for a stable diatomic species. Contours of the wavefunctions $\psi_0^e(\mathbf{r}, R)$ at several values of R are shown in Figure 5.2: this wavefunction is cylindrically symmetric about the internuclear axis.

The energy E_i^e which appears in equation (5.4) is called the potential energy because it is the potential energy that governs the motion of the nuclei. In other words, the Hamiltonian for the nuclear motion is

$$\mathscr{H}_n = T_n + E_i^e, \tag{5.5}$$

and this is a function only of the nuclear coordinates. We have labelled the electronic potential energy E_i^e by a suffix i to emphasize that each electronic state ψ_i^e, which is a solution of equation (5.4), has a corresponding potential energy curve (for a diatomic molecule) or multi-dimensional surface (for a polyatomic) E_i^e. The solutions of the Schrödinger equation

$$(T_n + E_i^e)\,\psi^{n(i)} = E^{(i)}\psi^{n(i)}, \tag{5.6}$$

define the nuclear wavefunctions $\psi^{n(i)}$ and the total energy $E^{(i)}$. The methods of solving equation (5.6) and a discussion of the nuclear wavefunctions and energies, both of which can be separated into vibrational and rotational parts, can be found in books on molecular spectroscopy.

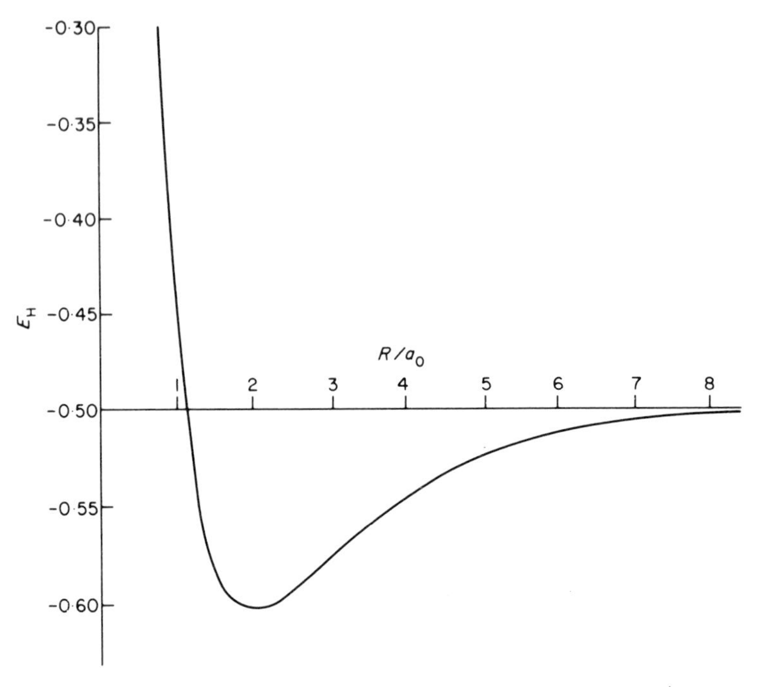

Figure 5.1 Electronic energy of the ground state of H_2^+ as a function of the internuclear distance.

The total wavefunction of the molecule is, in the Born–Oppenheimer approximation, a product of the electronic and nuclear parts. For example, the kth nuclear state of the ith electronic state has an energy $E_k^{(i)}$ and a wavefunction

$$\Psi_{ki} = \psi_i^e \, \psi_k^{n(i)}. \tag{5.7}$$

Within the approximation that the nuclear kinetic energy operator T_n has no effect on the electronic wavefunction ψ_i^e, this is a solution of the full Schrödinger equation for the molecule

$$\mathscr{H}\Psi_{ki} = E_k^{(i)} \Psi_{ki}. \tag{5.8}$$

As T_n is a differential operator this amounts to the assumption that ψ_i^e is a slowly varying function of the nuclear coordinates. Apart from special situations which are of no concern in this book, this is a good assumption.

Calculations have been made to find the errors introduced by the Born–Oppenheimer approximation. For H_2, which, having the lightest nuclei may be considered one of the least favourable cases, it has been calculated† that the error in the calculated dissociation energy is less than 100 J mol^{-1} which is about 0.03%

†W. Kolos and L. Wolniewicz, *J. Chem. Phys.*, **41**, 3663, 3674 (1964).

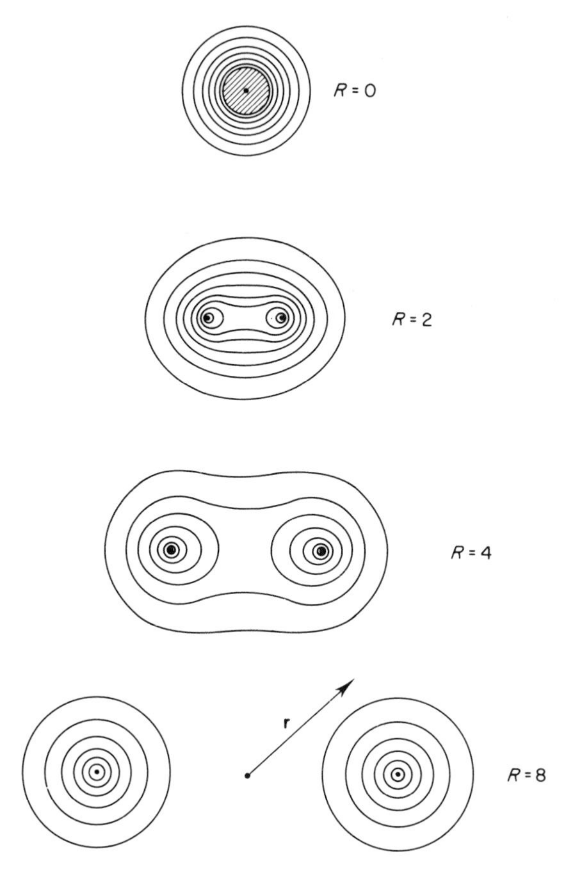

Figure 5.2 Contours of the ground state wavefunction of H_2^+, $\psi_0(\mathbf{r})$, for several values of the internuclear distance R/a_0. The outer contour is 0.05 and contours increase by steps of 0.05 towards the nucleus. For $R = 0$ the contours are too close together in the inner region to be shown clearly.

of the total dissociation energy. For other molecules such as error would be much smaller than the errors we make in solving the electronic Schrödinger equation and can be ignored.

The potential energy curves of most stable diatomic molecules in their ground states are well-known from spectroscopy and the curves for many excited states are also known. For polyatomic molecules our knowledge is much less and it must be realized that even for triatomic molecules the potential energy surface is a function of the three variables that are needed to specify the relative positions of the nuclei and hence a complete diagrammatic representation of such surfaces is difficult. Figure 5.3 shows contours of the potential energy surface that has been calculated

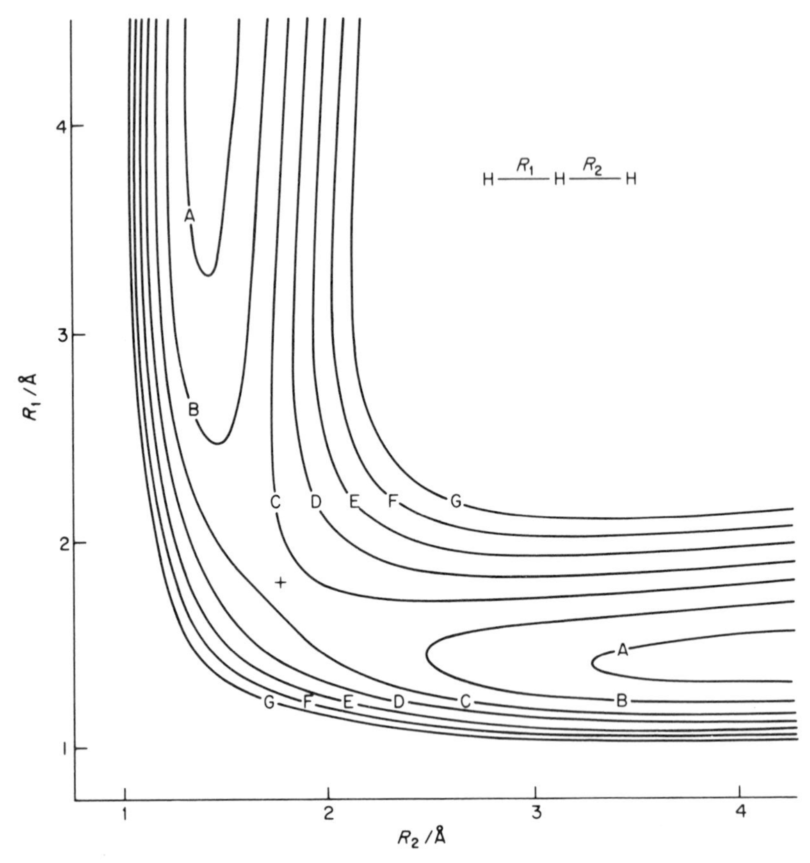

Figure 5.3 Contours of the potential energy surface for collinear configurations of H_3. Contour A has energy 0.1eV (9.65kJ mol^{-1}) greater than the energy of $H + H_2(r_e)$ and successive contours are at 0.2eV intervals. Contours above 1.3eV are not shown.

for the system H_3 with all nuclei on a line and which is important for an understanding of the elementary reaction

$$H + D_2 \rightarrow HD + D.$$

It is worth emphasizing that as the nuclear masses only enter the Hamiltonian through the nuclear kinetic energy operator, the electronic Hamiltonian is independent of nuclear mass and hence H_3 and HD_2, for example, have the same potential energy surface.

5.2. Electronic wavefunctions and observables

When we make a quantitative measurement in the physical sciences we record facts, like the trace of a pen recorder or the clicks of a geiger counter; these are our observations. The term 'observable' in quantum mechanics goes further than referring to that which we can actually see and encompasses any quantity that can in principle be measured. There is an important distinction between a wavefunction whose form depends on the choice of coordinates (variables) in which the system is described and which is not an observable and, for example, a dipole moment which has a value that should be independent of any technique for calculating it or measuring it. The dipole moment is therefore referred to as an observable although what is actually observed in an experiment to determine it may be a meter reading of a capacitance.

In Section 2.4 we described the relationship between the electronic wavefunction and the probability distribution in space for the electrons. There is a more general relationship between the wavefunction and any observable, which we must now examine. This relationship may be considered as one of the postulates of quantum mechanics; in other words its truth lies not in the fact that it may be proved from more fundamental postulates or axioms but that the results we obtain from it are found to agree with experimental measurements.

Every observable is characterized in quantum mechanics by an operator (**B** say)†. The average value of this observable ($\bar{b}$) in a state having a normalized wavefunction Ψ is given by the integral

$$\bar{b} = \int \Psi^* \, \mathbf{B} \, \Psi \, dv, \tag{5.9}$$

where the integration is taken over the complete space of all the variables on which Ψ depends.

There are simple rules for determining the appropriate operator for any observable. In classical mechanisms the observable can be written as a function of the position (x, y, z) and momentum (p_x, p_y, p_z) variables of the system. The corresponding quantum mechanical operator is obtained from such functions by replacing the momentum variables by the partial differential operators

$$\left(-i\hbar\frac{\partial}{\partial x}, -i\hbar\frac{\partial}{\partial y}, -i\hbar\frac{\partial}{\partial z}\right).$$

For systems in the presence of electric or magnetic fields there are other rules to be followed but we shall have no cause to use them in this book.

As an example of the use of expression (5.9) we take the average position along the coordinate x and the average momentum in this direction for an electron in an orbital whose wavefunction is ψ. The relevant expressions are

$$\bar{x} = \int \psi^* x \psi dv, \tag{5.10}$$

and

$$\bar{p}_x = \int \psi^* \left(-i\hbar\frac{\partial}{\partial x}\right) \psi dv. \tag{5.11}$$

†Symbolic operators (as opposed to explicit expressions for the operators) are generally written in bold type as here, and average values are denoted by a superior bar, as in $\bar{b}$.

In the first of these expressions x is simply a multiplying factor in the integrand and its position is not important. We can therefore write

$$\bar{x} = \int \psi^* \psi x \, \mathrm{d}v = \int Px \, \mathrm{d}v, \tag{5.12}$$

where P is the probability density for the wavefunction.

$$P = \psi^* \psi. \tag{5.13}$$

A similar relationship between $\bar{p}_x$ and the derivative of P cannot generally be established because the position of the operator in (5.11) indicates that only the wavefunction to the right of the integrand is differentiated.† This also emphasizes the fact that a knowledge of the probability density alone is not sufficient to enable us to calculate all observables. The wavefunction implicitly contains more information than the probability density.

The rules given above enable us to calculate the energy of a system from a knowledge of its wavefunction. The quantum mechanical operator for the energy, which is an observable, is the Hamiltonian. We noted in Chapter 2 that the replacement of momentum by a differential operator (2.23) allowed us to go from the classical Hamiltonian, which is the energy expressed as a function of coordinates and momenta, to the quantum mechanical Hamiltonian. The average, or expectation value of the energy is therefore, according to (5.9), given by

$$\bar{E} = \int \Psi^* \mathscr{H} \Psi \, \mathrm{d}v. \tag{5.14}$$

The importance of expression (5.14) lies in the fact that it provides a method of calculating an *approximate* energy from a wavefunction which is an *approximate* solution of the Schrödinger equation. For most systems of chemical interest approximate solutions are all that we can obtain.

If we write the Schrödinger equation (2.27) in the form

$$\frac{\mathscr{H}\Psi}{\Psi} = E, \tag{5.15}$$

then for an exact solution, the function which is on the left-hand side of this equation should be a constant. That is, if we evaluate $\mathscr{H}\Psi/\Psi$ at several points in space we should always obtain the same value. However, if Ψ is only an approximate solution to the Schrödinger equation then $\mathscr{H}\Psi/\Psi$ is not a constant but differs from one point to another. The question then arises as to how we should calculate the energy in this case. There have been some attempts to define the energy as the average of $\mathscr{H}\Psi/\Psi$ over all points in space, but the difficulty with this approach is that Ψ may be closer to the exact wavefunction at some points than at

†Consideration of the case $\psi = \exp(\mathrm{i}kx)$ will serve to illustrate the difference between

$$\int \psi^* \left(-\mathrm{i}\hbar \frac{\partial}{\partial x} \right) \psi \, \mathrm{d}v \quad \text{and} \quad -\mathrm{i}\hbar \int \frac{\mathrm{d}}{\mathrm{d}x} P \, \mathrm{d}v.$$

others. The implication is that we should weight this average more heavily at some points than at others, but no generally satisfactory procedure has been found for doing this.

In contrast to the above difficulty expression (5.14) has proved to be a reliable route to approximate energies as we shall see in this book. This leaves open the question of how we calculate approximate wavefunctions to insert in (5.14) and that topic will be considered in the next chapter.

It should be noted that expression (5.14) is consistent with the Schrödinger equation if Ψ is an exact solution. This follows by multiplying the Schrödinger equation on both sides by Ψ^* and integrating over the space of all the variables as follows:

$$\mathscr{H}\Psi = E\Psi, \tag{5.16}$$

therefore

$$\Psi^*\mathscr{H}\Psi = \Psi^*E\Psi, \tag{5.17}$$

therefore

$$\int \Psi^*\mathscr{H}\Psi \, dv = \int \Psi^*E\Psi \, dv. \tag{5.18}$$

Because for an exact solution of (5.16) E is a constant it may be taken outside the integral in (5.18). We then have

$$\int \Psi^*\mathscr{H}\Psi \, dv = E\int \Psi^*\Psi \, dv = E, \tag{5.19}$$

provided that Ψ is a normalized wavefunction.

For an n-electron wavefunction the probability density $\Psi^* \Psi$ is a function of 3n dimensions and it will also be a property of the spins of the electrons. Such complicated functions are of no general interest. It is necessary to convert them into one-electron densities, which are functions in 3-dimensional space and these can be compared with electron densities determined by X-ray scattering techniques. The procedure for this is simple if the electronic wavefunction is based upon the orbital approximation as described in Section 3.4, as the total electron density is then obtained by summing the contributions from each orbital.

Because an orbital is a one-electron function, each orbital ψ_i gives directly a one-electron density according to the expression

$$P_i(\mathbf{r}) = \psi_i^*\psi_i(\mathbf{r}), \tag{5.20}$$

and the total density is

$$P(\mathbf{r}) = \sum_i n_i P_i(\mathbf{r}), \tag{5.21}$$

where n_i is the number of electrons (0, 1, or 2) in orbital ψ_i, the maximum of two following from the Pauli exclusion principle. Figure 5.4 shows the radial electron density $r^2 R^2(r)$ calculated from Hartree–Fock orbitals of K^+ as a function of the radial coordinate r (c.f. Figure 3.4). The three peaks arise from the 1s electrons, the

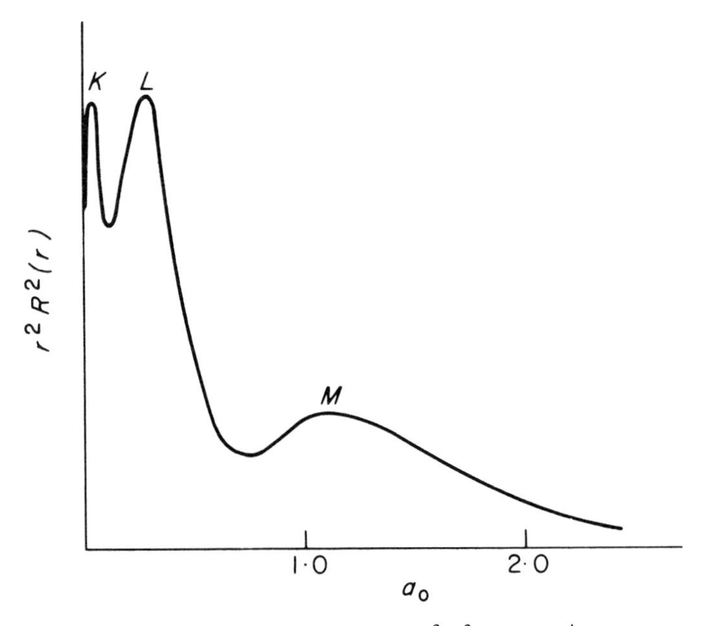

Figure 5.4 Radial electron density $r^2R^2(r)$ for K^+ calculated by the Hartree-Fock method.

2*s* and 2*p* electrons, and the 3*s* and 3*p* electrons: these are labelled *K*, *L*, and *M* shells respectively. For all but the heaviest atoms electron densities have been calculated which are accurate enough for the interpretation of most experimental data. In particular, from these densities the X-ray scattering amplitudes of atoms have been calculated and are used in the analysis of molecular structure by X-ray diffraction.

For molecules the X-ray diffraction experiment can be used to deduce electron densities and these in turn show how electrons change their positions when atoms form bonds. The resolution of the X-ray experiment and the assumptions inherent in the analysis lead to some uncertainty in the resulting electron density maps, but as can be seen from the example of Figure 5.5 there are plenty of data on which molecular electronic wavefunctions can be tested. This example, 1,2,3-tricyanocyclopropane,† shows that in small carbon ring compounds, which chemists generally associate with bond strain, there is a build-up of electron density on the outside of the triangle defined by the carbon nuclei rather along the line of nuclei which is typical of an unstrained bond.

One quantity of considerable interest to chemists that can be calculated directly from the electron density is the electric dipole moment. The dipole moment of an electron probability density $P(\mathbf{r})$ is the vector quantity

$$\mu_e = -e \int P(\mathbf{r})\mathbf{r}\, dv. \qquad (5.22)$$

†A. Hartman and F. L. Hirschfeld, *Acta Cryst.*, **20**, 80 (1966).

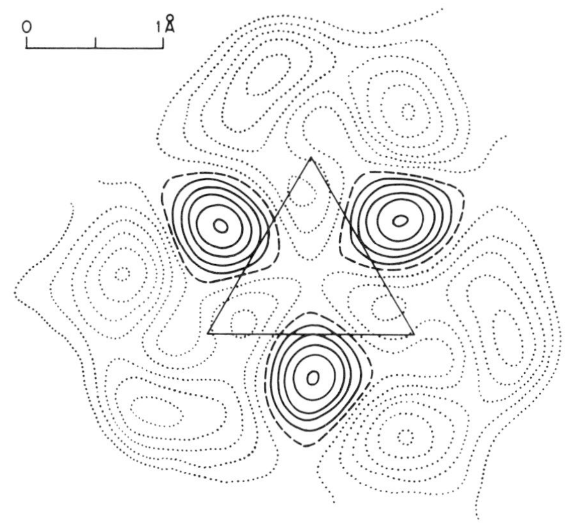

Figure 5.5 The difference in electron density between 1,2,3-tricyanocyclopropane and the summed electron densities of its component atoms. The difference is shown as contours in the plane of the ring. Positive differences are full lines, negative are dotted lines. The zero contour is shown as a broken line and the interval is $0.01e$ Å^{-3}.

The electron charge $-e$ is inserted to make the dimensions correct, but in atomic units e is unity. The negative sign is adopted to accord with the usual sign definition of a molecular dipole moment. The nuclei also contribute to the net moment with a contribution

$$\mu_{\mathrm{n}} = e \sum_k Z_k \mathbf{r}_k. \tag{5.23}$$

Table 5.1. A comparison of calculated and experimental dipole moments for some diatomic molecules [S. Green, *Adv. Chem. Phys.*, **25**, 179 (1974)]. In each case a positive dipole for AB implies A^+B^-. The experimental value for ClF is a recent measurement [B. Fabricant and J. S. Muenter, *J. Chem. Phys.*, **66**, 5274 (1977)]

Molecule	Orbital model	More exact calculation	Experimental value/D
LiH	6.001	5.814	5.828
HF	1.934	1.816	1.826
CO	0.280	−0.121	−0.112
CS	−1.56	−2.03	−1.97
ClF	1.096	0.839	0.85

The total dipole moment is then $\boldsymbol{\mu}_e + \boldsymbol{\mu}_n$ and for neutral molecules this is independent of the origin to which the vectors **r** are defined. Table 5.1 lists the dipole moments that have been calculated for some diatomic molecules. Calculations are shown within the SCF model (see Section 3.4 and 5.3), and for more exact wavefunctions. In most cases there is little difference and the agreement with experiment is good. An exception is the result for CO which shows that the orbital model gives the wrong sign (the sign can be determined by measuring the magnetic moments of rotational states in isotopically different molecules).

We shall look at other molecular properties that can be calculated by quantum mechanics later in this book, but before that we must take a further look at the concept of an orbital in a molecular context.

5.3. Molecular orbitals of diatomic molecules

In Section 3.4 we discussed the concept of an orbital for a many-electron atom and described the self-consistent-field method for calculating orbital wavefunctions. This concept can be extended to molecules without difficulty. A molecular orbital is the wavefunction of an electron in a molecule moving under the influence of the nuclear attraction and the average repulsion of all other electrons.

Molecular orbitals are in general more complicated mathematical functions than atomic orbitals because there is more than one nucleus contributing to the attractive part of the potential. In addition, it is not customary to make any spherical average of the electron density in defining the repulsive part of the potential as is the case with atoms. That is, the effective potential for an atomic orbital is spherically symmetric, that for a molecule is not.

In the next chapter we will examine a method which has general applicability for calculating the wavefunctions and energies of molecular orbitals. At this point, however, we give a brief discussion of the solution of the Schrödinger equation for the simple molecular system H_2^+. This one-electron-molecule plays a similar role in the field of molecular wavefunctions to that which the hydrogen atom plays in the field of atomic wavefunctions. The electronic Schrödinger equation can be solved exactly for H_2^+ and the wavefunctions have certain characteristic features which are general to the orbitals of other diatomic molecules and in some important respect to the orbitals of polyatomic molecules.

The potential of H_2^+ has axial symmetry, this is, it is independent of the angle φ which defines rotation about the internuclear axis. The molecular orbitals of H_2^+ can therefore be written as a product of a function of φ and a function of the two other variables r and ϑ as follows

$$\psi(r,\vartheta,\varphi) = \Omega(r,\vartheta)\,\Phi(\varphi), \tag{5.24}$$

the coordinates being defined in Figure 5.6. It is possible to make a further separation of the function Ω with a suitable choice of coordinates but that does not concern us here.

The angular part of the wavefunction Φ is the solution of a differential equation which is not dependent on the nuclear charges and hence these functions are the

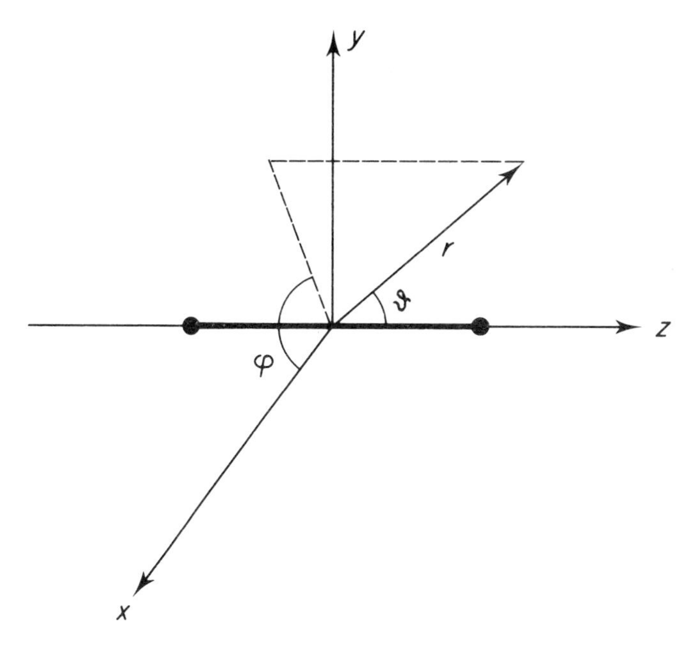

Figure 5.6 Coordinates used to construct the wavefunctions of H_2^+ in equation (5.24).

same for the orbitals of all diatomic molecules (there is an analogy with the angular functions (3.12) which are the same for all atomic orbitals). Their precise form is

$$\Phi(\varphi) = \begin{cases} \cos \lambda\varphi \\ \sin \lambda\varphi \end{cases}, \tag{5.25}$$

where λ is zero or an integer. For $\lambda = 0$ only one function (the cosine) exists and has a constant value, but for integer values of λ the two solutions (5.25) correspond to orbitals which are doubly degenerate, that is, their energies are equal.

The constant λ is analogous to the m quantum number of atomic orbitals and defines the angular momentum of electrons about the internuclear axis. We represent molecular orbitals by symbols which are the Greek equivalents of the spectroscopic symbols (s, p, d, etc.) used for atomic orbitals.

$$\begin{aligned} \lambda \quad &= 0, 1, 2, \ldots \\ \text{symbol} &= \sigma, \pi, \delta, \ldots . \end{aligned} \tag{5.26}$$

Figure 5.7 shows polar diagrams of σ, π, and δ orbitals for a diatomic molecule and we can see the close resemblance to some of the atomic orbital polar diagrams of Figure 3.5. This is an important resemblance which we make use of in the next chapter.

Figure 5.2 shows the contours of the function $\Omega\ (r, \vartheta)$ for the lowest energy state of H_2^+, which we have already looked at for other reasons. This function has

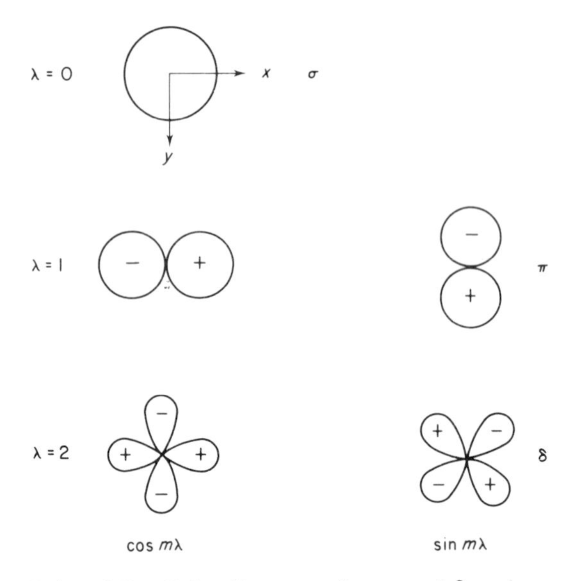

Figure 5.7 Polar diagrams of σ, π, and δ molecular orbitals of a diatomic molecule. The nuclei lie along the z axis (perpendicular to the plane of the paper).

cylindrical symmetry about the internuclear axis and it is therefore designated as a σ orbital. The orbital is shown for four different values of the internuclear distance R. At large distances the contours are spherically symmetric about each nucleus and this clearly corresponds to the dissociation into two atoms (with equal chance of finding the electron about each nucleus). At very short distances the orbitals approach spherical symmetry about the centre of the two nuclei, and for the limit in which the two nuclei coalesce (the so-called united–atom limit) we have the 1s orbital of He^+. In the intermediate region the contours are ellipsoidal.

The equilibrium internuclear distance of H_2^+ is $2\,a_0$, and Figure 5.8 shows contours of some of the lowest energy states at this distance. The orbitals have been given a second label g or u which, like σ, π, etc., defines its general characteristics. Orbitals labelled g are symmetric to inversion through the centre of the molecule. This means that in Cartesian coordinates with the molecular centre as orgin, the wavefunction at the point (x, y, z) is the same as at the point $(-x, -y, -z)$. The orbitals labelled u are antisymmetric, that is, the wavefunctions have different signs at these points. We shall have more to say about such labels when we discuss symmetry in Chapter 7. Finally, we add an integer which represents the order of increasing energy. This integer has some resemblance to the principal quantum number of atomic orbitals.

The potential energy curve for the ground state of H_2^+ is shown in Figure 5.1. It is interesting to compare this with the classical electrostatic energy of a proton and a hydrogen atom. If the proton is entirely outside the electron cloud of the

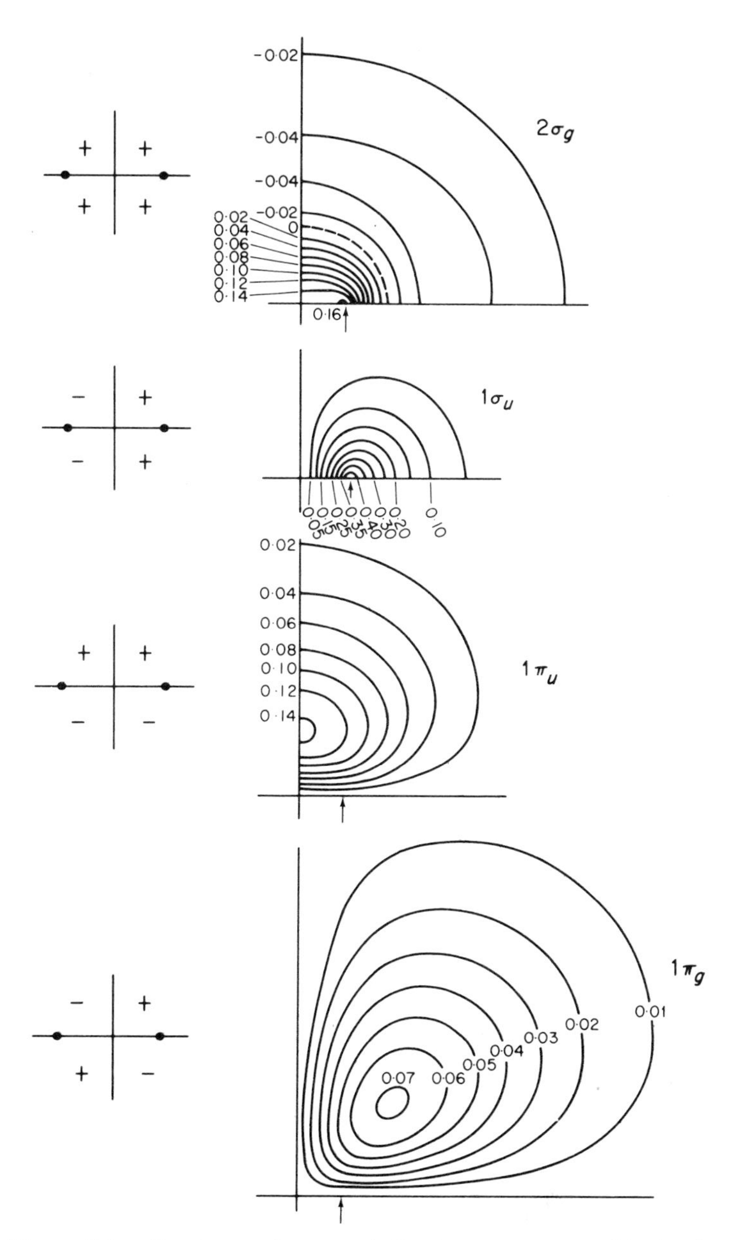

Figure 5.8 Contours of some of the lowest energy orbitals of H_2^+. Contours are shown for only one quadrant in the *xz* plane, the relative signs for all the quadrants being indicated to the left. The position of one nucleus is marked by an arrow.

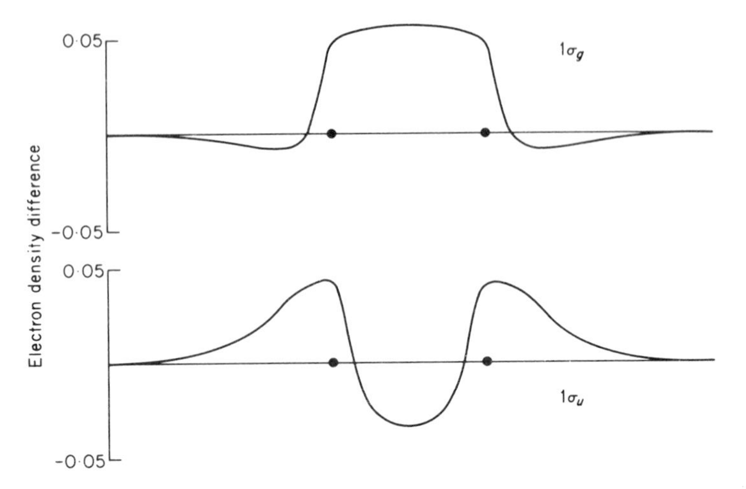

Figure 5.9 The difference between the electron densities of the H_2^+ $1\sigma_g$ and $1\sigma_u$ states and that appropriate to one electron shared equally between two hydrogen 1s orbitals. These differences are plotted along the line of the nuclei. The internuclear distance is $2a_0$.

hydrogen, the electrostatic energy is zero but as it penetrates the cloud there is a repulsion due to the fact that the nuclear–nuclear repulsion increases more rapidly than the nuclear–electron attraction. Thus we can confirm that the binding energy must come from the reorganization of the electrons as the two nuclei come together.

Figure 5.9 shows the difference between the electron density of the ground state of H_2^+ and the density appropriate to an electron shared equally between two hydrogen 1*s* orbitals. This illustrates the change in electron density arising from the formation of the chemical bond. It is seen that the electron has moved into the region between the two nuclei. This is the region where the electron experiences the greatest attractive potential from the nuclei. This decrease in the potential energy of the electrons is partly offset by an increase in the kinetic energy of the electrons, but overall it is the potential energy that wins and the total energy is therefore lowered. There is in fact a general theorem of both classical and quantum mechanics called the *Virial Theorem* from which it can be shown that for particles interacting by Coulomb forces, at equilibrium the total energy is one half the potential energy. Thus if the potential energy is more negative on bond formation, the total energy must be more negative also.

The lowest energy molecular orbital of H_2^+ is called a *bonding* orbital for obvious reasons. If we examine the next higher energy state, $1\sigma_u$, we find that its energy is *greater* than that of the separate species H and H^+ and we refer to this orbital as *antibonding*. These are terms which we shall use widely in the rest of the book. Figure 5.9 also shows the electron density difference for this antibonding orbital and we see that in this state electrons have moved out of the internuclear region and the origin of the repulsion is obvious.

The molecular orbitals of all diatomic molecules have the general form given by expression (5.24). Moreover, this expression is also applicable to linear polyatomic molecules such as carbon dioxide or acetylene because, within the SCF approximation, the effective potential acting on an electron in such molecules is independent of the angle φ. For diatomic molecules in which both nuclei are the same (homonuclear) the orbitals can also be labelled by the g or u suffixes.

In Chapter 4 we showed how the electronic structure of the ground states of atoms could be understood on the basis of Bohr's building up procedure. In this electrons are assigned successively to the lowest available orbital subject to the requirement of the Pauli exclusion principle that an orbital shall contain a maximum of two electrons. The same procedure can be followed for molecules but to do this we have to know the order on an energy scale of the molecular orbitals. This order can be established by experiment or by calculation. The method of calculation will be considered in the next chapter but it is convenient at this point if we describe the experimental technique because it brings out features of molecular wavefunctions which are more general than any model we may adopt for their calculation.

5.4. Photoelectron spectroscopy and Koopmans' theorem

We have already seen in Section 4.2 that the ionization potential of an atom is an important quantity because it measures the tendency of the element to show metallic character. For molecules it is equally important as it measures the ease with which a molecule can be oxidized, the essential step in oxidation being removal of an electron from the molecule. However, atomic or molecular ions like their neutral parents have many possible electronic energy levels and the process of removing an electron from an atom or molecule may leave the ion in its lowest (ground) state or in an excited state. Although the term ionization potential is generally used for the energy required to remove an electron from an atom or molecule in its ground state leaving the ion in its ground state, we can more generally define a kth ionization potential for a situation in which the ion is left in its kth electronic state.

There is a likely source of confusion between the above definition and one which is often used for atoms. The latter refers to the least energy required to remove the kth electron from the atom, $(k-1)$ electrons having already been removed. These two definitions refer to quite different processes.

The first and second ionization potentials of Ne according to the first definition given above are represented by the following schemes:

$$\mathrm{Ne}(1s^2 2s^2 2p^6) \rightarrow \mathrm{Ne}^+(1s^2 2s^2 2p^5) + \mathrm{e},\ I_1 = 21.6\ \mathrm{eV},$$

$$\mathrm{Ne}(1s^2 2s^2 2p^6) \rightarrow \mathrm{Ne}^+(1s^2 2s^1 2p^6) + \mathrm{e},\ I_2 = 48.5\ \mathrm{eV}.$$

The ground state of Ne^+ has one electron fewer in the $2p$ orbitals than the ground state of Ne, and the first excited state of Ne^+ has one electron fewer in the $2s$ orbital. According to the second definition, however, the second ionization potential is given by the scheme

$$Ne^{+}(1s^2\,2s^2\,2p^5) \rightarrow Ne^{2+}(1s^2\,2s^2\,2p^4) + e,\; I_1^{+} = 41.1 \text{ eV},$$

in which a second electron is removed from the $2p$ orbitals. To prevent confusion we will not use the second definition, but will specify such processes as the ionization potentials of the $(k-1)$th positive ion.

The most direct and widely applicable technique for measuring the ionization potentials of molecules is by photoelectron spectroscopy. In this a molecule in the gas phase is subjected to monochromatic radiation of frequency ν, whose photon energy, $h\nu$, is greater than the ionization potential to be measured. For ionization potentials below 20 eV the most common source is the 58.4 nm (21.2 eV) emission line from excited helium atoms. For higher ionization potentials, particularly those associated with the removal of inner-shell electrons, an X-ray source is used. This latter experiment can also be applied to molecules in the solid state.

By measuring the kinetic energy of the emitted electrons we can determine the ionization potential according to the energy conservation equation (c.f. 2.2)

$$h\nu = T_k + I_k \tag{5.27}$$

For a fixed frequency ν, electrons are ejected with different kinetic energies and if the number of electrons emitted in a given time is plotted as a function of T (or directly as $I_k = h\nu - T_k$) then this defines a spectrum. For helium and similar sources this technique is called ultra-violet photoelectron spectroscopy (or UPS) and for X-ray sources it is called X-ray photoelectron spectroscopy (XPS) or ESCA (electron spectroscopy for chemical analysis).

Photoelectron spectroscopy has proved to be a more valuable technique for molecules than for atoms because it has shown itself to be a unique tool for confirming the validity of the molecular orbital approach in valence theory.

Figure 5.10 shows the photoelectron spectra of N_2, O_2, and NO. There are for each molecule two distinct regions, one of several hundred electron volts and one of tens of volts. The latter region is divided at approximately 20 eV as the photon sources used to ionize the molecules are different above and below this. The region below 20 eV (excited by the He I line) has higher resolution than the remainder of the spectrum partly for instrumental reasons and partly due to the narrower band width of the ionizing radiation. The fine structure in this region is due to vibrational levels of the positive ion. In other words, ionization can not only leave the ion in different electronic states but each of these has a set of possible vibrational levels. The envelope of this vibrational pattern is determined by the Franck–Condon principle which is described in books on molecular spectroscopy.

Finally, it must be mentioned that there are some bands in a photoelectron spectrum which are not associated simply with the removal of an electron from a molecule but which can also involve excitation of other electrons in the molecule. The band marked A in the O_2 spectrum probably arises from such a process. Such bands have no relevance at the level to which we discuss photoelectron spectra in this book.

The link between the position of bands in photoelectron spectra and the energies

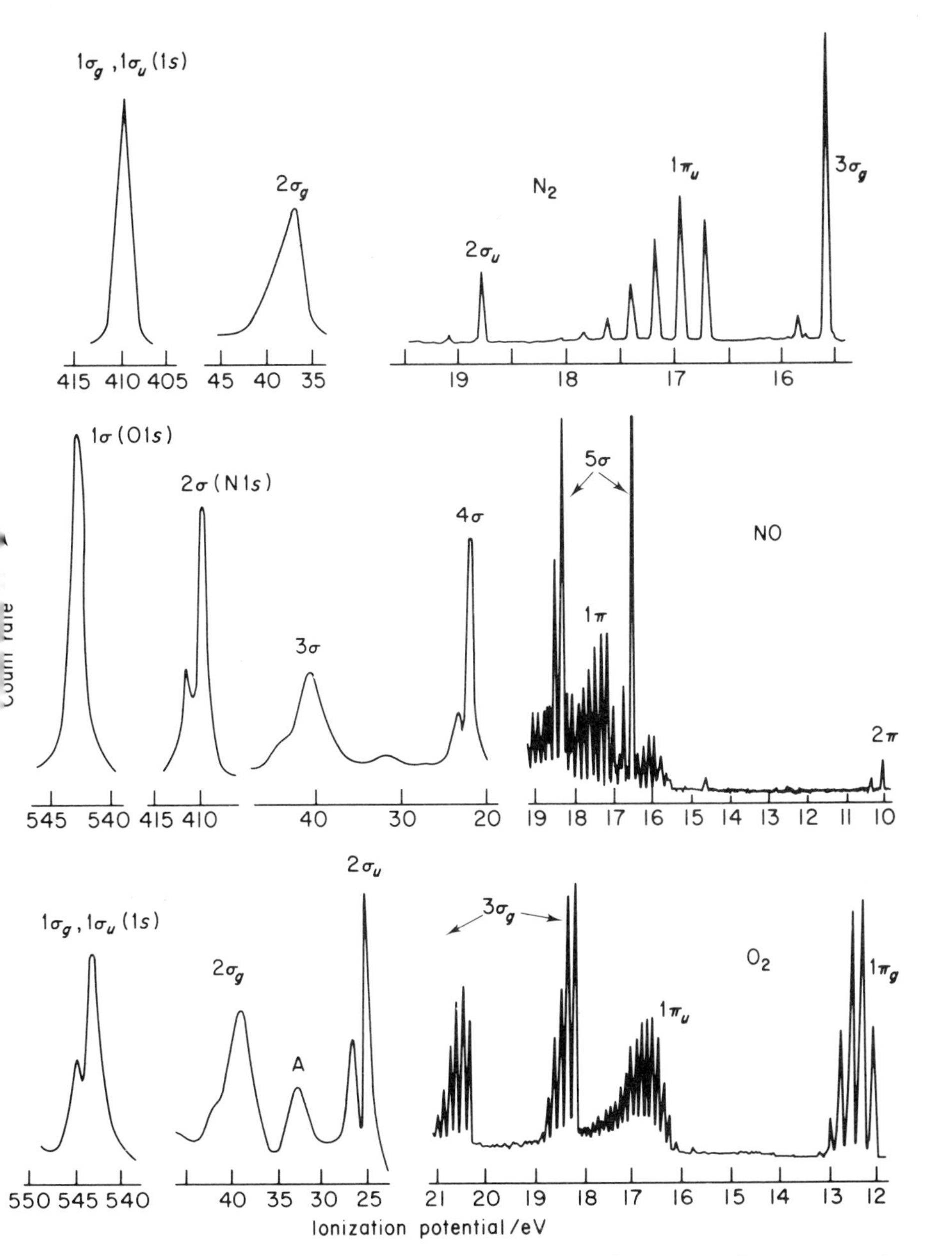

Figure 5.10 Photoelectron spectra of N_2, NO, and O_2. Spectra are shown as count rates (different scales for each segment of the spectra) against the ionization potential. The low ionization limits below 21eV were obtained with a HeI source (W. C. Price) and the region above with Mg $K\alpha$ irradiation (K. Siegbahn and coworkers).

of molecular orbitals is provided by a theorem first stated by Koopmans. This is that the orbital energy in the Hartree–Fock limit is the negative of the ionization potential for the removal of an electron from that orbital. The theorem applies strictly only to molecules in which all orbitals are either filled with two electrons or are empty (so-called closed shell systems) but this covers the majority of stable molecules. It also rests on the assumption that on ionization the remaining electrons do not adjust their positions to the new potential in the ion. In other words, the molecular orbitals for the positive ion are the same as those for the neutral molecule. This last assumption is clearly not exact but the errors inherent in it are not usually sufficiently large to invalidate the association between molecular orbital energies and ionization potentials expressed through Koopmans' theorem.

A corrollary of Koopmans' theorem is that the distribution of positive charge in the ion may be determined from the wavefunction of the molecular orbital from which the electron has been removed. If $\rho(\mathbf{r})$ is the total electron density function for the ground state of the neutral molecule, then the electron density in the ion obtained by removing an electron from ψ_k is

$$\rho(\mathbf{r}) - \psi_k^2(\mathbf{r}). \tag{5.24}$$

If this total electron density is broken down to contributions on individual atoms, then, after combining these with the positive nuclear charges, we obtain the distribution of the net positive charge on the individual atoms.

Several important conclusions can be drawn from Figure 5.10. Firstly, the two high-energy ionization potentials in NO are close to the corresponding potentials observed for N_2 and O_2 and by calculation we know that these are close to the ionization potentials for removal of an electron from the $1s$ orbitals of the free atoms. We conclude that the inner shell ($1s$) orbitals of atoms are little affected by bond formation. This supports the important chemical rationalization brought out in the periodic table, that the chemistry of the elements depends only on their outer shell or valence electrons.

As each of the atoms in N_2 has a $1s$ orbital there should be two molecular orbitals with ionization potentials approximately 410 eV. This follows by analogy with the situation for H_2^+ which was discussed in the last section. If we imagine a process of bringing the two atoms together from infinity, the two $1s$ atomic orbitals will gradually evolve into two molecular orbitals which will be labelled $1\sigma_g$ and $1\sigma_u$. However, the high energy band in the N_2 spectrum shows no evidence of two separate ionization potentials. Calculations on N_2 indicate that these two levels are separated by approximately 0.1 eV and this cannot be resolved with the present spectroscopic equipment.

The high-energy band of O_2, however, does show a resolution into two peaks with intensity ratio approximately 2:1. This separation (~1 eV) is too large to be due to the difference in energy of the $1\sigma_g$ and $1\sigma_u$ levels. It arises from the fact that the valence electrons of O_2 are not all spin-paired (this will be explained fully in Chapter 6) and there is a coupling between the unpaired electrons in the valence shell and the unpaired electron in the 1σ orbitals that arises after ionization. A similar spin coupling occurs for NO as there is an odd number of valence electrons (hence one whose spin in unpaired) for this molecule. Note, however, that it is only the high energy band at 410 eV that is split and this suggests that the unpaired electron in the valence shell of the molecule resides mainly on the nitrogen atom.

The low-energy ionization regions of the three spectra are more complicated but have been interpreted as due to ionization from different valence molecular orbitals. The assignments given in the figure are based partly on calculations and partly on the fact that the integrated area under a band is approximately proportional to the number of electrons in the orbitals contributing to that band. Thus a doubly occupied orbital should give a band twice the height of a singly occupied orbital. There is a splitting of the bands assigned as $3\sigma_g$ in the O_2 and 5σ in the NO spectra which is due to spin-exchange coupling as already described for the high-energy regions. We shall leave a more detailed discussion of the valence regions of these spectra until we have studied the form of the valence molecular orbitals in the next chapter.

Chapter 6

The LCAO Approximation to Molecular Orbitals

6.1. Justification of the LCAO approach

In the last chapter we examined the general form of the molecular orbitals of diatomic molecules. We also gave a more detailed description of H_2^+, this being a system for which an exact solution of the Schrödinger equation is possible. In this chapter we will consider the molecular orbitals of molecules in general. It must be emphasized at the start that our approach will be quite different from that taken for H_2^+.

We re-iterate the qualitative definition of an orbital: it is the wavefunction of an electron moving under the influence of the nuclear potential and the average repulsive potential of the other electrons. For a molecule with many atoms the nuclear potential itself is a complicated function in space and the potential of the electrons can be no simpler. Thus any attempts to express the effective potential explicitly as a function in three-dimensional space $V(x,y,z)$ and to solve the appropriate Schrödinger equation will be doomed to failure. Nevertheless, suppose that it could be done. The molecular orbitals would then be obtained as functions $\psi(x, y, z)$ for which there would be, in general, no separation into the product of separate functions of the three coordinates, as there is for atoms (equation 3.11) or linear molecules (equation 5.24).

The only way we could describe orbitals obtained in this way would be to present a table of ψ at discrete points in space. If we chose a course grid of ten points in each dimension then our table would comprise 10^3 values. If we had a finer grid (say 100 points in each dimension) then we would tabulate a million values. Such a method of communicating is clearly unrealistic for a general description of molecular orbitals.

An alternative method of representing a complicated function is to express it in terms of simpler functions whose tabulated values are well known. This is a standard technique in mathematics as illustrated by the Fourier expansion of a complicated but periodic function.

In one dimension the Fourier cosine expansion is

$$f(x) = \sum_{n=0}^{\infty} a_n \cos(2\pi nx/x_0) \tag{6.1}$$

x_0 being the length of the repeating unit. The cosine functions are referred to as the *basis* for the expansion. Thus, instead of tabulating $f(x)$ at different values of x we list the coefficients a_n of the expansion.

This procedure does not allow us to give an exact representation of $f(x)$ unless we list an infinite number of coefficients. However, we may obtain a satisfactory approximation to $f(x)$ by listing the first few coefficients (a_0, a_1, a_2, etc.) providing that the series a_n converges. This book is not the place to discuss the conditions under which this is true although such matters are of considerable importance for accurate calculations in quantum chemistry. We must, however, consider what are appropriate basis functions for the expansion of a molecular orbital. Although several bases have been examined there are few that have been found to be generally useful.

We note first that the basis must provide an expansion in three dimensional space and second that the function we have to expand is not periodic for a finite molecule, although an extension of the orbital model to polymers or to crystalline solids will introduce conditions of periodicity.

The problem we have to solve is expressed by the finite expansion

$$\psi(x, y, z) = \sum_n c_n \phi_n (x, y, z). \tag{6.2}$$

We want a basis set of standard functions ϕ_n for which a good representation of the molecular orbitals can be obtained by a relatively small number of terms, with the added requirement that any mathematical operations involved in the determination of the coefficients c_n can be carried out with reasonable speed using modern computers.

Our discussion of the problem has so far been rather abstract and expressed in mathematical terms. The *solution* of the problem has, however, been largely based upon physical arguments. The most widely used basis functions for the expansion of molecular orbitals are the atomic orbitals of the component atoms. In other words equation (6.2) is in this case an expansion of a molecular orbital as a *linear combination of atomic orbitals*: this is commonly referred to as the LCAO approximation.

The physical grounds for the LCAO approximation are two-fold. Firstly, in the limit that a molecule is pulled apart into its component atoms the set of molecular orbitals must go smoothly into the set of atomic orbitals of the atoms. Secondly, the effective potential for an electron in a molecule when it is close to one nucleus is so dominated by the potential due to that nucleus that all other potentials are negligible in comparison. It follows that the functional form of a molecular orbital near to a nucleus must be similar to the functional form of an atomic orbital of that atom.

The above features have already been noted in our discussion on H_2^+ in the last chapter. Figure 5.2 shows the contours of the lowest energy orbital ($1\sigma_g$) of H_2^+ as a function of the internuclear distance. At large distances this orbital looks like the separate atomic orbitals of the atoms and at shorter distances the inner contours of the molecular orbital around each nucleus are approximately spherical, that is approximately like those of a free atom. We note also in the limit of the coalescing nuclei ($R = 0$) that the molecular orbital is again atomic in nature.

The atomic characteristics of diatomic molecular orbitals are also illustrated by the nodal properties and other features of the higher energy orbitals of H_2^+ (Figures 5.7 and 5.8). For example, the $1\sigma_u$ orbital of H_2^+ is seen to have approximately spherical symmetry close to the nuclei.

There are two aspects in which molecular orbitals have proved useful in chemistry. The first is at a qualitative level in which the general shape or symmetry of an orbital and its approximate energy are sufficient to give an assignment of a spectrum or an interpretation of a chemical reaction. The second is a quantitative calculation of some molecular property to confirm or predict chemical observations. For quantitative work the emphasis is usually on obtaining the most accurate representation of the molecular orbitals by the expansion (6.2) and many basis atomic orbitals are taken in the expansion. For qualitative work the emphasis is on the shortest expansion which will give results satisfactory for the purpose in mind. In this book we emphasize mainly the qualitative aspect. It will be seen in the following discussion that qualitatively satisfactory results can be obtained by taking what is usually called the *minimal basis* of atomic orbitals. Roughly speaking these are the atomic orbitals that contain electrons in the ground states of the free atoms.

A minimal basis for the molecular orbitals of H_2^+ will be the two hydrogen $1s$ atomic orbitals. These will have wavefunctions (c.f. Table 3.1)

$$\phi_a = N \exp(-r_a), \quad \phi_b = N \exp(-r_b), \tag{6.3}$$

where N is a normalizing factor and r_a and r_b are the distances of the electron from the two nuclei, expressed in atomic units. The molecular orbitals of H_2^+ will, in this minimal basis, have the form

$$\psi = c_a\phi_a + c_b\phi_b. \tag{6.4}$$

We shall see in Section 6.4 that there are general techniques available to determine the coefficients in the LCAO expansion. In the case of the function (6.4) as it applies to H_2^+, however, we have no need to use general techniques as the coefficients can be determined by symmetry arguments.

Let us start by considering the probability density for an electron in the molecular orbital (6.4): This is

$$\psi^2 = c_a^2\phi_a^2 + c_b^2\phi_b^2 + 2c_ac_b\phi_a\phi_b. \tag{6.5}$$

Because the two nuclei in H_2^+ are equivalent the probability of finding an electron in the vicinity of one nucleus must be the same as for an equivalent point relative to the other nucleus. This result is clearly only obtained from the probability density (6.5) if the condition

$$c_a^2 = c_b^2 \tag{6.6}$$

is satisfied. If this were not the case then ψ^2 would be a lop-sided function with greater weight around one nucleus than around the other.

There are two solutions of the equation (6.6) for real values of the coefficients and these are

$$c_a = c_b, \tag{6.7}$$

and

$$c_a = -c_b. \tag{6.8}$$

Thus we can obtain from expression (6.4) two molecular orbitals of H_2^+ which make physical sense. From (6.7) we have

$$\psi_g = c_{a(g)}\,(\phi_a + \phi_b), \tag{6.9}$$

and from (6.8)

$$\psi_u = c_{a(u)}\,(\phi_a - \phi_b). \tag{6.10}$$

We have identified these two solutions by suffixes g and u which are the symmetry labels used to characterize the molecular orbitals of H_2^+ in Chapter 5. The coefficient c_a does not necessarily have the same value for the two solutions. The coefficients, c_a in (6.9) and (6.10), are just scale factors for the amplitude of ψ and we determine them by applying the normalization condition to the orbitals. This normalization condition (equation 2.29) applied to (6.5) is

$$\int \psi^2\, dv = \int (c_a^2\phi_a^2 + c_b^2\phi_b^2 + 2c_a c_b \phi_a \phi_b)\, dv = 1, \tag{6.11}$$

the integration being over the whole of the three dimensional space in which the electron is confined. Substituting (6.7) and (6.8) into (6.11) gives us the following conditions:

$$\int \psi_g^2\, dv = c_{a(g)}^2 \int (\phi_a^2 + \phi_b^2 + 2\phi_a\phi_b)\, dv = 1, \tag{6.12}$$

$$\int \psi_u^2\, dv = c_{a(u)}^2 \int (\phi_a^2 + \phi_b^2 - 2\phi_a\phi_b)\, dv = 1. \tag{6.13}$$

We have already defined the basis atomic orbitals ϕ_a and ϕ_b to be normalized functions, by (6.3), hence

$$\int \phi_a^2\, dv = \int \phi_b^2\, dv = 1. \tag{6.14}$$

If we now further introduce the symbol S_{ab} for the integral

$$S_{ab} = \int \phi_a \phi_b\, dv, \tag{6.15}$$

then the equations (6.12) and (6.13) become respectively

$$c_{a(g)}^2\,(2 + 2S_{ab}) = 1 \quad \text{or} \quad c_{a(g)} = \pm(2 + 2S_{ab})^{-1/2}, \tag{6.16}$$

and

$$c_{a(u)}^2\,(2 - 2S_{ab}) = 1 \quad \text{or} \quad c_{a(u)} = \pm(2 - 2S_{ab})^{-1/2}. \tag{6.17}$$

The choice of sign in these expressions for c_a is arbitrary as the overall sign of a wavefunction has no meaning; only ψ^2 has physical significance. It is therefore customary to take the positive sign in each case and, on substituting the expressions for c_a, (6.16) and (6.17), into (6.9) and (6.10) respectively, we have the following normalized molecular orbital wavefunctions

$$\psi_g = (2 + 2S_{ab})^{-1/2} (\phi_a + \phi_b), \tag{6.18}$$

$$\psi_u = (2 - 2S_{ab})^{-1/2} (\phi_a - \phi_b). \tag{6.19}$$

Examination of the contours of these wavefunctions shows that they are very similar to those of the exact $1\sigma_g$ and $1\sigma_u$ respectively, whose contours were shown in Figures 5.2 and 5.8. This is illustrated in Figure 6.1 where we show the values of the wavefunctions along the line of centres and compare them with the exact results. It is seen that in the $1\sigma_g$ state the LCAO wavefunction underestimates the true magnitude at the nuclei and in the $1\sigma_u$ it overestimates this. One can, in fact, get much better wavefunctions for H_2^+ of the LCAO form if the wavefunctions of the 1*s* atomic orbitals (equation 6.3) are modified to have a variable exponent as follows:

$$\phi = N \exp(-kr). \tag{6.20}$$

The contraction at the nucleus in the $1\sigma_g$ orbital can be reproduced by using a value of k which is greater than unity and the expansion in the $1\sigma_u$ orbital requires a value of k less than unity. We will see in Section 6.4 that there is a well defined procedure for determining the best value of k.

As we have determined approximate wavefunctions for the molecular orbitals of H_2^+ we can obtain energies which will be approximations to the exact energies by

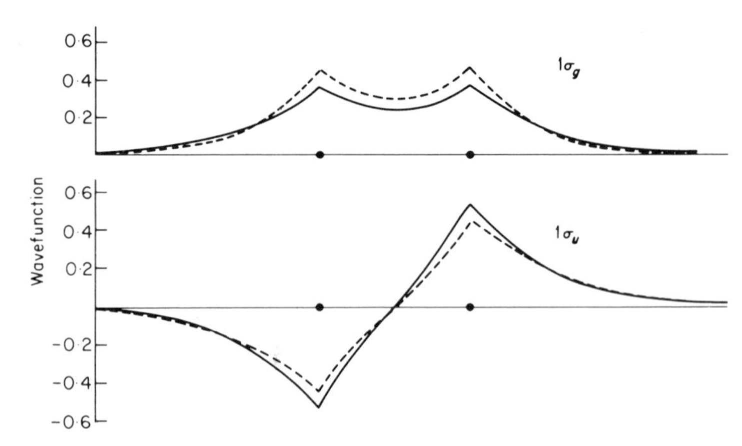

Figure 6.1 The LCAO wavefunctions of equations (6.18) and (6.19) (full lines) plotted along the line of the nuclei and compared with the exact wavefunctions for H_2^+. The bond length is the ground state equilibrium value, $2a_0$.

the use of expression (5.14). Inserting (6.18) and (6.19) into (5.14) gives us the orbital energies in the form

$$E_{g,u} = (2 \pm 2S_{ab})^{-1} \int (\phi_a \pm \phi_b) \mathscr{H} (\phi_a \pm \phi_b) \, dv, \tag{6.21}$$

the positive and negative signs referring to the $1\sigma_g$ and $1\sigma_u$ orbitals respectively.

The integral in (6.21) can be expanded to give four terms which can be abbreviated by the symbols

$$H_{aa} = \int \phi_a \mathscr{H} \phi_a \, dv, \quad H_{ab} = \int \phi_a \mathscr{H} \phi_b \, dv, \text{ etc.} \tag{6.22}$$

As ϕ_a and ϕ_b are equivalent atomic orbitals for the two atoms it follows that $H_{aa} = H_{bb}$ and $H_{ab} = H_{ba}$. Expression (6.21) can there fore be written in the form

$$E_{g,u} = (1 \pm S_{ab})^{-1} (H_{aa} \pm H_{ab}). \tag{6.23}$$

The Hamiltonian for H_2^+ consists of the kinetic energy operator, the potential of attraction to the nuclei, and the potential of repulsion between the two nuclei. In atomic units, and using the symbols of equation (3.46), this is

$$\mathscr{H} = -\frac{1}{2}\nabla^2 - \frac{1}{r_a} - \frac{1}{r_b} + \frac{1}{R}. \tag{6.24}$$

The integrals H_{aa} and H_{bb}, and S_{ab} defined earlier by (6.14), can in this case be evaluated analytically and expressed as functions of the internuclear distance R. The results are as follows:

$$S_{ab} = e^{-R}\left(1 + R + \frac{R^2}{3}\right) \tag{6.25}$$

$$H_{ab} = e^{-R}\left(\frac{1}{R} - \frac{1}{2} - \frac{7R}{6} - \frac{R^2}{6}\right), \tag{6.26}$$

$$H_{aa} = -\frac{1}{2} + e^{-2R}\left(1 + \frac{1}{R}\right). \tag{6.27}$$

Figure 6.2 shows how these integrals and the energies E_g and E_u vary as a function of R. H_{ab} and S_{ab} decay exponentially to zero as R becomes large but H_{aa} tends to the value $-0.5E_H$ which is the energy of the hydrogen 1*s* orbital. It can be seen from this figure and expression (6.23), that H_{ab} is the important function that leads to the minimum in the E_g potential energy curve at about $2a_0$.

Functions such as S_{ab}, H_{ab}, and H_{aa} play an important role in MO theory. Even in cases where their functional form is not known exactly we shall rely on a knowledge of their approximate form to reach conclusions about the strength of chemical bonds. Figure 6.2 therefore has a wider relevance than for calculations on H_2^+.

The overlap integral, in particular, is an important quantity. In the first place it

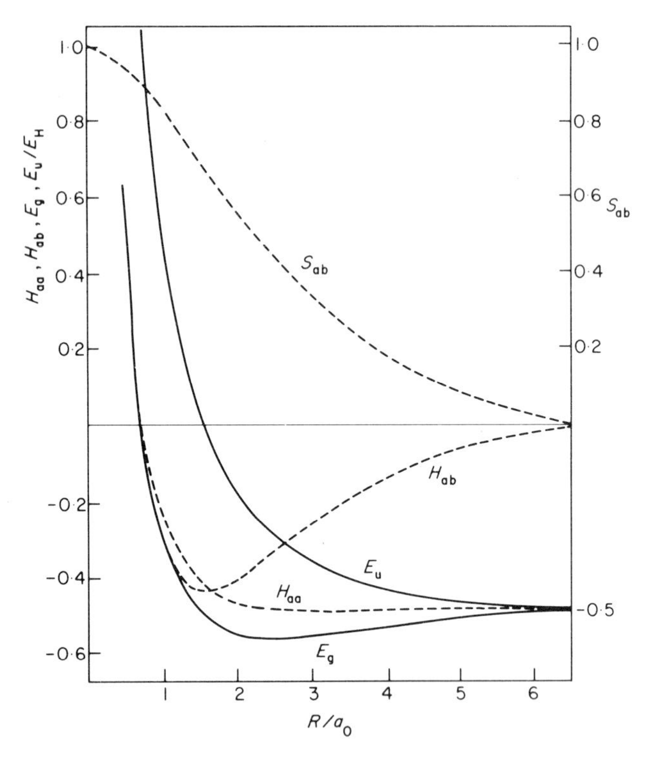

Figure 6.2 The functions S_{ab}, H_{aa} and H_{ab} and the LCAO MO energies plotted against internuclear distance.

is always an easy integral to calculate whereas the integrals involving the Hamiltonian are often very difficult. In the second place we note from either (6.25) and (6.26) or from Figure 6.2, that the important Hamiltonian integral H_{ab} has an R dependence that is quite similar to that of S_{ab} (but with opposite sign) for values of R greater than the equilibrium bond length. It is a common assumption in qualitative MO theories that H_{ab} is proportional to S_{ab}.

If S_{ab} is zero because there is no region of space where both ϕ_a and ϕ_b have non-negligible values (we say in this case that ϕ_a and ϕ_b do not overlap) then on the same grounds H_{ab} must be zero also. If S_{ab} is zero because the product $\phi_a\phi_b$ has positive and negative regions which cancel in their contribution to the integral, then H_{ab} is also either zero or small depending on the precise nature of the Hamiltonian. Situations in which integrals such as S_{ab} are zero are shown in Figure 6.3 and, as we shall see, these have an important role in the theory of the chemical bond.

The physical arguments we have given in favour of the LCAO approximation of molecular orbitals and the proved success in the case of H_2^+ justifies our further examination of the approximation for other molecules. We therefore turn our attention in the next two sections to a more extensive examination of the molecular orbitals of homonuclear diatomic molecules.

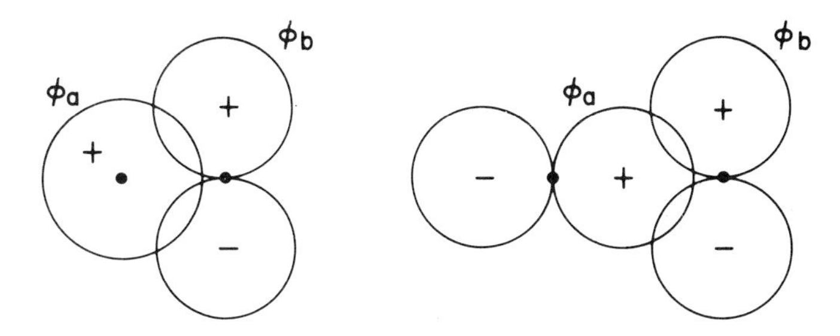

Figure 6.3 Two situations in which the overlap integral S_{ab} is zero because of the cancellation of positive and negative contributions from $\phi_a\phi_b$ to the integral.

6.2. Application of LCAO MO theory to the molecules Li_2 to F_2

In this section we will describe the LCAO MO interpretation of the properties of the homonuclear diatomic molecules formed from first row atoms. It will be seen that there is a strong similarity between the way we used the pattern of atomic orbital energies and the building-up principle to describe the electronic structure of the atoms Li – F and the way we use the pattern of molecular orbital energies and the building-up principle to describe the structure of the molecules Li_2 – F_2.

The molecules Li_2, Be_2, B_2, C_2, N_2, O_2, and F_2 have widely different properties. For example, N_2 is a very stable molecule. Its dissociation energy, D_0 = 942 kJ mol^{-1}, is greater than that of any other diatomic molecule except CO. On the other hand, Li_2 is rather unstable (D_0 = 107 kJ mol^{-1}) and is only detected in 1% concentration in the vapour of the metal at its boiling point. Be_2 has never been detected even as a transient species.

The molecules also differ in their magnetic properties. The molecules B_2 and O_2 are paramagnetic, that is they are attracted into a magnetic field, whereas the others are diamagnetic; they are repelled from a magnetic field. We shall see that this arises from the fact that the spins of the electrons are not all paired in the paramagnetic molecules.

The traditional structural formulae for dinitrogen, dioxygen, and difluorine are N≡N, O=O and F–F. Such formulae satisfy the normal valencies of 3, 2, and 1 for N, O, and F respectively, and account qualitatively for the relative bond strengths of the molecules. However, such formulae give no indication that O_2 is paramagnetic and the other two diamagnetic, nor for other facts such that ionization of O_2 leads to a stronger bond, whereas ionization of N_2 leads to a weaker bond. These facts as we shall see are readily explained by molecular orbital theory.

The building up (aufbau) principle that we have already mentioned was introduced by Bohr to explain the structure of the Periodic Table; this was described in Chapter 4. In the context in which we shall use it to explain molecular structure it can be stated as follows. The lowest energy state of a molecule is obtained by allocating electrons in turn to the lowest available molecular orbitals subject to the requirement of the Pauli exclusion principle that not more than two electrons (which must have opposite spins) occupy any one orbital.

To illustrate the application of this principle we will first consider the systems H_2^+, H_2, He_2^+, and He_2. We have already seen that the first molecular orbitals of H_2^+ are designated $1\sigma_g$ and $1\sigma_u$ in order of increasing energy, and in the minimal basis LCAO model their wavefunctions are given by combinations of the two $1s$ atomic orbitals, equations (6.18) and (6.19) respectively. These will also be the lowest energy molecular orbitals of H_2.

Because the ground state of He has two electrons in a $1s$ orbital, the minimal basis LCAO molecular orbitals for He_2^+ and He will also have the form given by (6.18) and (6.19). However, the wavefunction of the $1s$ orbital ϕ will not be the function (6.3) but it can be approximated quite well by the function (6.20) with an appropriate value for k. From Table 3.2 it can be seen that k (ζ) for He_2 would be 1.687 5.

Applying the building-up principle to the four species H_2^+, H_2, He^+, and He_2 we arrive at the configurations for their ground states shown in Table 6.1. We have noted in Section 5.3 that the $1\sigma_g$ orbital has a lower energy than that of the $1s$ atomic orbitals from which it is composed and hence electrons in this orbital make the molecule stable to dissociation. $1\sigma_g$ is called a bonding orbital. On the other hand $1\sigma_u$ has a higher energy than that of its component $1s$ orbitals and it is therefore destabilizing or antibonding. We show in Table 6.1 the difference between the number of bonding and antibonding electrons.

We see that on passing from H_2^+ to H_2 we double the number of bonding electrons and this is accompanied by a reduction in the bond length and an increase in the dissociation energy. Moreover, when the next electron is added to give He_2^+, this, being in an antibonding orbital, leads to an increase in bond length and a decrease in dissociation energy. Finally He_2, which would have equal numbers of bonding and antibonding electrons, is not a known species in its ground state. A general rule, although not one we prove rigorously by qualitative arguments, is that if there is no excess of bonding over antibonding electrons then the molecule is not stable.

We turn now to the molecules formed by the atoms Li to F. In their ground states these atoms all have electrons in $2s$ atomic orbitals, and the atoms B to F also have electrons in $2p$ orbitals. The minimal basis for the LCAO approximation must

Table 6.1. Properties of the species H_2^+, H_2, He_2^+, and He_2

	Electron configuration	No. bonding −no. antibonding electrons	Equilibrium bond length/ Å	Dissociation energy/ kJ mol^{-1}
H_2^+	$(1\sigma_g)^1$	1	1.06	256
H_2	$(1\sigma_g)^2$	2	0.74	432
He_2^+	$(1\sigma_g)^2(1\sigma_u)^1$	1	1.08	~300
He_2	$(1\sigma_g)^2(1\sigma_u)^2$	0	—	—

therefore include $2s$ and $2p$ atomic orbitals as well as $1s$, so that the expansion (6.2) for a molecule like F_2 will in general consist of ten terms; $1s$, $2s$, $2p_x$, $2p_y$, and $2p_z$ for each atom. This is a relatively complicated situation on which to base qualitative arguments. There are, however, two rules which allow us to simplify the problem.

The first rule is that two atomic orbitals will both make large contributions to a molecular orbital (their LCAO coefficients will be large) only if their atomic orbital energies are very similar. The second rule is that two atomic orbitals for which the overlap integral is zero will not make large contributions to the same molecular orbital. Both of these rules will be given a firmer basis in the next section.

From the first rule it is a very good approximation to ignore the mixing between the inner $1s$ orbitals and the valence $2s$ and $2p$ orbitals. Also it is a reasonably good approximation, for the molecules we are considering, to ignore the mixing between the $2s$ and $2p$ orbitals. (The energy difference between the $2s$ and $2p$ orbitals has been given in Table 3.3.) It follows that the first two molecular orbitals of the molecules Li_2 and F_2 have a similar form to equations (6.18) and (6.19) where ϕ_a and ϕ_b are the $1s$ orbitals of the two atoms. The $2s$ orbitals will also combine to give molecular orbitals

$$
\begin{aligned}
2\sigma_g &= 2s_a + 2s_b, \\
2\sigma_u &= 2s_a - 2s_b,
\end{aligned}
\tag{6.28}
$$

where we have used the symbol $2s_a$ to represent the wavefunction of a $2s$ orbital of atom A. Each of these wavefunctions has to be multiplied by a factor such as (6.16) or (6.17) for normalization.

We can surmise from the arguments already presented in the case of H_2^+, that $2\sigma_g$ will have a lower energy and $2\sigma_u$ will have a higher energy than the $2s$ atomic orbitals from which they are constructed.

The mixing of the $2p$ orbitals can be simplified on the basis of the second rule if we refer these orbitals to a z axis which is along the line of nuclear centres, with x and y axes perpendicular to this. As shown in Figure 6.3 the overlap integral between a $2p_z$ orbital of one atom and a $2p_x$ or $2p_y$ orbital of the other will be zero. The only non-zero overlap integrals between the p orbitals will be S_{xx}, S_{yy}, and S_{zz}.

Following the same symmetry arguments used in constructing the H_2^+ molecular orbitals, we conclude that the molecular orbitals derived from linear combinations of the $2p$ orbitals on the two atoms have the following form:

$$
\begin{aligned}
1\pi_g(x) &= 2p_{xa} - 2p_{xb}, \quad 1\pi_g(y) = 2p_{ya} - 2p_{yb}, \\
1\pi_u(x) &= 2p_{xa} + 2p_{xb}, \quad 1\pi_u(y) = 2p_{ya} + 2p_{yb}, \\
3\sigma_u &= 2p_{za} - 2p_{zb}, \\
3\sigma_g &= 2p_{za} + 2p_{zb}.
\end{aligned}
\tag{6.29}
$$

The orbitals have been labelled by the symbols used for the orbitals of H_2^+ (see

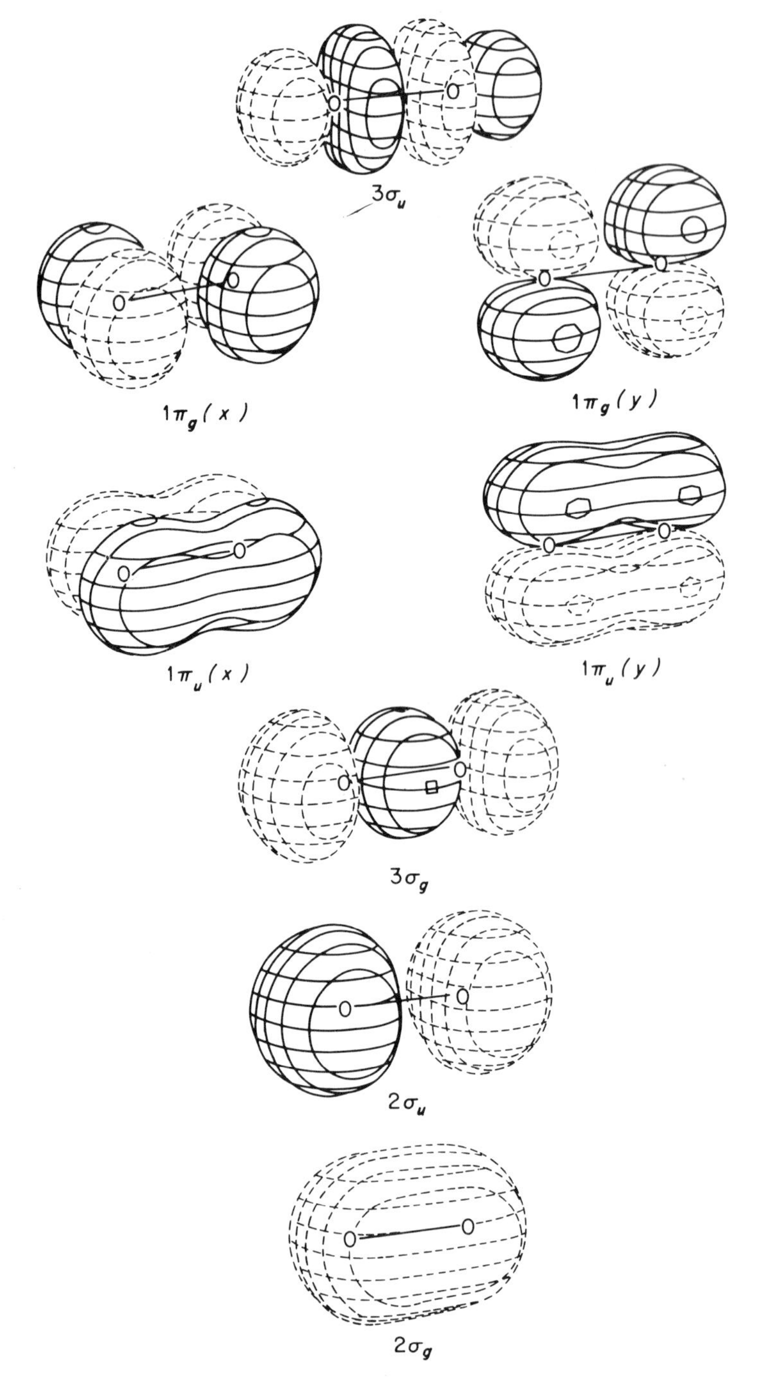

Figure 6.4 Contour surfaces of the valence shell molecular orbitals of the ground state of O_2.

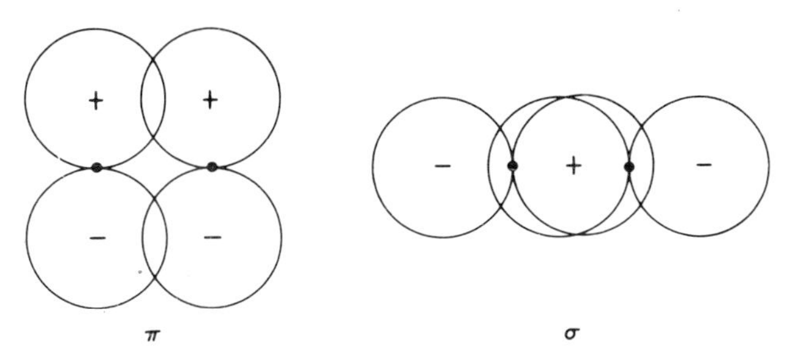

Figure 6.5 The overlap of $2p$ atomic orbitals to form σ and π molecular orbitals.

Figure 5.8). The $\pi(x)$ and $\pi(y)$ orbitals are equivalent apart from their direction in space and hence they must have the same energy. Figure 6.4 shows 3-dimensional representations of the valence orbitals of O_2 that have been produced by Jorgensen and Salem† from wavefunctions of Krauss. These show the contour of the wavefunction at $\pm$ 0.1 atomic unit (represented by its sections parallel to the xz and yz planes), the positive and negative lobes of the orbitals being represented by broken lines and by full lines respectively. It can be seen that the σ_g and π_u orbitals have greater amplitude in the region between the two nuclei than do the σ_u and π_g: the former are bonding orbitals and the latter antibonding

Some decision about the relative energies of these valence orbitals can be based upon the type of atomic orbital from which they are derived and the magnitude of the overlap integral between the two atomic orbitals. Thus the molecular orbitals formed from the $2s$ orbitals will have a lower energy than those from the $2p$. The overlap integral between the two $2p_z$ orbitals in O_2 is approximately 0.3, whereas the overlap between the $2p_x$ (or $2p_y$) orbitals is approximately 0.15. This can be anticipated by examining the amount of overlap in schematic diagrams such as Figure 6.5. We therefore expect the $3\sigma_g$ orbital to be more bonding than the $1\pi_u$, and the $3\sigma_u$ to be more antibonding than $1\pi_g$. This is supported by calculations. The change in energy of the orbitals that occurs on bond formation can therefore be represented schematically as in Figure 6.6.

The qualitative arguments we have presented for the order of the molecular orbitals for the homonuclear diatomic molecules Li_2 to F_2 include one questionable assumption, which is that the $2s$ and $2p$ atomic orbitals do not contribute to the same molecular orbital. The data in Table 3.3 shows that the energy difference between the $2s$ and $2p$ orbitals increases across the Periodic Table and hence our assumption is more justified for F_2 than for Li_2. In fact, from photoelectron spectroscopy we can deduce the ordering of the molecular orbitals and for F_2 and O_2 this confirms the qualitative correctness of the diagram shown in Figure 6.6. For the molecules Li_2 to N_2, however, this order is not in agreement with experiment.

†W. J. Jorgensen and L. Salem, *The Organic Chemists Book of Orbitals*, Academic Press, New York, 1973.

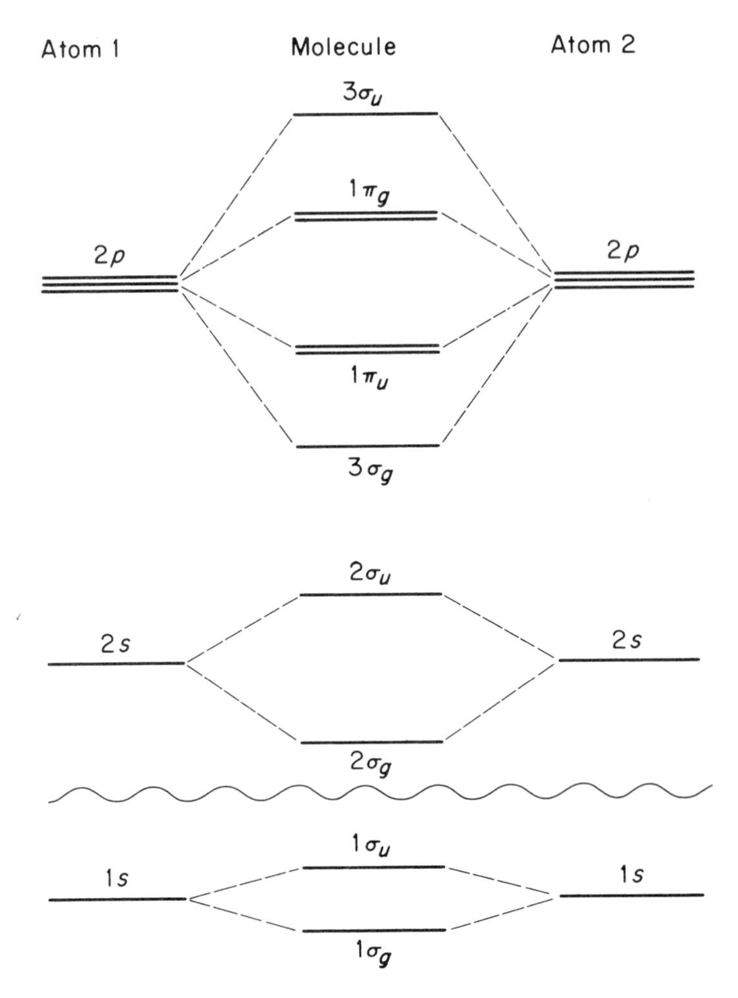

Figure 6.6 Relationship between the energies of the atomic orbitals and the energies of the molecular orbitals for O_2 and F_2.

It is clear from diagrams such as those shown in Figures 6.3 and 6.5 that there is a non-zero overlap integral between a $2s$ orbital of one atom and a $2p_z$ orbital of the other. It follows that $2s$ and $2p_z$ (but not $2p_x$ or $2p_y$) orbitals will contribute to the same molecular orbital. This will modify the wavefunctions of the σ orbitals we have already given in equations (6.28) and (6.29) so that the $2\sigma_g$ orbital, for example, will have a wavefunction of the form

$$2\sigma_g = c_1(2s_a + 2s_b) + c_2(2p_{za} + 2p_{zb}). \tag{6.30}$$

The coefficient c_2 will be rather small for F_2 but quite large for Li_2. The effect of this $2s$–$2p_z$ mixing is primarily to lower the energy of the $2\sigma_g$ and $2\sigma_u$ and to raise

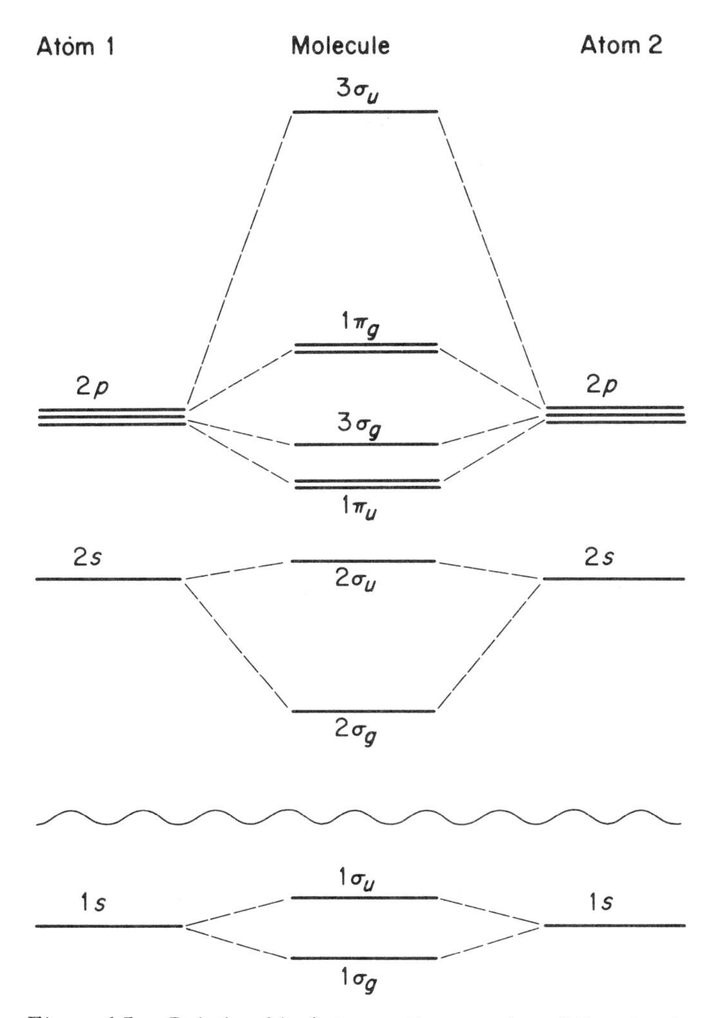

Figure 6.7 Relationship between the energies of the atomic orbitals and the energies of the molecular orbitals for Li_2 to N_2.

that of $3\sigma_g$ and $3\sigma_u$. The important outcome is that for the molecules Li_2 to N_2 the $3\sigma_g$ orbital comes *above* $1\pi_u$ on the energy scale, as shown in Figure 6.7.

On the basis of the energy diagrams shown in Figures 6.6 and 6.7 and the classification of the orbitals as bonding or antibonding, we can rationalize the relative strengths of the bonds in the molecules Li_2 to F_2, and we can also understand the magnetic properties of these molecules. Table 6.2 shows the ground state electron configurations of the molecules deduced by applying the building up principle to the orbital energy diagrams.

The electron configuration is simply a description of the number of electrons in any orbital and does not imply any particular spin of these electrons. If an orbital

Table 6.2. Summary of the properties of the ground states of the homonuclear diatomic molecules Li_2 to F_2. The dissociation energies are subject to experimental uncertainty to varying degrees: 1 kJ mol^{-1} would be typical in most cases. [Be] stands for the electron configuration of Be_2.

Molecule	Electron configuration	Excess of bonding electrons	Number of unpaired electrons	Bond length/ Å	Dissociation energy/ kJ mol^{-1}
Li_2	$(2\sigma_g)^2$	2	0	2.67	107
Be_2	$(2\sigma_g)^2(2\sigma_u)^2$	0	0	–	–
B_2	$[Be](1\pi_u)^2$	2	2	1.59	291
C_2	$[Be](1\pi_u)^4$	4	0	1.24	590
N_2^+	$[Be](1\pi_u)^4(3\sigma_g)^1$	5	1	1.12	841
N_2	$[Be](1\pi_u)^4(3\sigma_g)^2$	6	0	1.09	942
O_2^+	$[Be](3\sigma_g)^2(1\pi_u)^4(1\pi_g)^1$	5	1	1.12	644
O_2	$[Be](3\sigma_g)^2(1\pi_u)^4(1\pi_g)^2$	4	2	1.21	494
F_2	$[Be](3\sigma_g)^2(1\pi_u)^4(1\pi_g)^4$	2	0	1.44	154

has its maximum occupation of two electrons then these must, by the Pauli exclusion principle, have m_s spin quantum numbers + ½ and – ½ respectively and there will be no net spin angular momentum and hence no net magnetic moment from these spins. However, if there are two electrons in a degenerate pair of π orbitals then these electrons might both be in the same orbital, in which case they have opposite spins, or they may be one in $\pi(x)$ and one in $\pi(y)$. In the latter situation they may, or may not, have opposite spins. There is, therefore, more than one arrangement of spins for so-called incomplete shells of electrons and hence more than one energy state.

There is a rule which was first stated for atoms by Hund and which has been seen to apply equally well to molecules. The rule is that for an incomplete shell of electrons the lowest energy state is the one in which the maximum number of electrons have the same spin. In other words, for the configuration $(1\pi_u)^2$ of B_2 the lowest energy state is one in which the two electrons are in different orbitals with the same value of the spin quantum number: these two electrons are said to be unpaired. Such a state has a net spin angular momentum and hence a non-zero magnetic moment: it is therefore a paramagnetic molecule. The species B_2, N_2^+, O_2^+, and O_2 can be seen from Table 6.2 to have this property.

The paramagnetism of B_2 confirms that the orbital pattern of Figure 6.7 rather than that of Figure 6.6 applies to this molecule. If this were not the case then the ground state configuration would be $[Be](3\sigma_g)^2$, which would be a spin-paired diamagnetic state.

The trends in the dissociation energies and relative bond lengths of the molecules reflects the excess number of bonding electrons in the molecule. Be_2, which has no excess of bonding electrons, is not known as a stable species. It is to be noted, in particular, that ionization of N_2 reduces the number of bonding electrons and leads to a less stable molecule whereas ionization of O_2 by removing an anti-bonding electron leads to an ion which is more stable to dissociation and has a shorter bond length.

The family of molecules should be completed by adding Ne_2. The two electrons additional to those of F_2 will go into the $3\sigma_u$ orbital and the excess bonding electrons will then be zero. Ne_2 is not known as a stable molecule in its ground state.

The satisfactory description we have given of the properties of these molecules by MO theory is an encouragement to look at other systems. To do this, however, we require some method of determining the coefficients in the LCAO expansion (6.2) when these are not determined by symmetry. The theorem which enables us to do them is arguably the most important theorem of quantum mechanics for chemists and we shall state and prove it in the next section.

6.3. The variation theorem and secular equations

The variation theorem is simple to state and far reaching in its application in qunatum chemistry. *If the expectation value of the energy is calculated from an approximate solution of the Schrödinger equation according to expression (5.14) then this energy is always greater than the exact ground state energy for that Hamiltonian.*

If the energy is calculated from (5.14) with the exact ground state wavefunction then, as we showed in expressions (5.16)–(5.19), the result is the exact ground state energy (E_0). The variation theorem can therefore be expressed by the relationship

$$\bar{E} = \int \Psi \mathscr{H} \Psi \, dv > E_0. \tag{6.31}$$

There is one famous example in which $\bar{E}$ was calculated to be below the value of E_0 deduced from experiment. In 1968 accurate calculations on H_2 by Kolos and Wolniewicz† gave a dissociation energy for H_2 which was 0.05kJ mol^{-1} greater than the then accepted experimental value. As the energy of two separate hydrogen atoms is known precisely (see Chapter 3) an overestimate of the dissociation energy could only mean that the calculated energy of H_2 was below that deduced from experiment. A re-examination of the electronic spectrum of H_2 by Herzberg‡ later established a new dissociation limit in agreement with the calculation, within experimental uncertainty.

The calculation by Kolos and Wolniewicz not only emphasized the power of the variation theorem, but showed that accurate solutions of the Schrödinger equation for molecules would give results in agreement with experiment. There is therefore no question that the equation is of the correct form to describe molecular properties, at least for situations in which relativistic effects are unimportant, although to achieve solutions with the accuracy often achieved in experiments is for most cases not yet possible.

†W. Kolos and L. Wolniewicz, *J. Chem. Phys.*, **49**, 404 (1968).
‡G. Herzberg, *J. Mol. Spectrosc.*, **33**, 147 (1970).

A rigorous proof of expression (6.31) requires a greater emphasis on the mathematical foundations of quantum mechanics than is appropriate for this book. It rests firstly on the fact that the wavefunction for an approximate solution of the Schrödinger equation can be written as a linear combination of the *exact* wavefunctions. This is not easy to prove but appears in the light of the discussion surrounding expansions such as (6.1) and (6.2), to be a reasonable statement. We express it formally as

$$\Psi = N \sum_s a_s \Psi_s, \tag{6.32}$$

where Ψ_s are the exact wavefunctions.

Expression (6.31) applies only to normalized wavefunctions and hence we must choose the value of N in (6.32) accordingly. The following condition must be satisfied

$$\int \Psi^2 \,\mathrm{d}v = N^2 \sum_r \sum_s a_r a_s \int \Psi_r \Psi_s \,\mathrm{d}v = 1. \tag{6.33}$$

The second fact we require is that the integral of a product of two wavefunctions which are different solutions of the Schrödinger equation is zero:

$$\int \Psi_r \Psi_s \,\mathrm{d}v = 0 \quad (r \neq s). \tag{6.34}$$

This is known as the *orthogonality* condition.

The word orthogonal is used in several contexts in mathematics. In the field of vectors it is used for two vectors which are mutually perpendicular: two vectors **a** and **b** are orthogonal if their scalar product **a** · **b** is zero. With this condition the two vectors are independent in the sense that the projection of one vector on the direction of the other is zero. In other words, there is no component of one vector in the other. The term orthogonal has a similar connotation for wavefunctions, implying that they are independent functions.

We can combine (6.34) with the normalization condition to give the expression

$$\int \Psi_r \Psi_s \,\mathrm{d}v = \delta_{rs}, \tag{6.35}$$

where δ_{rs} is called the Kronecker delta and has the value 0 if $r \neq s$ and 1 if $r = s$. Expression (6.35) is usually called the *orthonormality* condition for wavefunctions.

If we apply the condition (6.35) to equation (6.33) we have

$$N^2 \sum_s a_s^2 = 1, \tag{6.36}$$

whence

$$N = \left(\sum_s a_s^2 \right)^{-1/2}. \tag{6.37}$$

A corollary of the orthogonality condition (6.34) is that Hamiltonian integrals such as

$$\int \Psi_r \mathscr{H} \Psi_s \, dv, \tag{6.38}$$

will be zero if $r \neq s$. This is only necessarily true if Ψ_r and Ψ_s are *exact* solutions of the Schrödinger equation as we see from the following proof.

From (2.27) we can simplify (6.38) as follows:

$$\int \Psi_r \mathscr{H} \Psi_s \, dv = \int \Psi_r E_s \Psi_s \, dv = E_s \int \Psi_r \Psi_s \, dv. \tag{6.39}$$

The second step is possible because E_s is a constant and it may therefore be taken outside the integration. Applying the condition (6.35) we have the required result

$$\int \Psi_r \mathscr{H} \Psi_s \, dv = \delta_{rs} E_s. \tag{6.40}$$

After this brief mathematical interlude we may proceed to a proof of the variation theorem. Inserting (6.32) into (6.31), and using expression (6.37) for the normalizing constant, we are required to prove

$$\bar{E} = \left(\sum_s a_s^2 \right)^{-1} \left[\sum_r \sum_s a_r a_s \int \Psi_r \mathscr{H} \Psi_s \, dv \right] \geqslant E_0. \tag{6.41}$$

This simplifies using (6.40) to

$$\bar{E} = \left(\sum_s a_s^2 \right)^{-1} \left[\sum_s a_s^2 E_s \right] \geqslant E_0, \tag{6.42}$$

or

$$\bar{E} - E_0 = \left(\sum_s a_s^2 \right)^{-1} \left[\sum_s a_s^2 (E_s - E_0) \right] \geqslant 0. \tag{6.43}$$

The proof now follows from the fact that a_s^2 is a positive quantity and $E_s - E_0$ is either zero ($s = 0$) or positive because E_0 is the energy of the ground state. Thus all the terms in the bracketted expression are positive, or zero in the special case that all a_s are zero, except a_0 (which means that Ψ is Ψ_0).

There is a straightforward extension of the variation theorem to excited states which we state but do not prove. If $\tilde{\Psi}_i$ is orthogonal to a set of approximate or exact wavefunctions for all lower energy states, and it is normalized, then $\bar{E}_i$ is greater or equal to the exact energy E_i, that is

$$\bar{E}_i = \int \tilde{\Psi}_i \mathscr{H} \tilde{\Psi}_i \, dv \geqslant E_i. \tag{6.44}$$

The variation theorem is used to obtain approximate wavefunctions in the following manner. A function is selected whose general form is, by experience, capable of providing a satisfactory representation of the wavefunction, but which contains within it parameters whose values have yet to be determined. The energy is then calculated by expression (5.14) as a function of these parameters and the 'best' wavefunction is taken as the one for which the calculated energy is a minimum. This energy will then be as close to the exact energy as can be obtained with the type of function selected, and the parameters which give this energy define the optimum wavefunction.

Unfortunately an energy minimizing criterion does not imply that other properties calculated from the resulting wavefunction according to expression (5.9) will necessarily be closest to the exact values of these properties. We hope, and in many cases find, that energy-minimized wavefunctions give the best predictions for other properties but there is no theorem to prove that this is the case.

As an example of the application of the variation theorem which we have already discussed we have the radial wavefunctions of atomic orbitals taken as the Slater functions (3.35). The exponents (ζ) of these functions are determined as those values which minimize the energies of the ground states of the atoms, and selected optimum exponents calculated by this recipe have been given in Table 3.2.

In some cases it is possible to obtain a simple expression for the energy in terms of variational parameters in the wavefunction and this expression can then be differentiated and the resulting expression equated to zero to find the minimum in the usual way. In many cases, however, this is not possible and the energy must then be obtained for selected numerical values of the parameters and a curve-fitting criterion applied to determine the energy minimum. This might be called a numerical as opposed to an analytical method of minimizing the energy. There is one type of variational parameter for which the minimum can always be found by analytical means and that is when the parameters to be varied are a set of linear expansion coefficients such as the coefficients c_n of expression (6.2). As this procedure is central to the determination of the LCAO coefficients of molecular orbitals, we will go through the necessary derivation in detail.

The task is to find the coefficients in the expansion

$$\psi = N \sum_n c_n \phi_n, \tag{6.45}$$

which minimize the energy

$$\bar{E} = \int \psi \mathscr{H} \psi \, \mathrm{d}v, \tag{6.46}$$

with the restriction that ψ shall be normalized. Expression (6.45) will typically be the LCAO expansion of a molecular orbital or the expansion of the many electron wavefunction in terms of orbital product functions of the type we shall examine in Chapter 8. We emphasize that the set of functions ϕ_n are not in general eigenfunctions of the same Hamiltonian nor individually are they necessarily eigenfunctions of any Hamiltonian. Thus conditions such as (6.34) and (6.40) will not in general hold for the functions ϕ.

The normalization of (6.45) is imposed through the equation

$$N^2 \sum_m \sum_n c_m c_n \int \phi_m \phi_n \, \mathrm{d}v = 1. \tag{6.47}$$

In (6.15) we introduced the symbol for the overlap integral

$$S_{mn} = \int \phi_m \phi_n \, dv, \tag{6.48}$$

and this is now generalized to include cases in which ϕ is any function, not only an atomic orbital. Equation (6.47) therefore leads to a normalization condition

$$N = \left[\sum_m \sum_n c_m c_n S_{mn} \right]^{-\frac{1}{2}}. \tag{6.49}$$

Writing the Hamiltonian integrals as (c.f. 6.22)

$$H_{mn} = \int \phi_m \mathscr{H} \phi_n \, dv, \tag{6.50}$$

we can substitute (6.45) into (6.46) and with the use of (6.49) obtain

$$\bar{E} = \left[\sum_m \sum_n c_m c_n S_{mn} \right]^{-1} \left[\sum_m \sum_n c_m c_n H_{mn} \right]. \tag{6.51}$$

Expression (6.51) can be re-written as

$$\sum_m \sum_n c_m c_n (H_{mn} - \bar{E} S_{mn}) = 0, \tag{6.52}$$

and the requirement is that $\bar{E}$ shall be minimized for variations in all the coefficients c_n. If we differentiate (6.52) with respect to c_n keeping all other c_m constant, we obtain the result

$$2 \sum_m c_m (H_{mn} - \bar{E} S_{mn}) - \left[\sum_m \sum_n c_m c_n S_{mn} \right] \frac{\partial \bar{E}}{\partial c_n} = 0, \tag{6.53}$$

where the factor of 2 occurs because c_m occurs both as $c_m \, c_n$ and $c_n \, c_m$ in the double summation. Notice that both S_{mn} and H_{mn} are independent of the coefficients but $\bar{E}$ is a function of the coefficients.

At a minimum we have $(\partial \bar{E} / \partial c_n) = 0$ hence (6.53) becomes

$$\sum_m c_m (H_{mn} - \bar{E} S_{mn}) = 0. \tag{6.54}$$

There is one such equation for each coefficient c_n and they are known in mathematics as the *secular equations*. The set of equations is satisfied by the trivial solution that all the coefficients are zero, but this solution is of no interest: a non-trivial solution exists provided that the following determinant (known as the secular determinant) is zero

$$| H_{mn} - \bar{E} \, S_{mn} | = 0. \tag{6.55}$$

This is a standard result in the theory of simultaneous equations and we shall shortly prove it for the simple case of two variable coefficients.

If the determinant (6.55) is expanded it leads to an Nth degree polynomial in $\bar{E}$ (if there are N variable coefficients) and this has N solutions which may be ordered as follows: $\bar{E}_0, \bar{E}_1, \ldots \bar{E}_{N-1}$. If each of these solutions is substituted in turn into the secular equations (6.54) the coefficients appropriate to each solution may be found. The lowest energy $\bar{E}_0$ leads to a wavefunction which is a variational approximation to the ground state and the higher energies are variational approximations to excited states.

Before we turn to a practical application of the above equations we will make an algebraic analysis of a 2-variable problem. This will enable us to draw some general conclusions about the way in which atomic orbitals combine to form molecular orbitals. In particular we can establish the two rules that were used in Section 6.2 to determine the mixing of $2s$ and $2p$ atomic orbitals in the molecular orbitals of homonuclear diatomic molecules.

We consider a basis of two functions (ϕ_a, ϕ_b) which are atomic orbitals of neighbouring atoms and which interact to give molecular orbitals of the form

$$\psi = c_a\phi_a + c_b\phi_b. \tag{6.56}$$

The secular equations (6.54) in this case are

$$c_a(H_{aa} - \bar{E}S_{aa}) + c_b(H_{ab} - \bar{E}S_{ab}) = 0, \tag{6.57}$$

$$c_b(H_{ba} - \bar{E}S_{ba}) + c_b(H_{bb} - \bar{E}S_{bb}) = 0. \tag{6.58}$$

Without any loss of generality the basis functions can be taken as normalized ($S_{aa} = S_{bb} = 1$). Another simplification arises from the fact that the H and S integrals are symmetric in their indices. This is obvious for the overlap integrals as the order of the two functions in an integral like (6.48) is unimportant. It is not always obvious for the H integrals in view of the fact that H contains differential operators, but it can be proved that this is so for the Hamiltonian as a special class of operator. The secular equations therefore have the form

$$c_a(H_{aa} - \bar{E}) + c_b(H_{ab} - \bar{E}) = 0. \tag{6.59}$$

$$c_a(H_{ab} - \bar{E}S_{ab}) + c_b(H_{bb} - \bar{E}S) = 0. \tag{6.60}$$

If a value for $\bar{E}$ is known then either (6.59) or (6.60) allows us to deduce the ratio $c_a : c_b$. From (6.59)

$$\frac{c_a}{c_b} = -\frac{H_{ab} - \bar{E}S_{ab}}{H_{aa} - \bar{E}}. \tag{6.61}$$

From (6.60)

$$\frac{c_a}{c_b} = -\frac{H_{bb} - \bar{E}}{H_{ab} - \bar{E}S_{ab}}. \tag{6.62}$$

The ratio must clearly be the same in both cases hence equating the left-hand sides of (6.61) and (6.62) and re-arranging we obtain the condition

$$(H_{aa} - \bar{E})(H_{bb} - \bar{E}) = (H_{ab} - \bar{E}S_{ab})^2. \tag{6.63}$$

This condition is identical to that provided by equating the secular determinant

(6.55) to zero

$$\begin{vmatrix} H_{aa} - \bar{E} & H_{ab} - \bar{E}S_{ab} \\ H_{ab} - \bar{E}S_{ab} & H_{bb} - \bar{E} \end{vmatrix} = 0, \tag{6.64}$$

if one follows the standard rule for expanding a determinant.

Rearranging (6.63) leads to the following quadratic equation in $\bar{E}$:

$$\bar{E}^2(1 - S_{ab}^2) - E(H_{aa} + H_{bb} - 2H_{ab}S_{ab}) + (H_{aa}H_{bb} - H_{ab}^2) = 0, \tag{6.65}$$

The roots of this can be given explicity in terms of the H and S integrals, but the expressions are a little complicated. We will look at the simpler situation in which S_{ab} can be neglected, as the conclusions we arrive at are valid for most practical values of the overlap integral (remember that the absolute magnitude of S_{ab} can never be greater than unity).

Taking $S_{ab} = 0$ in (6.65) gives a quadratic equation whose roots are

$$\bar{E} = ½[(H_{aa} + H_{bb}) \pm [(H_{aa} - H_{bb})^2 + 4H_{ab}^2]^{½}]. \tag{6.66}$$

Two limiting cases are of particular interest. Firstly, if $H_{aa} = H_{bb}$ (the situation in homonuclear diatomic molecules for two atomic orbitals of the same type) then the two solutions are

$$\bar{E} = H_{aa} \pm H_{ab}. \tag{6.67}$$

In other words, the two energies are disposed either side of H_{aa} and separated by $2H_{ab}$. This leads to the symmetrical splitting between bonding and antibonding molecular orbitals typified by Figure 6.6.

The second situation is if $|H_{ab}| \ll |H_{aa} - H_{bb}|$. In this case the square root in (6.66) can be expanded by the binomial theorem in powers of the ratio

$$\frac{H_{ab}}{H_{aa} - H_{bb}}, \tag{6.68}$$

to give from the leading terms the two solutions

$$\bar{E} = H_{aa} + \frac{H_{ab}^2}{H_{aa} - H_{bb}}, \tag{6.69}$$

and

$$\bar{E} = H_{bb} - \frac{H_{ab}^2}{H_{aa} - H_{bb}}. \tag{6.70}$$

In this case the energies of the molecular orbitals are shifted from the atomic orbital energies by

$$\frac{H_{ab}^2}{H_{aa} - H_{bb}}, \tag{6.71}$$

one level going up in energy and one going down. Moreover if $H_{aa} > H_{bb}$ then it is the level whose energy is approximately H_{aa} that goes up and that whose energy is approximately H_{bb} that goes down: the energy shift acts as though the two levels are repelling one another. Clearly if the ratio (6.68) is very small, either by the fact that H_{ab} is small or $|H_{aa} - H_{bb}|$ large, then the energies are shifted by only a small amount. In this case if we substitute (6.69) or (6.70) back into (6.62) to find the ration of the coefficients we find that this ratio is very small or very large. In other words, the wavefunction (6.56) is dominated by one of the atomic orbitals. This result justifies our assumption in section (6.2) that we can ignore the mixing of two atomic orbitals whose energies are very different or for which the Hamiltonian interaction term is small or zero. It is a result that can be proved more generally by perturbation theory, a topic that will be dealt with in Chapter 11.

In expressions (6.31)–(6.71) we have consistently used a bar over the variational energy to indicate that it is not an exact eigenvalue of the Hamiltonian. If the variational wavefunction is completely flexible, and this may in principle be arranged by having an expansion such as (6.45) of infinite length, then $\bar{E}$ will equal the exact energy. If this is the case then we can obtain the secular equations (6.54) by a shorter and simpler derivation than that already given.

If we substitute the expansion (6.45) into the Schrödinger equation

$$(\mathscr{H} - E)\psi = 0, \tag{6.72}$$

we obtain

$$\sum_n c_n(\mathscr{H} - E)\phi_n = 0. \tag{6.73}$$

This equation is no simpler to solve than (6.72) but we can convert it into a set of simultaneous equations by multiplying (6.73) from the left by each of the basis functions in turn and integrating over the variables. If we multiply by ϕ_m and integrate we get

$$\sum_n c_n(H_{mn} - ES_{mn}) = 0 \tag{6.74}$$

where the symbols have the same meaning as in (6.48) and (6.50). These equations (one for each m) are identical in form with the secular equations except that E appears instead of $\bar{E}$. This derivation of (6.74) holds only if (6.72) is satisfied and in practice this will not in general be so unless the expansion is infinite. However, we will for brevity drop the bar over the energy when we use the secular equations in the rest of this book whilst recognizing that the energies we obtain by solution of the secular equations will not be exact eigenvalues of the Hamiltonian but some variational approximation to them.

6.4. Molecular orbitals of heteronuclear diatomic molecules

As an illustration of the use of the secular equations to determine molecular orbital coefficients we will consider two simple heteronuclear diatomic molecules LiH and HF. LiH is the simplest heteronuclear molecule of any chemical importance. As it has four electrons we are faced immediately with making some allowançe for the repulsion between them. As an exercise the one-electron system HeH^{2+} would be simpler and amenable to an exact solution, but the calculation would not have much chemical significance.

As a basis for the molecular orbitals we take the hydrogen 1*s* and the lithium 1*s* and 2*s* atomic orbitals. If we were aiming for an accurate representation of the molecular orbitals we would certainly include the lithium 2*p* orbitals in the basis, but we choose the smaller basis to simplify the analysis.

The hydrogen 1*s* orbital has the wavefunction and energy

$$1s_{\mathrm{H}} = \pi^{-\frac{1}{2}} \exp(-r), \quad E(1s_{\mathrm{H}}) = -0.5E_{\mathrm{H}}. \tag{6.75}$$

For the lithium 1*s* orbital we take a Slater-type function such as (3.35) which when normalized is

$$1s_{\mathrm{Li}} = (\zeta_1^3/\pi)^{\frac{1}{2}} \exp(-\zeta_1 r). \tag{6.76}$$

The lithium 2*s* orbital we take to be a noded function as (3.38) which is orthogonal to $1s_{\mathrm{Li}}$ and a good representation of the Hartree–Fock 2*s* atomic orbital

$$2s_{\mathrm{Li}} = \lambda_2(\zeta_2^5/3\pi)^{\frac{1}{2}}\, r \exp(-\zeta_2 r) - \lambda_1\, 1s_{\mathrm{Li}}. \tag{6.77}$$

A variational calculation on the lithium atom in which ζ_1, ζ_2, λ_1, and λ_2 are treated as variational parameters (that is the energy is minimized with respect to all of these parameters) leads to the result

$$\zeta_1 = 2.69, \quad \zeta_2 = 0.64, \quad \lambda_2 = 1.10, \quad \lambda_1 = 0.18, \tag{6.78}$$

and the energies of the two orbitals are

$$E(1s_{\mathrm{Li}}) = -2.45E_{\mathrm{H}}, \quad E(2s_{\mathrm{Li}}) = -0.19E_{\mathrm{H}}. \tag{6.79}$$

The overlap integrals between the three basis functions at the experimental internuclear distance of LiH (3.0 a_0) are as follows:

$$S(1s_{\mathrm{H}}, 1s_{\mathrm{Li}}) = 0.096, \quad S(1s_{\mathrm{H}}, 2s_{\mathrm{Li}}) = 0.463, \quad S(1s_{\mathrm{Li}}, 2s_{\mathrm{Li}}) = 0. \tag{6.80}$$

We see that there is only a small overlap between the lithium 1*s* and the hydrogen orbital and in addition we note that the lithium 1*s* orbital has a much lower energy than the other two. From the analysis of the last two sections we conclude that the dominant effect on bringing the two atoms together will arise from the mixing of the hydrogen 1*s* and lithium 2*s* orbitals. We will therefore consider the molecular orbitals of the valence electrons as a linear combination of two basis functions

$$\psi = c_{\mathrm{a}}\phi_{\mathrm{a}} + c_{\mathrm{b}}\phi_{\mathrm{b}} \tag{6.81}$$

where

$$\phi_{\mathrm{a}} = 2s_{\mathrm{Li}} \text{ and } \phi_{\mathrm{b}} = 1s_{\mathrm{H}}.$$

The effective one-electron Hamiltonian for the valence electrons consists of the kinetic energy operator $(-\frac{1}{2}\nabla^2)$ and the following potential energy terms:

1. the attraction to the proton ($-1/r_{\mathrm{H}}$);
2. the attraction to the lithium nucleus ($-3/r_{\mathrm{Li}}$);
3. the repulsion by the two lithium $1s$ electrons;
4. the repulsion by the other valence electron.

The third of these terms is a little complicated but can be calculated knowing the wavefunction of the lithium $1s$ orbital. By the laws of electrostatics we know that the potential outside a spherical shell of charge is the same as if the charge were placed at the centre, and inside a spherical shell the potential is a constant equal to the charge in that shell. Thus, far from the nucleus the potential experienced by a valence electron from the two $1s$ electrons is ($2/r_{\mathrm{Li}}$) and this monotonically increases to the limiting value +2 at the lithium nucleus. If we combine this repulsive potential with term 2, which is the attraction to the lithium nucleus, we obtain the effective attractive potential of the ion Li^+, and this is shown in Figure 6.8.

It is more difficult to calculate term 4, the repulsion by the other valence electron because we do not yet know the wavefunction of the valence molecular orbital. As we have explained in Chapter 3 when discussing the orbitals of many-electron atoms, this problem is solved by making a first guess at the wavefunctions (that is at c_{a} and c_{b}) and calculating an approximate potential. The Hamiltonian integrals can then be calculated from this potential and the solution of the resulting secular equations will give us a better approximation to the coefficients. On repeating this cycle several times we converge on the final result.

In practice, for this simple two-basis problem, the SCF procedure converges quite rapidly on the final solution whatever the initial assumption made about the

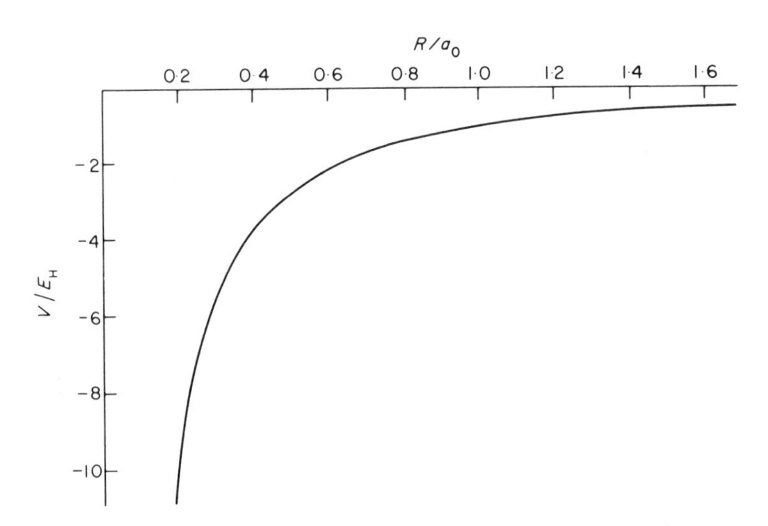

Figure 6.8 The effective attractive potential (V) of Li^+.

coefficients c_a and c_b. We shall show only the final step in this iterative process. The Hamiltonian matrix elements have been calculated to have the values (in atomic units)

$$H_{aa} = -0.102, \quad H_{bb} = -0.267, \quad H_{ab} = -0.223, \tag{6.82}$$

Noting the values of the overlap integral (equation 6.80) this leads to the secular equations (6.59) and (6.60):

$$\begin{aligned} c_a(-0.102 - E) + c_b(-0.223 - 0.463\,E) &= 0, \\ c_a(-0.223 - 0.463\,E) + c_b(-0.267 - E) &= 0, \end{aligned} \tag{6.83}$$

and expansion of the secular determinant gives the following quadratic equation in E

$$0.785\,6\,E^2 + 0.162\,5\,E - 0.022\,5 = 0. \tag{6.84}$$

The roots of this equation are $E = -0.302$ and 0.095.

The lowest energy state will have two electrons in the orbital of energy $-0.302E_H$. To find the coefficients of this orbital we substitute this value for the energy back into the secular equations (6.83) and this gives us the ratio

$$c_a/c_b = 0.416 \tag{6.85}$$

The absolute values of the coefficients are determined by the condition that the molecular orbital be normalized, and this leads, from equation (6.47), to

$$c_a^2 + c_b^2 + 2c_a c_b S_{ab} = 1. \tag{6.86}$$

The solution of equations (6.85) and (6.86) gives us the molecular orbital

$$\psi = 0.333\phi_a + 0.801\phi_b. \tag{6.87}$$

We emphasize again that if the potential is calculated from this orbital then the Hamiltonian matrix elements of equation (6.82) are obtained.

From the wavefunction (6.87) we can calculate the distribution of the two valence electrons in the molecule as follows:

$$2\psi^2 = 0.222\phi_a^2 + 1.283\phi_b^2 + 1.067\phi_a\phi_b. \tag{6.88}$$

The electron density has contributions from each of the atomic orbitals and a cross term which is called the overlap density. For a simple picture it is usual to divide up this overlap density equally between the two atomic centres by the replacement

$$\phi_a\phi_b \rightarrow \frac{S_{ab}}{2}(\phi_a^2 + \phi_b^2). \tag{6.89}$$

This condition conserves the total density as can be seen by integrating both sides over all space to give the result S_{ab}. After substituting (6.89) into (6.88) we obtain the so-called *Mulliken population*

$$2\psi^2 = 0.469\phi_a^2 + 1.531\phi_b^2. \tag{6.90}$$

We see that there is a greater electron population of the hydrogen orbital (ϕ_b) than of the lithium orbital (ϕ_a). The net charge in the molecule is calculated by combining the electron charge with the nuclear charges. The electron population on the lithium atom is 2.469 (remembering the two 1*s* electrons) hence the net charge is $+0.531e$. The net charge on the hydrogen atom is $+1-1.531 = -0.531e$.

The above result is consistent with our chemical knowledge of LiH as a species in which there is a dipole moment in the sense Li^+H^-. The experimental value for this is 5.8 D (Table 5.1) and our calculation gives 4.0 D.

We will contrast the above calculation on LiH with an analysis of the molecular orbitals of HF. To simplify the analysis we make the approximation that the fluorine 2*s* orbital is non-bonding and contains, in the ground state, two electrons. This is an extension of the approximation made for LiH that the lithium 1*s* orbital is non-bonding. It is a less justified approximation both on the criterion of energy and the magnitude of the overlap integrals between the 2*s* and the hydrogen 1*s* orbitals but we shall see it leads to results which are qualitatively similar to those from more exact calculations.

The basis for the calculations will be the fluorine 2*p* orbitals and the hydrogen 1*s* orbital. Following the procedure used for the homonuclear diatomics we choose axes such that the 2*p* orbitals may be classified as $2p\sigma$ and $2p\pi$ (doubly-degenerate). The latter will also be non-bonding as there are no atomic orbitals of the hydrogen of π symmetry in our basis. The orbital energy diagram is therefore as shown in figure 6.9.

To make the calculation we will use the zero-overlap approximation which led to expression (6.66). Instead of attempting a calculation of the H integrals we will approximate the difference between H_{aa} and H_{bb} by the difference in ionization potentials of the orbitals of the free atoms. In other words, we are making use of Koopmans' theorem and also assuming that the effect on these atomic energies due to the presence of the other atom is not large. The 2*p* energy deduced from atomic spectral data is -18.7 eV.† To arrive at this value it is necessary to average over several states of the atom and ion which have the same atomic configuration. The hydrogen 1*s* energy is -13.6 eV.

The π orbital energy will in our model be the same as the atomic 2*p* energy except for the effect of polarity in the HF bond. Our chemical experience tells us that the molecule has an electron distribution which is dipolar in the sense H^+F^-. The negative charge on the fluorine will lead to an electrostatic destabilization of the π electrons and hence a reduction of the ionization potential relative to that of the free atom. The photoelectron spectrum of HF‡ shows that the π ionization potential is 16.0 eV, so that this destabilization is 2.7 eV. This may be thought large, but it should be noted that the repulsion energy of two electrons 1 Å apart is 14 eV.

The σ ionization potential of HF is 19.1 eV, and the difference between this and

†As the parameters for this calculation are deduced from photoelectron spectroscopy it is convenient to use electron volts as energy units.
‡ C. R. Brundle, *Chem. Phys. Letters.*, 7, 317 (1970).

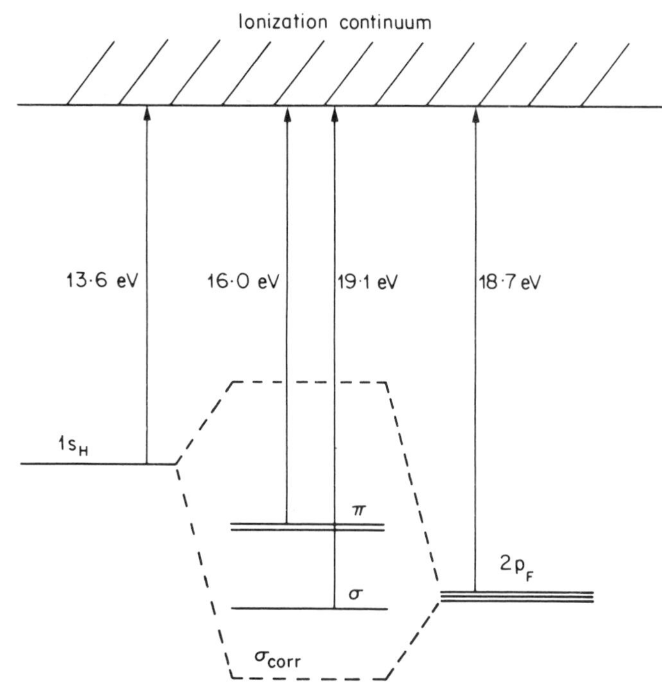

Figure 6.9 Orbital energy diagram for HF. σ_{corr} is the energy of the bonding σ orbital corrected for electron repulsion by the amount deduced from the energy of the π orbitals.

the π ionization potential (3.1 eV) will, in our model, arise from the interraction between the $2p\sigma$ orbital and the hydrogen $1s$ orbital. The electrostatic repulsion, referred to above, we assume to be the same for σ and π electrons on the fluorine. Thus before we correct for this repulsion, the energy of the bonding σ molecular orbital would be 19.1 + 2.7 = 21.8 eV. We now ask what value of the interaction Hamiltonian element H_{ab} will reproduce this molecular orbital energy. Let us first calculate this from expression (6.69). Taking the suffixes a and b to refer to the fluorine and hydrogen orbitals respectively we find

$$-21.8 = -18.7 + \frac{H_{ab}^2}{-18.7 + 13.6} \tag{6.91}$$

whence

$$|H_{ab}| = 4.0 \text{ eV}. \tag{6.92}$$

The sign of H_{ab} is not determined by this analysis and will in fact depend on our choice of sign for the $2p$ orbital (whether the positive lobe points towards or away from the hydrogen atom). For positive overlap integrals between the two atomic orbitals, H_{ab} will be negative.

The validity of expression (6.69) rests on the assumption that $|H_{ab}|$ is considerably smaller than $|H_{aa}-H_{bb}|$ but we have deduced from the above equations that this is not the case. We must therefore go back to expression (6.66) and obtain from this the value of H_{ab} which leads to the adjusted σ orbital energy of -21.8 eV. Substituting the appropriate quantities gives

$$-21.8 = \tfrac{1}{2}[-18.7 - 13.6 - [(-18.7 + 13.6)^2 + 4H_{ab}^2]^{1/2}], \tag{6.93}$$

whence

$$|H_{ab}| = 5.0 \text{ eV}. \tag{6.94}$$

We note that this is a slightly larger value than was deduced from the perturbation expression (6.69).

We are now in a position to calculate the wavefunction of the bonding σ molecular orbital by substituting this value of H_{ab} and the stabilization energy into the secular equations. From (6.59), with the zero overlap approximation and the choice of negative sign for H_{ab}, we have

$$c_a(3.1) + c_b(-5.0) = 0, \tag{6.95}$$

whence

$$\psi = 1.6\phi_a(2p\sigma) + \phi_b(1s_H). \tag{6.96}$$

As anticipated this molecular orbital has a greater amplitude on the fluorine than on the hydrogen. The two electrons in this orbital are $(1.6/2.6)^2 \times 100 = 62\%$ on the fluorine and 38% on the hydrogen so that the net charge on the hydrogen is $1 - 2 \times 0.38 = +0.24e$. Accurate calculations using the minimal basis and including the fluorine $2s$ orbitals in the molecular orbital give a net charge of $+0.15e$.

The above calculation on HF makes no pretence to accuracy and has had very limited objectives. The input data for the calculation have been the ionization potentials of the atoms and the molecule and the output has been the predicted electron distribution in the molecule. In calculations of this type the aim should be to maximize the ratio of output to input data and by considering one molecule in isolation from similar molecules as we have done it is difficult to get this ratio high. When we come to discuss similar calculations for organic molecules in Chapter 9 it will be seen that such calculations can have considerable predictive value.

6.5 The *ab-initio* and empirical approaches

Our discussion of the molecular orbitals of LiH and HF in the last section has shown two different approaches which are widely used in valence theory. In the LiH calculation we started with basis functions defined explicitly by algebraic functions and a Hamiltonian for the molecular orbitals defined by the kinetic energy and potential energy operators for the electrons. The integrals required to

† M. Krauss, = *J. Chem. Phys.*, **28**, 1021 (1958).

set up the secular equations could then be calculated explicitly and after carrying through the SCF procedure molecular orbitals were derived. Although we made an approximation in assuming that the lithium 1*s* orbitals were not involved in the valence molecular orbitals, at no time did we make use of any experimental data concerning the atoms or molecule to deduce values for the H and S integrals. Such a calculation is referred to as *ab initio* or approximate *ab initio* in the case that some approximations are made in calculating the H and S integrals.

Our discussion of HF was in a completely different vein. We took the energies of the basis functions from experimental atomic energy levels and estimated the interaction term H_{ab} by using observed ionization potentials of the molecule. Such a calculation is referred to as empirical meaning 'based upon experiment'. In a strict sense the whole of quantum mechanics is empirical in so far as its foundations rest on experimental observations, but in valence theory we use the word to contrast such calculations with *ab-initio* calculations in which no specific atomic or molecular experimental data are used.

Empirical valence theories have had a long period of development and have played an important role in the development of transition metal chemistry, solid state physics, and the chemistry of unsaturated organic hydrocarbons. These topics will be discussed in later chapters. In recent years, with the advances in computational techniques, the empirical methods have given way to some extent to *ab-initio* calculations. However, their place within the total scheme of chemistry is assured by the contributions they have made to chemical ideas and concepts, and it is for this reason that they will receive considerable attention in this book.

6.6 The isoelectronic principle

In the Lewis theory of valence the electrons in a molecule are associated with individual atoms but in bonding situations some are shared between two atoms. We shall see in Chapter 14 that this picture has been developed in quantum mechanical language into *Valence Bond* (VB) theory. In MO theory, in contrast, there is no identification of electrons with individual atoms. The emphasis is instead on molecular orbitals which in general are diffused over the whole molecule. The connection with atomic orbitals has been made only with the LCAO expansion which can be considered as a mathematical convenience having strong physical justification.

Because MO theory places no emphasis on where the electrons in a molecule originate from, it is ideally structured to interpret the isoelectronic concept which is widely used in chemistry. This is that molecules with similar nuclear configurations and the same number of electrons often have many similar properties. This is particularly true of physical properties such as ionization potential but less true of chemical reactivity.

Examples of the isoelectronic concept will be found later in the book, for example when comparing the properties of benzene and pyridine. It is opportune to develop the idea at this point because of the examples provided by the diatomic molecules of the first row elements.

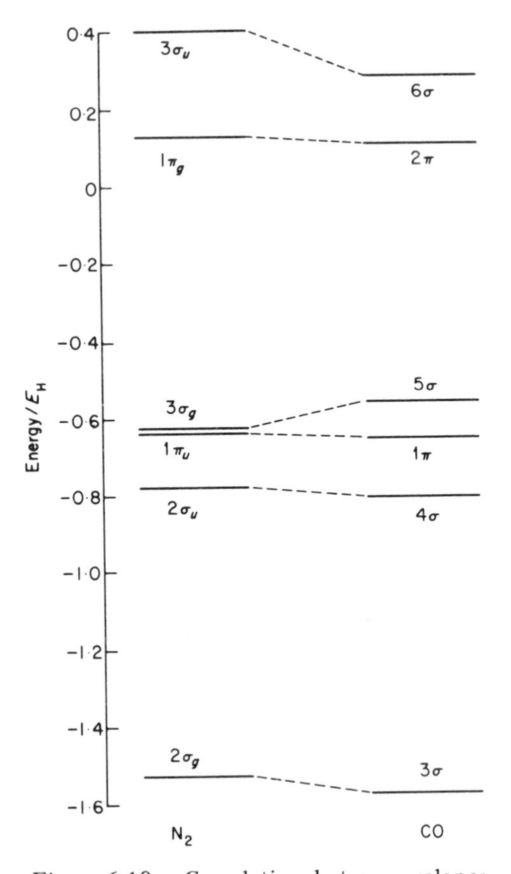

Figure 6.10 Correlation between valence orbital energies of N_2 and CO (L. C. Snyder and H. Basch, *Molecular Wave Functions and Properties*, Wiley-Interscience, New York, 1972).

Two molecules that are isoelectronic are N_2 and CO. The contrast in the chemical reactivity of CO and comparative unreactivity of N_2 hide their other similarities. Moreover, recently prepared complexes such as

trans-Re Cl [XY] [P Me_2 Ph]$_4$

where XY = N_2 or CO show that N_2 and CO can have similar properities when they occur as ligands in transition metals complexes.

The valence molecular orbitals of N_2 and CO have a 1:1 relationship and their energies are similar as shown by calculations or by their photoelectron spectra. The situation is summarized in Figure 6.10. The molecular orbitals of CO cannot be classified by the g, u symbols for inversion symmetry that are used for the homonuclear diatomic molecules (see Figure 5.8) because the coefficients of the

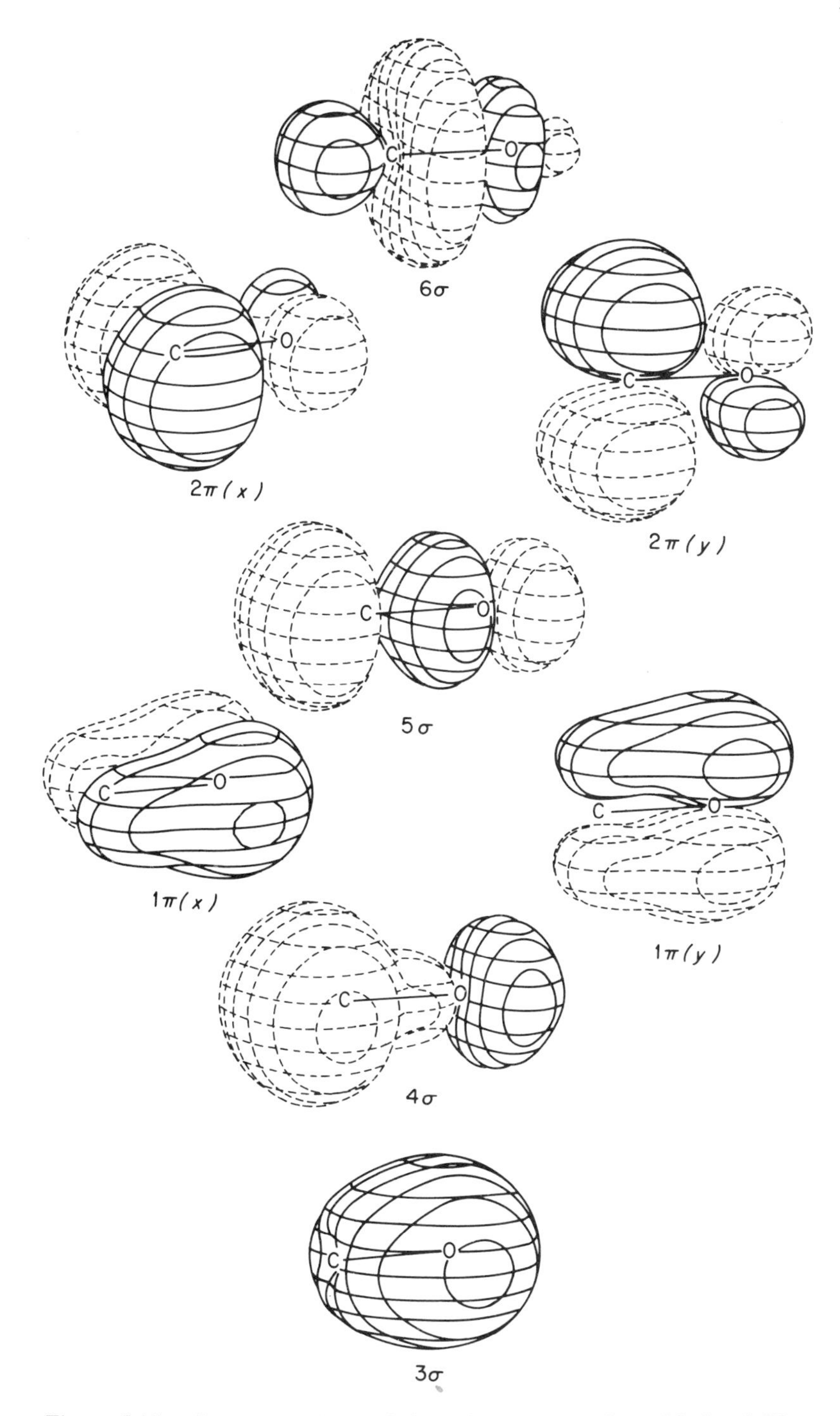

Figure 6.11 Representations of the valence molecular orbitals of CO (Jorgensen and Salem, see Figure 6.4).

carbon and oxygen atomic orbitals in any molecular orbital will not have the same magnitude. Hence the lowest energy valence orbital is the third σ orbital (the two inner shell orbitals being σ) rather than the second σ_g orbital.

Because oxygen is a more electronegative atom than carbon, that is electrons are more strongly bound to oxygen, the bonding molecular orbitals of CO (e.g. 3σ, 1π) tend to have a greater weighting of oxygen atomic orbitals than of carbon orbitals, and the anti-bonding (e.g. 4σ, 2π) a greater weighting of carbon atomic orbitals than of oxygen orbitals. This is shown by the 3-dimensional representations of the valence orbitals of CO shown in Figure 6.11.

In the ground state of CO the molecular orbitals up to and including 5σ are all filled with electrons. By analogy with N_2 we would consider this a triply bonded situation which could be represented by the classical valence structure $C^{-}\equiv O^{+}$. However, orbital 5σ is not as strongly bonding for CO as $3\sigma_g$ is for N_2, and a doubly bonded structure C=O in which the carbon is divalent might be more appropriate. A comparison of the force constants for the bonds of N_2, CO, and O_2, which are 2 294, 1 902, and 1 177 N m^{-1} respectively,† is in accord with a structure intermediate between these extremes.

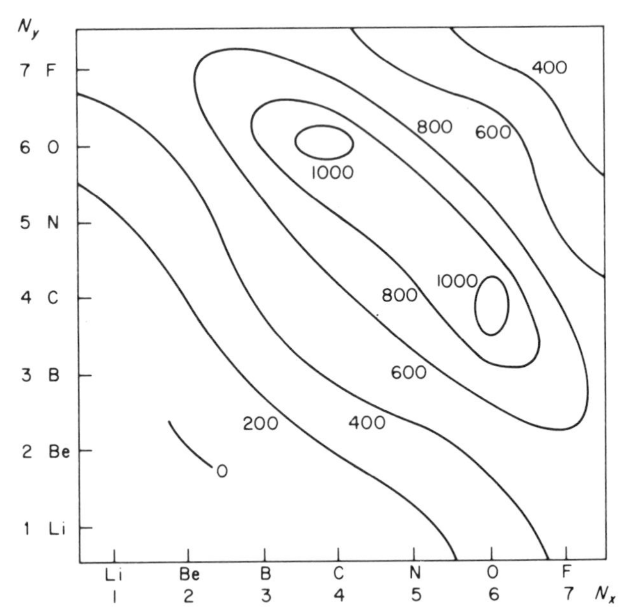

Figure 6.12 The dissociation energies (contours marked in kJ mol^{-1}) of the first row diatomic molecules $D_e(N_x, N_y)$ considered as a continuous function of the numbers of valence electrons of the separate atoms. (The authors are indebted to W. F. Sheehan for pointing out this correlation.)

† 10^2 Nm^{-1} = 1 mdyne Å^{-1}.

One of the most striking indications of the isoelectronic principle can be seen from the dissociation energies of the first row diatomic molecules.

If one imagines these as the grid values of a continuous function $D_e(N_x,N_y)$ where N_x and N_y are the number of electrons of the two atoms, then the contours of this surface, as sketched in Figure 6.12, are predominantly parallel to the lines $N_x + N_y$ = constant. The dissociation energies decrease slightly as the difference between N_x and N_y increases (the increase on going from N_2 to CO being an exception) but the approximate constancy for isoelectronic molecules is the dominant pattern of Figure 6.12.

6.7. Inert-gas configurations and the formal oxidation state

In the early theories of valence developed by G. N. Lewis and others the dominant concept was that atoms exchanged or shared electrons so as to adopt electron configurations which had special stability or inertness to further chemical reaction. The rule of eight proposed separately by Kossel and Lewis in 1916 was the most important guide to stability, it being recognized that the inert gases all had an outer shell of eight electrons. Later, Lewis changed the emphasis from the rule of eight to the rule of two, with his concept of the electron-pair bond.

For transition metal compounds the relevant inert gases (Kr, Xe, Rn) have an outer shell of eighteen electrons if the *d* electrons are included in this number, and, as we shall see, the eighteen electron rule, first postulated by Sidgwick, has had some success in rationalizing the formulae of known compounds.

A chemist normally uses the adjective stable for a molecule that is unreactive under normal laboratory conditions. That is, it does not dissociate or polymerize or react with water or oxygen at normal temperatures. Advances in laboratory techniques, using dry-boxes, vacuum lines, and low temperatures have meant that many compounds can now be prepared and their properties investigated which are not stable under the above defined conditions. The spectroscopist can go further and examine molecules or ions in very small concentrations, or species which have a very short lifetime, using techniques such as flash photolysis. A spectroscopist might consider a molecule stable if its potential energy surface (as defined in Section 5.1) has a minimum. Thus H_2^+ is spectroscopically stable, as we have seen in Figure 5.1, yet it is not a chemically stable species. The 2, 8, or 18-electron rules are therefore only meant to apply to chemically stable molecules.

One measure of chemical stability is the total energy of a molecule (or system of molecules) relative to the energies of molecules into which it may be converted. For example, *ab-initio* SCF MO calculations on some hydrides of boron, carbon, and nitrogen give the following energies for complexation:†

	kJ/mol^{-1}		kJ/mol^{-1}
$BH_3 + BH_3 \rightarrow B_2H_6$	86	$CH_3 + CH_3 \rightarrow C_2H_6$	284
$BH_3 + NH_3 \rightarrow BH_3.NH_3$	89	$CH_4 + NH_3 \rightarrow CH_4.NH_3$	2.

†W. A. Lathan, L. A. Curtiss, W. J. Hehre, J. B. Lisle, and J. A. Pople, *Progr. Phys. Org. Chem.*, **11**, 175 (1974); J. D. Dill, P. V. R. Schleyer, and J. A. Pople, *J. Am. Chem. Soc.*, **97**, 3402 (1975).

The species BH_3 and CH_3, which do not satisfy the octet rule are considerably stabilized by dimerization or by combining with a stable species (e.g. $BH_3 + NH_3$), whereas there is almost no decrease in energy on complexing the two species (CH_4 and NH_3) which do obey the octet rule. Although it is known that calculations of this type will not always give accurate complexation energies, due to the fact that the SCF method does not give an exact solution to the Schrödinger equation, the striking difference in these energies is in accord with the inert-gas rules.

Chemical stability depends not only on whether there is a point on the potential energy surface of lower energy, but whether such points can be reached without going over a large potential energy barrier. For example CH_3CN has a calculated energy which is 72 kJ mol^{-1} lower than that of its isomer CH_3NC‡ and yet CH_3NC is chemically stable. There must therefore be a high barrier ($> kT$) between the CH_3NC minimum and the CH_3CN minimum on the potential energy surface. We refer to CH_3NC as kinetically stable but thermodynamically unstable.

Other features of MO calculations that are relevant to chemical stability are the energies of the highest occupied and the lowest empty molecular orbitals, commonly referred to as the HOMO and LUMO energies respectively. If the HOMO energy is high then it is easy to remove an electron from that orbital and the molecule will be easily oxidized. If the LUMO energy is low then it will readily accept an electron from another species and the molecule will easily be reduced. In free radicals the highest occupied orbital is only half full, and this therefore fills the role of both HOMO and LUMO, thus one of these criteria for instability is certain to be satisfied.

In general, compounds which satisfy the eight-electron rule have low HOMO energies and high LUMO energies as is illustrated by the data of Table 6.3. BH_3 has a very low LUMO energy and it is for this reason that the species forms complexes with electron donors such as NH_3 which can be considered as providing the two electrons needed to complete the boron octet.

To check whether an atom possesses an inert gas structure one requires an accounting procedure to allocate electrons in a molecule to particular atoms. This is most conveniently done through the concept of the *formal* oxidation state. The

Table 6.3. The HOMO and LUMO energies of some small molecules

	BH_3	NH_3	CH_4
LUMO	$0.057\,0E_H$	$0.235\,8E_H$	$0.304\,8E_H$
HOMO	$-0.493\,9E_H$	$-0.412\,5E_H$	$-0.514\,7E_H$

‡L. C. Snyder and H. Basch, *Molecular Wave Functions and Properties*, Wiley-Interscience, New York, 1972.

adjective formal indicates that this does not always agree with a realistic view of the true oxidation state of the atom.

The oxidation state of a free atom or ion is a number equal to the net charge of the ion. Thus neutral atoms are in oxidation state zero, M^+ has oxidation state +1, M^- has oxidation state −1, etc. A word of warning: atomic spectroscopists use, for example, the symbol Fe I for the spectrum of the neutral iron atom, Fe II for the spectrum of Fe^+, etc., so there is a possible confusion between the spectroscopists roman numeral and the chemists oxidation state. The formal oxidation state of an atom in a molecule is the oxidation state deduced according to certain rules for assigning electrons to atoms or groups in the molecule.

Let us now consider the following known complex ions of cobalt:

$$[Co(CO)_4]^-, \quad [Co(CN)_4]^{4-}, \quad [Co(CN)_3CO]^{2-},$$
$$[CoCl_4]^{2-}, \quad [CoF_6]^{3-}, \quad [CoF_6]^{2-}.$$

To determine the formal oxidation state of the metal we must assign charges to the ligand atoms or groups (CO, CN, Cl, F). This is done by adopting those charges that bring the ligand up to a stable structure according to the inert-gas rules. Thus Cl and F are assigned one negative charge. CO, being a stable species for which a structure can be written satisfying the inert-gas rules of each atom, is taken as neutral (see Section 6.6) and CN is assigned one negative charge to make it isoelectronic with CO. In terms of these formal charges the complex ions listed above can be represented as follows:

$$[Co^{-1}(CO)_4], \quad [Co(CN^-)_4], \quad [Co^{+1}(CN^-)_3(CO)],$$
$$[Co^{+2}(Cl^-)_4], \quad [Co^{+3}(F^-)_6], \quad (Co^{+4}(F^-)_6].$$

The charges on the colbalt, which are taken as the formal oxidation state of the metal, therefore run from −1 to +4.

Unfortunately the above method of defining the formal oxidation states of the transition metal in a complex is not without ambiguity. For example, in hydride complexes the hydrogen could be considered as H^+ or H^-: for metal–hydrogen bonds it is always taken as H^-. A more difficult common ligand is NO. This can either be considered as NO^+ (isoelectronic with N_2 or CO) or as NO^- (isoelectronic with O_2), and both are used. The very wide range of observed infrared NO stretching frequencies in complexes (1 100 – 1 900 cm^{-1}) indicates a wide range of bonding types for this ligand.

Low oxidation states of a transition metal are stabilized by neutral ligands, particularly the so-called π bonding ligands like CO, and high oxidation states are stabilized by atoms with a high electron affinity such as F and O (which is formally written O^{2-}.

Changes in oxidation state of a free atom or a complex can be measured directly by the oxidation-reduction potential. For example, E° is 0.6 V (relative to a standard hydrogen electrode) for the process

$$(MnO_4)^- + e^- \rightleftharpoons (MnO_4)^{2-}.$$

However, changes in the formal oxidation state of an atom in an ion cannot be directly measured, e.g.

$$Mn^{7+} + e^- \rightleftharpoons Mn^{6+},$$

as the free atomic ions may not exist under any conditions.

That formal oxidation states are indeed formal and have no relation to real atomic charges can be confirmed by MO calculations and by experiment. Calculations show that transition metals in complexes are in fact nearly neutral. In MnO_4^-, for example, the net charge on the Mn has been calculated to be approximately +1.3 and not the formal oxidation state of +7.

One experimental method of inferring the net charge on an atom in a molecule is by XPS spectroscopy (see Section 5.4). The inner-shell ionization potentials of atoms in molecules of the lighter elements (e.g. carbon) have been found to correlate linearly with calculated net charges on the atoms.† For complexes of the transition metals there is generally poor correlation between such ionization potentials and the formal oxidation state as can be seen from Figure 6.13. The spread of ionization potentials for a given formal oxidation state can be larger than the difference in the mean value from one oxidation state to another. Although there is still some uncertainty in the way such data should be related to net atom charges for transition metals, there is general agreement that formal oxidation state is not a reliable measure of the net atom charge.

Having defined the formal oxidation state we are in a position to examine the validity of the 18-electron rule for transition metal complexes. The area of greatest success is in the carbonyl and nitrosyl complexes. The known mononuclear carbonyl complexes of the first transition series are $Cr(CO)_6$, $Fe(CO)_5$, and $Ni(CO)_4$. The metals are, as we have explained, all to be considered as in the zeroth oxidation state, and the number of 4*s* and 3*d* electrons that they posses will be (c.f. Table 4.1), 6, 8, and 10 for Cr, Fe, and Ni respectively.

Although CO is not generally considered a strong electron donor (or a so-called Lewis base) it does form complexes such as $BH_3.CO$ in which, to satisfy the eight-electron rule for the boron, it must be considered to donate two electrons. If we count two electrons from each CO towards the effective electron configuration of the metal in the transition metal carbonyls then we have for each of the three molecules listed a total of 18 electrons.

The absence of mono-nuclear carbonyl complexes of Mn and Co, both of which have an odd number of valence electrons in their zeroth oxidation states, gives strength to the 18-electron rule. These elements form binuclear carbonyl complexes, $Mn_2(CO)_{10}$ and $CO_2(CO)_8$ in which there is a metal-metal bond, and this, being a two-electron bond, satisfies the 18-electron rule for both metal atoms.

Mononuclear carbonyl complexes of Mn and Co can be formed by changing the oxidation state of the metal to +1, as in $Mn(CO)_5Cl$ and $Co(CO)_4H$, or to −1, as in $Co(CO)_3NO$. In the latter compound the NO is considered as a two-electron donor

†K. Siegbahn *et al.*, *ESCA: Atomic, Molecular and Solid State Structure Studies by means of Electron Spectroscopy*, Almquist and Wiksells, Uppsala, 1967.

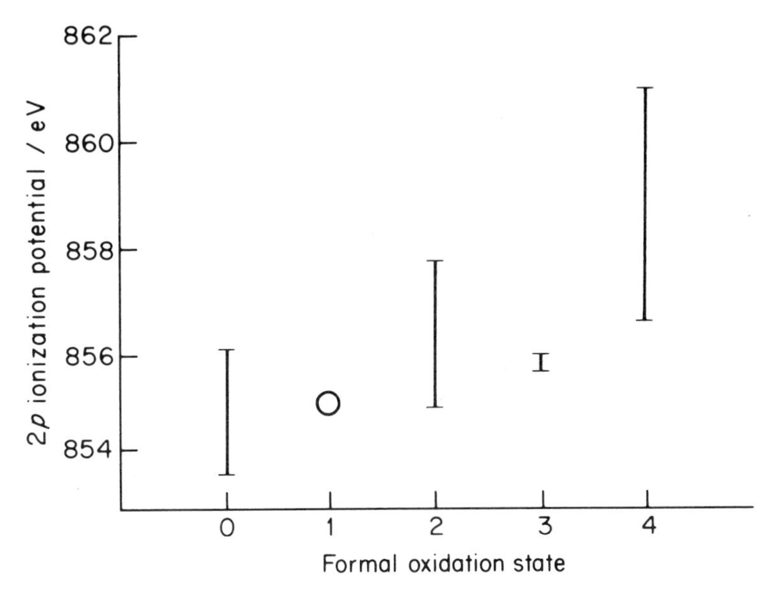

Figure 6.13 Relationship between the $2p$ ionization potential and the formal oxidation state of some Ni complexes. Only one example for Ni^{+1} has been measured. Data from C. A. Tolman, W. M. Riggs, W. J. Linn, C. M. King, and R. C. Wendt, *Inorg. Chem.*, **12**, 2770 (1973); L. O. Pont, A. R. Siedle, M. S. Lazarus, and W. L. Jolly, *Inorg. Chem.*, **13**, 483 (1974).

NO^+ or a three-electron donor NO. Calculations give a net charge of +0.27 on the NO group.†

The above successes of the 18-electron rule should not hide the failures. For example, $V(CO)_6$ exists, counting 17 electrons for the metal, although it is less stable than the other compounds we have mentioned, and it also readily accepts an electron to form $[V(CO)_6]^-$.

Many stable ammino complexes exist which do not satisfy the rule {e.g. $[Cr(NH_3)_6]^{3+}$, $[Ni(NH_3)_6]^{2+}$}, and for the common square-planar complexes of Ni such as $[Ni(CN)_4]^-$ the metal count is 16 electrons.

In summary, therefore, the inert-gas rules introduced in the early theory of valence have been shown by the test of time to be useful and to be in accord with MO theory when tested. MO calculations on the exceptions to the rules should be of interest when these become available.

6.8. The ionic bond

Although in elementary valence theory ionic and covalent bonds are described in quite different ways. G. N. Lewis in 1916 had already proposed that the

†M. Barber, J. A. Connor, M. F. Guest, M. B. Hall, I. H. Hillier, and W. N. E. Meredith, *J. C. S. Faraday Discussions*, **54**, 219 (1972).

electron-pair bond should encompass a range of bond types extending on the one hand to the extremely polar and on the other to the non-polar. This view is also contained in modern valence theory, and we should add further that the polarity of a bond depends also on the distance apart of the two atoms.

There is no neutral diatomic molecule whose ground state potential curve will lead to dissociation to ions rather than into two neutral atoms. This follows from the fact that the energy required for the process

$$A + B \rightarrow A^{+} + B^{-},$$

with A and B infinitely separated, is equal to the ionization potential of A less the electron affinity of B. The element with the lowest ionization potential is Cs (3.893 eV) and the element with the greatest electron affinity is Cl(3.613 eV). Thus Cs + Cl has a lower energy than $Cs^{+} + Cl^{-}$ at infinite separation by 0.280 eV (27 kJ mol^{-1}).

The Coulomb attraction between oppositely charged ions will compensate for the above energy difference when the ions are brought together. This is equal to $-1\,389(R/\text{Å})^{-1}$ kJ mol^{-1} so that $Cs^{+} + Cl^{-}$ will have the same energy as Cs + Cl at 51 Å. For all effective purposes we can say that CsCl is ionic for the whole of its ground state potential curve. If we carry out the same calculation for LiF, however, we find

$$\frac{1\,389}{R(\text{Å})} = I_{\text{Li}} - A_{\text{F}} = (5.390 - 3.448) \times 96.487 \text{ kJ mol}^{-1},$$

from which

$$R = 7.4 \text{ Å}.$$

Hence for greater bond lengths than this the ground state of LiF would represent two neutral atoms rather than ions.

Figure 6.14 shows the result of *ab-initio* calculations on the two lowest energy potential curves of LiF.† In a first approximation the two curves can be labelled as ionic and covalent, and these cross at approximately 6 Å (broken curves in Figure 6.14). The crossing point is not in exact agreement with the above estimate because the wavefunctions are not exact. A normal SCF calculation would lead to curves that cross which are often referred to as diabatic potentials, as distinct from adiabatic potentials that satisfy the Born–Oppenheimer Hamiltonian (see Section 5.1). From calculations that go beyond the SCF approximation the curves are found not to cross, as is shown in Figure 6.14 (solid curves). Such a calculation involves writing the wavefunction as a sum of two or more SCF-type wavefunctions each of which has a different allocation of electrons to SCF orbitals (i.e. different electron configurations). Thus for the calculations we are describing there will be an SCF configuration representing the ionic state of LiF and one representing the covalent state and the total wavefunction is a mixture of these, the mixing ratio being roughly 1:1 in the region of 6 Å.

†L. R. Kahn, P. J. Hay, and I. Shavitt, *J. Chem. Phys.*, **61**, 3530 (1974).

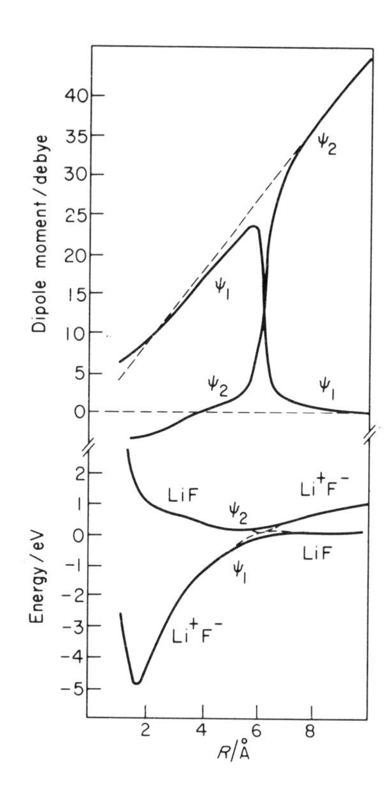

Figure 6.14 The *ab-initio* calculated potential energy curves and dipole moments of the lowest two states of LiF.

The calculated dipole moments of the two states of LiF bring out clearly this configuration mixing (or *configuration interaction*). The covalent state has a very small dipole moment as it is non-polar, and this must become zero at large internuclear distances. The ionic state has a dipole moment that is proportional to eR for large R: this follows from equations (5.22) and (5.23) by taking the probability density for the ionic state as the sum of two spherical atomic densities. In the region of the crossing point the dipole moments of the two adiabatic states change suddenly from the ionic to the covalent extremes.

We take this opportunity to point out that the non-crossing of diatomic potential energy curves of the same symmetry type that is illustrated in Figure 6.14, is a result of a general theorem proposed originally by Teller. A proof of this theorem will be left until Section 12.2.

Although we have shown that even the most polar molecule will dissociate into neutral atoms in the gas phase, in polar solvents we know that molecules like LiF dissociate spontaneously into ions. The reason for this is that the ions have a much greater solvation energy than do the neutral atoms or the undissociated molecule. This solvation energy is more than sufficient to overcome the 5 eV gas phase dissociation energy. The solvation energy itself can be considered as primarily electrostatic in origin,

Chapter 7
Symmetry and Orbitals

7.1. Symmetry constraints on molecular orbitals

One feature of valence theory which has become quite clear in the last two chapters is that even for quite a simple molecule the calculation of molecular wavefunctions is a difficult problem for which we can expect no more than an approximate solution. If this is the situation for simple molecules the problem for complicated molecules might well be thought hopeless. Fortunately the position is not as bad as it may seem and one of the redeeming features is the existence of molecular symmetry. The more symmetrical that a molecule is, the more easily we can reach some conclusions about its wavefunctions without attempting a solution of the Schrödinger equation.

Although the number of highly symmetrical molecules might in the total scheme of chemistry be considered small they have among their number the progenitors of some of the most important chemical families. For example, the molecules methane, ethylene, acetylene, and benzene characterize much of the physics and chemistry of hydrocarbons, and octahedral and tetrahedral complexes illustrate much of transition metal chemistry.

We have already seen how molecular symmetry may be used in valence theory. In Section 5.3 we first met the symbols g and u which were used to distinguish orbitals which are, respectively, symmetric and antisymmetric with respect to the operation of inversion in the centre of symmetry of a homonuclear diatomic molecule. We met these symbols again in Chapter 6 (equations 6.6 to 6.10) where we saw that orbitals carrying the g suffix obeyed equation (6.7) and those with the u suffix followed equation (6.8). That is, the relative signs of the coefficients with which different atomic orbitals appear in an expression for a molecular orbital may be symmetry-determined. This is true for all the elements of symmetry possessed by a molecule, not just centres of symmetry, and it is the details of this generalization which form the subject of this chapter.

To take full advantage of symmetry in solving molecular problems one should make use of a branch of mathematics known as group theory. This is a subject which is dealt with fully in more advanced texts than this, but we will find it convenient to use some of the simpler parts of the subject and some of the conventional symbols.

We can look at the symmetry operations of molecules in two ways. *Either* they are operations which change the molecule to an equivalent orientation, so that after the operation every point in the molecule is coincident with an equivalent point (or the same point) of the molecule in its original orientation *or* they are operations which leave the Hamiltonian unchanged. The first of these definitions is easier to comprehend on a physical basis, but the second is more general and emphasizes the important point that the symmetry does depend on what Hamiltonian is being considered. Thus an approximate or model Hamiltonian for a molecule may have a higher symmetry than the true Hamiltonian and this may be very useful in enabling us to calculate approximate wavefunctions for molecules which do not themselves have a very high symmetry.

To see in detail how molecular symmetry may be exploited we shall use arguments very similar to those used for H_2^+ in Chapter 6 (equations 6.6 to 6.10) but we shall apply them to a slightly more complicated molecule, water. Figure 7.1 shows the water molecule and four points (denoted by stars and numbers) at which the electron density must be identical. These four points are inter-related by the symmetry operations of the water molecule which are four in number. Three of them are indicated in Figure 7.1. A rotation of π radians about the axis denoted C_2 in Figure 7.1 interchanges the two hydrogen nuclei and leaves the oxygen nucleus unchanged. Thus this operation changes the molecule into a configuration which is

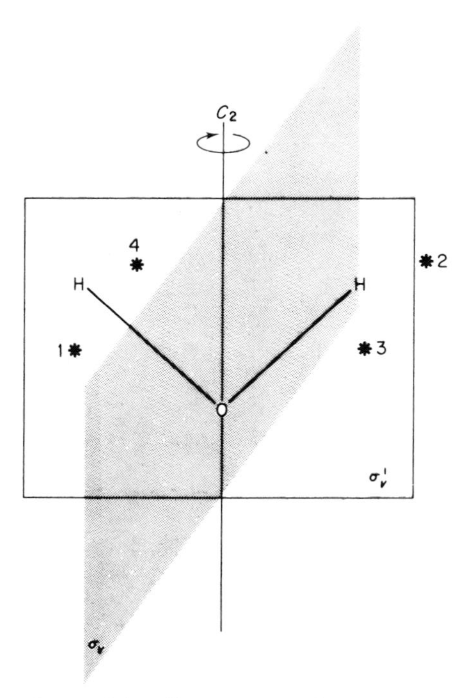

Figure 7.1 Symmetry elements for the water molecule. The numbers indicate four points at which the electron density is identical.

indistinguishable from the starting configuration. The electron densities at points 1 and 2 must be identical (as must the electron densities at points 3 and 4 because they are also interchanged by this *twofold* $2\pi/\pi$ rotation operation. Rotation operations are conventionally denoted by the capital letter C together with a suffix to indicate the order of the rotation: C_n is a rotation by an angle ϑ such that $n = 2\pi/\vartheta$, ϑ, being in radians.

A second symmetry operation indicated in Figure 7.1 is that denoted as σ_v which is reflection in a plane perpendicular to the molecular plane and containing the 2-fold axis. That it is not physically possible to carry out this operation is not important – the operation serves to express a relationship which must exist between points within the molecule. Thus in Figure 7.1 the electron density at point 1 must be identical to that at point 3. Similarly, points 2 and 4 must have identical densities. But we already know (from the existence of the C_2 operation) that points 1 and 2 and points 3 and 4 have identical densities. It follows that the electron densities at points 1, 2, 3, and 4 must all be identical. The operation of reflection in a mirror plane is denoted by the symbol σ. When the corresponding mirror plane contains the rotation axis C_n of highest n value (the so-called principal axis) then the mirror plane (and corresponding operation) are denoted σ_v, v standing for vertical. The mirror plane would indeed be vertical if a molecular model were held with the principal rotation axis vertical.

So far we have directly related by symmetry operations the electron density at point 1 with that at points 2 and 3. It is also directly related to that at point 4 by the existence of a second vertical mirror plane distinguished from the first by denoting it σ_v', which is the plane containing the three nuclei.

We stated earlier that the water molecule has four symmetry operations and so far we have mentioned only three. The fourth is trivial but important and is called the identity operation, it is the operation of leaving the molecule alone and is denoted by E (sometimes I). This operation may seem redundant but the need for it arises from the fact that using group theory one can construct an algebra associated with the operations of the water molecule. We therefore need the identity operation when we express the fact that a C_2 operation followed by another C_2 operation is equivalent to leaving the molecule unchanged. This may be expressed algebraically as $C_2 . C_2 = E$.

If we turn now to a molecular orbital description of the electron density in the water molecule we know that the total density is a sum of contributions from each of the occupied molecular orbitals (equation 5.21). We can therefore expect that the symmetry relationships between the electron density at different points in space which we have recognized above should also apply to the electron density contributed by each individual molecular orbital. It follows therefore that each of the molecular orbitals ψ of the water molecule must satisfy the condition

$$\psi^2(1) = \psi^2(2) = \psi^2(3) = \psi^2(4), \tag{7.1}$$

where ψ^2 (1) stands for the molecular orbital electron density at the point 1.

The equalities expressed by (7.1) are satisfied by wavefunctions which have the property

$$\psi(1) = \pm\, \psi(2) = \pm\, \psi(3) = \pm\, \psi(4). \tag{7.2}$$

There are 2 x 2 x 2 = 8 combinations of sign contained in (7.2) but not all of these are acceptable. The reason for this is that although the points 1,2,3, and 4 are quite arbitrary points they are related to one another by the four symmetry operations. Thus if we have a wavefunction with the property

$$\psi(1) = +\psi(2), \tag{7.3}$$

then, as this is true an arbitrary point 1, it must be true for all pairs of points that are related by the C_2 operation. However, 3 and 4 are another pair of symmetry related points so that if (7.3) is true then so also is

$$\psi(3) = +\psi(4). \tag{7.4}$$

Thus it is possible to allow from the set (7.2) the combination

$$\psi(1) = +\psi(2) = +\psi(3) = +\psi(4), \tag{7.5}$$

or

$$\psi(1) = +\psi(2) = -\psi(3) = -\psi(4), \tag{7.6}$$

both of which are consistant with (7.3) and (7.4); but it is not possible to allow

$$\psi(1) = +\psi(2) = +\psi(3) = -\psi(4), \tag{7.7}$$

which is inconsistant with (7.3) and (7.4). Likewise if we take the possibility that

$$\psi(1) = -\psi(2), \tag{7.8}$$

then it must also be true that

$$\psi(3) = -\psi(4), \tag{7.9}$$

and the combinations which satisfy these two conditions are

$$\psi(1) = -\psi(2) = +\,\psi(3) = -\psi(4), \tag{7.10}$$

or

$$\psi(1) = -\psi(2) = -\psi(3) = +\psi(4). \tag{7.11}$$

The four combinations (7.5), (7.6), (7.10), and (7.11) are the only ones of the eight possible in (7.2) which satisfy the symmetry conditions as applied to arbitrary points in space. A more succinct way of expressing these relationships is in terms of the effect of the symmetry operations on the wavefunctions rather than on the points. For example, if all four symmetry operations leave a wavefunction ψ unchanged, and we can express this mathematically as

$$\begin{aligned} E\psi &= +1\psi, \\ C_2\psi &= +1\psi, \\ \sigma_v\psi &= +1\psi, \\ \sigma_v'\psi &= +1\psi, \end{aligned} \tag{7.12}$$

then such a wavefunction satisfies the condition (7.5) for an arbitrary point 1. We have artificially included the factors +1 on the right-hand side of (7.12) so that we can characterize wavefunctions which have this property by the set of four numbers (1,1,1,1). Likewise if we have a wavefunction which is unchanged by E or C_2 but which changes sign under the operations σ_v and $\sigma_v{}'$ then we can express this property by

$$\begin{aligned} E\psi &= +1\psi, \\ C_2\psi &= +1\psi, \\ \sigma_v\psi &= -1\psi, \\ \sigma_v'\psi &= -1\psi, \end{aligned} \qquad (7.13)$$

and such a wavefunction is characterized by the set of numbers (1,1,–1,–1). A wavefunction of this type will satisfy (7.6) for an arbitrary point 1.

By similar reasoning to that above we can characterize two other types of wavefunction by the sets (1,–1,1,–1) and (1,–1,–1,1) which are generalizations of (7.10) and (7.11) respectively. The four sets of numbers which characterize the symmetries of physically acceptable wavefunctions for any molecule having the same symmetry elements as water are collected in Table 7.1. The numbers (1 or –1) are called the characters appropriate to each type of wavefunction and, as we shall see, each set is assigned a label according to standard conventions. Before giving a more detailed description of such character tables, and in particular before showing how they may be used to simplify problems in valence theory, we will describe the symmetry operations for molecules having other symmetry elements.

Table 7.1. The table of characters which describe the symmetries of physically acceptable wavefunctions of the water molecule

E	C_2	σ_v	σ_v'
1	1	1	1
1	1	–1	–1
1	–1	1	–1
1	–1	–1	1

7.2. Symmetry operations, point groups and character tables

In the preceding section we introduced three kinds of symmetry operations for the water molecule (E,C,σ) and in earlier chapters we encountered a fourth which was the inversion operation: this is represented by the symbol i. There is one other

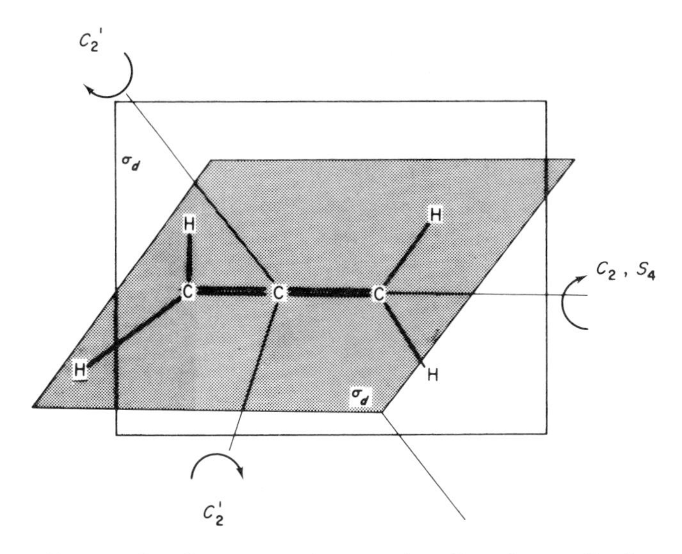

Figure 7.2 Symmetry elements for allene (group D_{2d}).

that we have not yet met and that is the so-called improper rotation. Such operations are denoted by the symbol S_n and comprise two parts; firstly a rotation by the angle $2\pi/n$ and then a reflection in a mirror plane which is perpendicular to the axis about which the rotation was carried out. A good example of such a rotation-reflection axis is the S_4 axis possessed by a molecule of allene. This is illustrated in Figure 7.2 from which the detailed form of the operation can be seen: first, a rotation by $2\pi/4$ about the axis containing the carbon atoms (hence the suffix 4) followed by a reflection in a plane perpendicular to this axis which passes through the central carbon atom. It sometimes happens that the C_n rotation and the reflection are themselves, independently, symmetry operations of the molecule. Sometimes they are not and the two components of the S_4 operation of allene are in this latter class.

To complete this review of symmetry operations, we have to enumerate the various suffixes which can occur on the symbol σ used to denote a mirror plane. These suffixes depend on the relationship between the mirror plane and rotation axes in the molecule. As we have seen, when the mirror plane contains the axis of highest rotation symmetry, i.e. is vertical with respect to it, a suffix v is used as in σ_v. Similarly, when the mirror plane is perpendicular to the axis of highest rotational symmetry, i.e. is horizontal with respect to it, the suffix h is used as in σ_h. Finally, when a mirror plane contains the principal axis of rotation (i.e. is of σ_v type) but in addition bisects the angles between two 2-fold rotation axes which are themselves perpendicular to the principal axis, than a suffix d is used as in σ_d. This suffix stands for dihedral, the 2-fold axes being described as dihedral axes.

The set of all symmetry operations for a molecule (or any other body) is called a symmetry group or point group. The number of operations in the group is called the order of the group. Groups must satisfy certain conditions. In particular, the effect of operating in succession with any two operations of the group must always be equivalent to the effect of some single element in the group. We note in the case of the water molecule that the σ_v' operation is equivalent to C_2 followed by σ_v. This condition provides a method (sometimes tedious) of checking that we have a complete group of operations. If we form all products of operations and if there is any result which is not equivalent to an operation already within the group, then the list of operations must be extended to include the additional single equivalent operation.

The result of applying successive operations of the group can be expressed algebraically by an equation of the type

$$BA = C. \tag{7.14}$$

This means that the effect of first applying the operation A and then applying B is equivalent to the application of a single operation C. It is important to note the order in which the operations are carried out. The result of carrying out first B and then A may, or may not, be C. In cases when

$$AB = BA, \tag{7.15}$$

we say that the operations commute, and when

$$AB \neq BA, \tag{7.16}$$

the operations do not commute. Groups which contain non-commuting operators have important properties not possessed by groups for which all operators commute.

Figure 7.3 illustrates the fact that for the ammonia molecule the operations of rotation about the principal axis by $2\pi/3$ followed by reflection in one of the vertical mirror planes, are not commuting operations.

A second condition that must be satisfied for a group is that the inverse of every element must itself be a member of the group. Successive operations by an element A and by its inverse, which we can represent as A^{-1} must leave the system unchanged, and we can express this fact mathematically as

$$A^{-1} A = E \tag{7.17}$$

We can see that as A^{-1} is an element of the group, then according to the rule (7.17) every group must contain the identity operation E. The symmetry groups of molecules are called point groups because all the symmetry elements, that is such quantities as a centre of symmetry, a rotation axis, an improper rotation axis, or a mirror plane, as a molecule may possess simultaneously pass through at least one point in space. An important class of groups that do not have this property is that which describes the symmetries of crystal lattices. These are called space groups and we give a brief discussion of them in Chapter 10.

There are three further points to note about symmetry operations. Firstly, just as we distinguished the two reflection operations in the case of the water molecule by applying a prime to one of the (σ_v and σ_v'), so it is general practice to distinguish symmetry operations of the same type, but corresponding to in-

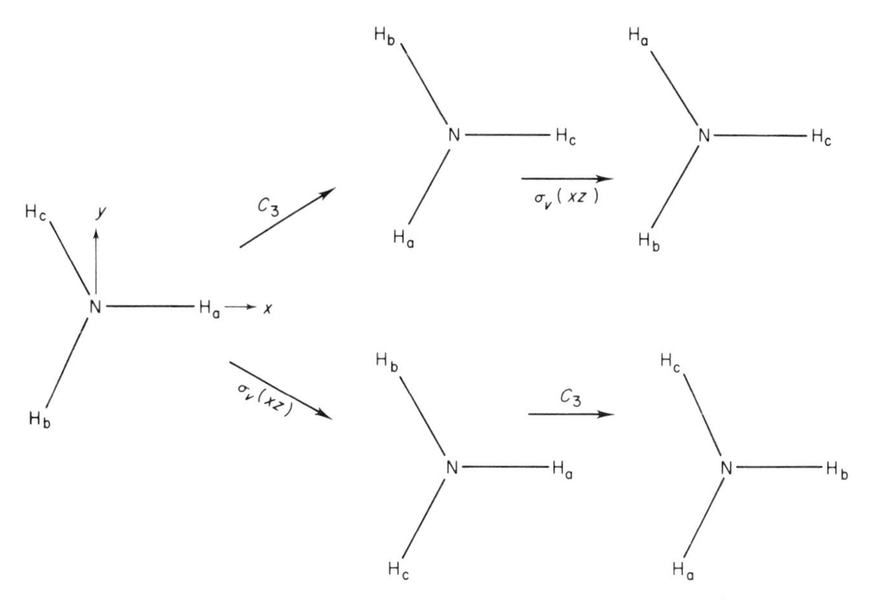

Figure 7.3 To show that the operations of C_3 (rotation clockwise about the z axis by $2\pi/3$) and σ_v (reflection in the vertical xz plane) do not commute.

equivalent symmetry elements, by primes. Alternatively one uses labels to show the relationship between the symmetry element and Cartesian axes in the molecule. Thus $C_2(x)$ would indicate a 2-fold rotation about the x axis, $\sigma_v(xz)$ indicates that the reflection plane is the xz plane.

Secondly, we note that there may be several operations for a molecule which are exactly of the same type. For example, it can be seen in Figure 7.3 that there are three σ_v planes for ammonia and any labels we might give to distinguish these three are quite arbitrary. Symmetry operations of this type are said to be of the same class and are always grouped together in the character table because the characters are always the same for members of a class. Thus the character table for the symmetry group of ammonia contains a column headed $3\sigma_v$ under which the character for each of the σ_v operations is to be found.

A more rigorous definition of class is provided by the fact that A and B are of the same class if an element X of the group can be found such that

$$X A X^{-1} = B. \tag{7.18}$$

Evidently to find which operations are in the same class as A one simply lets X be each element of the group in turn.

Thirdly, we note that some sets of elements of a group which are less than the complete group satisfy the rules we have above. Such sets are called a sub-group. For example, the elements E and C_2 alone of Table 7.1 are a sub-group of the group $(E, C_2, \sigma_v, \sigma_v')$. However, the set E, C_2, σ_v is not a sub-group because we have already noted that $\sigma_v C_2 = \sigma_v'$, so that one of our rules is not satisfied by this set.

The importance of sub-groups is that one may find it sufficient to classify wavefunctions by their behaviour under the operations of a sub-group rather than their behaviour under the operations of the full group.

Groups :of symmetry operations are labelled by standard symbols which are usually related to the elements of the group. For example, the group of the water molecule with elements $E, C_2, \sigma_v, \sigma_v'$ is labelled C_{2v}. That of ammonia, with elements E, $2C_3$, and $3\sigma_v$ (note that elements in the same class are grouped together) is called the C_{3v} group.

The label used to describe a group usually contains an indication of the multiplicity of the axis of highest symmetry – the two examples above illustrate this. The general rules are as follows:

1. If the molecule has a C_n rotational axis, but no other element of symmetry, it belongs to the C_n group.†

2. If, in addition to a C_n rotational axis the molecule has n σ_v mirror planes, it belongs to the C_{nv} group.

3. If, in addition to a C_n rotational axis the molecule has a σ_h mirror plane, but no other element of symmetry, it belongs to the C_{nh} group.

4. If, in addition to a C_n rotational axis there are n C_2 axes perpendicular to the C_n axis, it belongs to the D_n group.

5. If, in addition to a C_n rotational axis and n C_2 axes perpendicular to this, the molecule has a σ_h mirror plane, it belongs to the D_{nh} group.

6. If, in addition to a C_n rotational axis and n C_2 axes perpendicular to this, it has n mirror planes which bisect these 2-fold axes (these will be σ_d mirror planes) but no σ_h plane, it belongs to the D_{nd} group.

7. If the molecule has a C_n rotational axis and this axis is also the element of a S_{2n} operation, but there are no other symmetry elements except possibly i (which is equivalent to S_2), then the group is S_{2n}.

8. Linear molecules either have or do not have a mirror plane perpendicular to the C_∞ axis. If they do, they belong to the $D_{\infty h}$ group; if they do not, they belong to the $C_{\infty v}$.

9. Some molecules possess no axis of symmetry but do have a mirror plane; they belong to the C_s group.

10. Molecules with no axis or plane of symmetry but which possess a centre of symmetry belong to the C_i group.

11. Some high-symmetry groups carry quite obvious but non-systematic labels.

†A source of confusion in symmetry group theory is that the same symbol can be used for different things. Thus C_2 may stand for a 2-fold axis (the element) for a 2-fold rotation (the operation) or for the group whose operations are E and C_2. We shall also see that E can stand for the identity operation or for a doubly degenerate symmetry species. These symbols are a matter for international convention and such conventions are not necessarily perfect.

Table 7.2. Molecules exemplifying important point groups

Molecule	Point group
H_2CO	C_{2v}
PF_3	C_{3v}
trans – $C_2H_2Cl_2$	C_{2h}
C_6H_6 (benzene)	D_{6h}
eclipsed – C_2H_6	D_{3h}
staggered – C_2H_6	D_{3d}
HOCl	C_s
HCl	$C_{\infty v}$
C_2H_2	$D_{\infty h}$
CH_4	T_d
SF_6	O_h

Thus the group of the octahedron (which is also that of the cube) is O_h, that of the tetrahedron is T_d, and that of the icosahedron, I_h.

Table 7.2 lists some molecules and their point groups to illustrate the above rules. Table 7.3 details the operations of some of the more important point groups.

Once the point group of a molecule has been identified, one can use the techniques of symmetry group theory to simplify problems encountered in valence theory or molecular spectroscopy. The information required to do this is contained in the character table. Character tables are standard tables whose symbolism has been agreed by international convention.

Table 7.4 shows the character table for the C_{2v} group which we have already briefly discussed. There are four distinct parts of the table. On the top line are listed the symbol of the group and the symmetry operations. The identity is always listed first and the rotation about the principal axis (if any) second. The characters are the block of numbers in columns under symmetry operations. Each row of numbers represents the behaviour of a different type of function, we will call it a symmetry species, and it is identified by a symbol given in the lefthand column.†

Table 7.3. Operations of some important point groups

Group	Operations
D_{2h}	$E, C_2(z), C_2(y), C_2(x), i, \sigma(xy), \sigma(xz), \sigma(yz)$
D_{6h}	$E, 2C_6, 2C_3, C_2, 3C_2', 3C_2'', i, 2S_3, 2S_6, \sigma_h, 3\sigma_d, 3\sigma_v$
D_{2d}	$E, 2S_4, C_2, 2C_2', 2\sigma_d$
T_d	$E, 8C_3, 3C_2, 6S_4, 6\sigma_d$
O_h	$E, 8C_3, 6C_2, 6C_4, 3C_2, i, 6S_4, 8S_6, 3\sigma_h, 6\sigma_d$

†A more precise terminology is to call each set of characters an *irreducible representation* of the group. We will not discuss representation theory in this book, although it will be convenient to refer to irreducible representations later in this chapter.

Table 7.4. C_{2v} Character table

C_{2v}	E	C_2	σ_v	σ'_v	
A_1	1	1	1	1	z, x^2, y^2, z^2
A_2	1	1	−1	−1	R_z, xy
B_1	1	−1	1	−1	x, R_y, xz
B_2	1	−1	−1	1	y, R_x, yz

Thus one can refer to wavefunction of B_1 symmetry (or just a B_1 state) of water. The significance of the right hand column will be discussed later.

Table 7.5 shows the character table for the C_{3v} group, that is the point group appropriate to a molecule like ammonia. We see that it is more complicated than the C_{2v} table by having characters other than 1 or −1. All groups with the operations C_n or S_n where $n \geqslant 3$ have this complication (and incidentally these are the groups whose elements do not commute), and as it has an important relationship to the degeneracy of wavefunctions for this type of molecule we must examine the point in some detail. We remind the reader that degenerate wavefunctions are two or more with the same energy. To see how such degeneracy must arise we consider the following problem.

Table 7.5. C_{3v} Character table

C_{3v}	E	$2C_3$	$3\sigma_v$	
A_1	1	1	1	$z, x^2 + y^2, z^2$
A_2	1	1	−1	R_z
E	2	−1	0	$(x, y), (R_x, R_y), (x^2 - y^2, xy), (xz, yz)$

Suppose we take the ammonia molecule and extend the three NH bond lengths to very large but equal values whilst maintaining the bond angles as they are in the normal molecule. The extended molecule will then retain its C_{3v} symmetry. The molecular orbitals of ammonia must gradually evolve into the atomic orbitals of the component atoms as we pull the atoms apart, following the arguments presented in equation (6.1), and three of these must be the $2p$ orbitals of the nitrogen atom. Let us choose Cartesian coordinates such that z is along the C_3 axis. There is no unique obvious choice for the x and y axes but let us arbitrarily take x to be eclipsing one N–H bond as shown in Figure 7.3. We can now examine the effect of the C_{3v} symmetry operations on the three $2p$ orbitals. It is easy to see that $2p_z$ is unchanged by all the operations of the group so that its character must be (1,1,1) and it is clearly a wavefunction of A_1 symmetry.

The situation for $2p_x$ and $2p_y$ is much more complicated. For example, rotating the $2p_x$ orbital by $2\pi/3$ sends it into a function which is a linear conbination of the original $2p_x$ and $2p_y$ orbitals as shown in Figure 7.4. The effect of the rotation can be expressed mathematically as

$$C_3\ 2p_x = -\frac{1}{2} \cdot 2p_x - \frac{\sqrt{3}}{2} \cdot 2p_y \tag{7.19}$$

where the coefficients are determined by applying the rules for resolving a vector into its components. Likewise the effect of C_3 on $2p_y$ is expressed by

$$C_3\ 2p_y = \frac{\sqrt{3}}{2} \cdot 2p_x - \frac{1}{2} \cdot 2p_y. \tag{7.20}$$

It is clear that the effect of C_3 on $2p_x$ is not either $+1.2p_x$ or $-1.2p_x$ so that the appropriate characters for the $2p_x$ wavefunctions are not ± 1.

The operations of the group on either $2p_x$ or $2p_y$ lead to new functions which are mixtures of the original functions. We can only describe the effect of these operations by considering $2p_x$ and $2p_y$ as a pair. We have already noted that it is arbitrary how we choose the x and y axes for such a system, and the wavefunctions for one choice of axes are related to the wavefunctions for another choice of axes by some linear transformation like (7.19). Such a situation only makes sense if we conclude that the $2p_y$ and $2p_x$ orbitals are degenerate in C_{3v} symmetries.

We have implied but not stated explicitly in Chapter 3 that there is some freedom in defining the wavefunctions for degenerate states. We described two forms for the wavefunctions of atomic orbitals (except s orbitals), one real (e.g. equations 3.13 – 3.15) and one complex (e.g. equation 3.26), the two being related by the transformation (3.25). We can make the position clear in the following way.

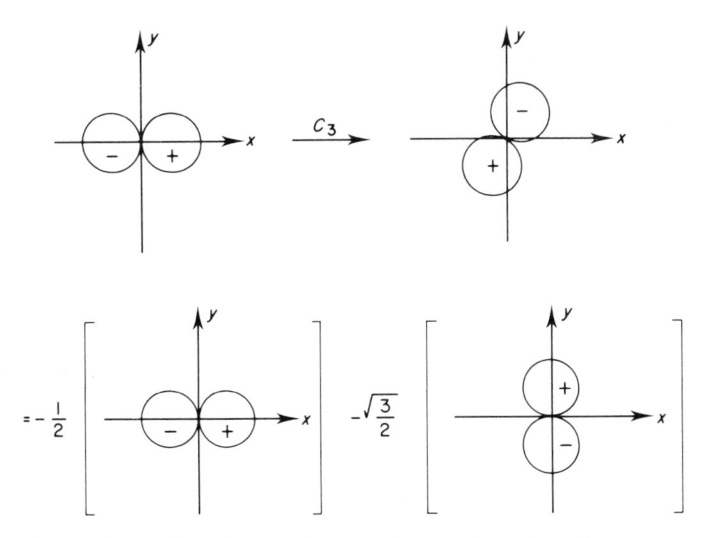

Figure 7.4 The effect of a rotation C_3 (clockwise) on a p_x orbital. The resolution of the orbital into two components along the original x and y directions is shown below.

Suppose we have two solutions of the Schrödinger equation (2.27)

$$\mathscr{H}\Psi_a = E_a\Psi_a, \tag{7.21}$$

and

$$\mathscr{H}\Psi_b = E_b\Psi_b. \tag{7.22}$$

If we multiply (7.21) by a constant k and (7.22) by a constant l and add the two, we arrive at the equation

$$\mathscr{H}(k\Psi_a + l\Psi_b) = kE_a\Psi_a + lE_b\Psi_b. \tag{7.23}$$

We now ask if $k\Psi_a + l\Psi_b$ is a solution of the Schrödinger equation. It will only be a solution if

$$\mathscr{H}(k\Psi_a + l\Psi_b) = E(k\Psi_a + l\Psi_b), \tag{7.24}$$

where E is a constant. However, the right-hand sides of (7.23) and (7.24) are identical only if

$$E = E_a \quad \text{and} \quad E = E_b, \tag{7.25}$$

which is only possible if $E_a = E_b$. We therefore conclude that if $E_a \neq E_b$ we *cannot* obtain new solutions of the Schrödinger equation by taking linear combinations of Ψ_a and Ψ_b, but if $E_a = E_b$ then *any* linear combination of Ψ_a and Ψ_b is a solution.

We can therefore understand the situation for the p orbitals of the nitrogen atom which we have discussed above. If the $2p_x$ and $2p_y$ orbitals are degenerate then any linear combination of them is also a solution of the Schrödinger equation, and the operations of the C_{3v} group on the degenerate pair can therefore be expressed by some constant multiplier of the degenerate pair. In other words, a character which describes the effect on the degenerate pair can be obtained.

A mathematical expression for the operations of the group on degenerate functions can only be given in the context of matrix algebra. For those readers with a knowledge of this we note that expressions (7.19) and (7.20) can be combined to give

$$C_3\begin{pmatrix}2p_x\\2p_y\end{pmatrix} = \begin{pmatrix}-\frac{1}{2} & -\frac{\sqrt{3}}{2}\\ \frac{\sqrt{3}}{2} & -\frac{1}{2}\end{pmatrix}\begin{pmatrix}2p_x\\2p_y\end{pmatrix}, \tag{7.26}$$

and the sum of the diagonal elements (trace) of the 2 x 2 matrix on the right hand side is −1, which is the character of the E symmetry species in Table 7.5.

It is obvious that the identity operation must leave all components of a degenerate set of wavefunctions unchanged so that we can write an equation analagous to (7.26) for the identity which is

$$E\begin{pmatrix}2p_x\\2p_y\end{pmatrix} = \begin{pmatrix}1 & 0\\0 & 1\end{pmatrix}\begin{pmatrix}2p_x\\2p_y\end{pmatrix}, \tag{7.27}$$

and the trace of the matrix is 2. We therefore see that the degeneracy of a symmetry species is given by the character under the identity operation.

The symbol E (sometimes with a suffix or a superfix) is always used for doubly degenerate species of the symmetry group, and T is used for the triply degenerate species that occur in O_h, T_d, or I_h. For this last group (rarely needed) four and five-fold degenerate species also exist. The symbols A and B are retained for non-degenerate species, and in the $C_{\infty v}$ and $D_{\infty h}$ groups the symmetry species are identified by the angular momentum symbols, Σ, Π, Δ, etc. which we have already met in their lower case form (σ, π, δ) in Chapter 6.

The right-hand side of each character table contains some additional information which is useful for problems involving symmetry. This is a list of functions which have the symmetry of that particular irreducible representation. The symbol z, for instance, against a symmetry species means that anything which behaves like the molecular z axis under the operations of the group, transforms as this species. Translation of the molecule along the z axis (sometimes denoted T_z) and a p_z orbital of an atom situated on the z axis are two examples. It is convention to take the principal axis as the z axis.

A d orbital, such as $3d_{xy}$, on an atom located on the axis of highest rotational symmetry has the symmetry properties of the species against which xy appears in the character table. Rotation of the entire molecule about a coordinate axis (usually denoted R_x, R_y, and R_z) are separately listed and are useful for spectroscopic analysis.

7.3. Direct products

We have seen in section 6.3 that to derive molecular orbitals using the LCAO approximation we are faced with the problem of evaluating integrals such as the overlap integral (6.48)

$$S_{mn} = \int \phi_m \phi_n \, dv, \tag{7.28}$$

and the Hamiltonian integral (6.50)

$$H_{mn} = \int \phi_m \mathcal{H} \phi_n \, dv. \tag{7.29}$$

A considerable simplification may arise if we can use symmetry arguments to choose basis functions ϕ_i for the LCAO expansion such that many of these integrals are zero. We have already made use of this procedure in Section 6.2 when determining the molecular orbitals of the homonuclear diatomic molecules.

We assume that we know how ϕ_m and ϕ_n, individually, transform and ask how the product, $\phi_m \phi_n$ transforms. We shall consider integration of the product later.

Let us take a specific example. Consider the C_{2v} point group and the case in which ϕ_m has B_1 symmetry and ϕ_n B_2 symmetry. The behaviour of the two functions under the operations of the C_{2v} point group can be deduced from the characters in Table 7.4 and is summarized in Table 7.6.

Table 7.6. Behaviour of the functions ϕ_m (B_1) and ϕ_n (B_2) and their product, under the operations of the C_{2v} group. The result of operating on each function and their product is listed under each operation

	E	C_2	σ_v	σ_v'
ϕ_m:	ϕ_m	$-\phi_m$	ϕ_m	$-\phi_m$
ϕ_n:	ϕ_n	$-\phi_n$	$-\phi_n$	ϕ_n
$\phi_m\phi_n$:	$\phi_m\phi_n$	$\phi_m\phi_n$	$-\phi_m\phi_n$	$-\phi_m\phi_n$

It is clear from the last row of Table 7.6 that the product $\phi_m\phi_n$ transforms like a function whose characters are (1,1,–1,–1) which is the A_2 symmetry species of the C_{2v} group. This result depends only on the symmetry of the functions ϕ_m and ϕ_n and in no way on their detailed form. It must therefore be true for any product of a B_1 and a B_2 function.

It should be obvious that the characters for the product are obtained by multiplying the characters for the separate functions. Such a product is called the *direct product* of two symmetry species (and is written in this case $B_1 \times B_2$). It is invariably true that the direct product of two symmetry species is either itself a symmetry species or, for some products involving degenerate species, is a sum of symmetry species of the group.

By forming direct products of pairs of other symmetry species of the C_{2v} group the table of direct products given in Table 7.7 can be generated.

To show how we obtain the direct product of two degenerate species we will consider the direct product $E \times E$ of the C_{3v} group (see Table 7.5). Applying the rules given above this gives the result

$$\begin{array}{lccc} & E & 2C_3 & 3\sigma_v \\ E \times E: & 4 & 1 & 0\ . \end{array} \tag{7.30}$$

It is clear that these are not the characters of any one symmetry species of the C_{3v} group. They are said to be the characters of a representation which is *reducible* and the process of reduction involves finding a sum of the characters of the symmetry

Table 7.7. The direct products of the C_{2v} group

C_{2v}	A_1	A_2	B_1	B_2
A_1	A_1	A_2	B_1	B_2
A_2	A_2	A_1	B_2	B_1
B_1	B_1	B_2	A_1	A_2
B_2	B_2	B_1	A_2	A_1

Table 7.8. Showing that the sum of the characters of A_1, A_2, and E gives the characters of the direct product $E \times E$

	E	$2C_3$	$3\sigma_v$
A_1	1	1	1
A_2	1	1	−1
E	2	−1	0
	4	1	0

species (the irreducible representations) which add up to the characters of the reducible representation. That this can be done in this case is shown by Table 7.8.

Formally we write

$$E \times E = A_1 + A_2 + E. \tag{7.31}$$

It can also be shown that this reduction is unique: every reducible representation has unique component irreducible representations for the same reason that a vector can be uniquely resolved into its component orthogonal unit vectors.

Our reason for discussing the direct product of two symmetry species was to draw some conclusions about the integrals (7.28) and (7.29). The integral over all space of any function that has the symmetries A_2, B_1, or B_2 of the C_{2v} group must be zero. The reason is that for every general point in space where the function is positive there is one other equivalent point where the function is also positive by two other equivalent points where it is negative. In other words, the integral can be divided into four parts which must be equal in magnitude (the four quadrants of xy space) and the integral over two of these must cancel the integral over the other two. For example, if an A_2 function is everywhere positive in the $x > 0$, $y > 0$ quadrant then it is also positive in $x < 0$, $y < 0$ but negative in both $x > 0$, $y < 0$ and $x < 0$, $y > 0$.

We conclude that:

> 'An integral over all space of a function transforming as a non-totally-symmetric symmetry species (i.e. all species except the one for each group whose characters are all +1) is identically zero.'

A second rule can be established for direct products

> 'The direct product of two symmetry species generates the totally symmetric species only when the two are the same symmetry species'.

It can be seen that this is true for the C_{2v} group in Table 7.7 and can easily be confirmed for other groups.

We are therefore in a position to be able to state that the overlap integral (7.28) is zero if the two functions ϕ_m and ϕ_n (which may be atomic orbitals or combinations of atomic orbitals) belong to different symmetry species.

For the Hamiltonian integral (7.29) we must consider a triple product of functions. This is not difficult because the Hamiltonian is unchanged by all the operations of the group. We have earlier given a definition of a symmetry operation as one which leaves the Hamiltonian unchanged, but it may be useful to note that as the Hamiltonian is only an operator equivalent of the energy of the molecule and clearly the energy cannot change in sign or magnitude by a symmetry operation, then the Hamiltonian cannot change either. Thus the symmetry species of the function $\phi_m \mathscr{H} \phi_n$ must be the same as the symmetry species of $\phi_m \phi_n$. Thus if the overlap integral is zero, *as a result of symmetry* the Hamiltonian integral must be zero also. Of course, there may be situations under which the overlap integral is zero for some reason other than symmetry, and it does not follow in these cases that the Hamiltonian integral is also zero.

We therefore see that we can always simplify the LCAO MO problem by either choosing Cartesian axes such that the atomic orbitals belong to distinct symmetry species or by taking the atomic orbitals in simple combinations that do so transform as symmetry species. Other problems in quantum mechanics that require the evaluation of Hamiltonian integrals can be simplified in a like manner. Thus symmetry is helpful in quantum mechanical problems but it is not a substitute for solving the Schrödinger equation. On the other hand, if we ignore symmetry when solving the Schrödinger equation we do so at the cost of increased computational effort. Even if we accept this increased effort it is a mistake to ignore the elegance that the use of symmetry can bring into the analysis of a problem. The next section in which we will apply some of the rules we have established should illustrate this point.

7.4. The molecular orbitals of linear and bent AH_2 molecules

As a simple illustration of the application of symmetry to valence problems, which goes beyond the treatment of diatomic molecules in Chapter 6, we will discuss the orbitals of molecules of general formula AH_2 where A is a first row atom. Water, with an equilibrium bond angle of 105°, is one member of the family and other species which are unstable, but which have been detected spectroscopically, are BH_2, CH_2, and NH_2. BeH_2 would be expected to be a linear but it is known only in the solid state where it has a polymeric structure.

Rather than consider a particular molecule we will look at the family as a whole and see how the molecular orbitals change as the bond angle is varied from 180° to 90°. We start first with linear configurations.

Linear H A H belongs to the group $D_{\infty h}$ whose character table is given in Table 7.9. As this group is different from the C_{2v} and C_{3v} groups considered in detail in the last section by being of infinite order, we will first make some explanatory comments about its character table.

Any rotation by an angle φ about the intermolecular axis leaves the nuclei unchanged and such an operation is represented by the symbol C_∞^φ. For all angles

Table 7.9. $D_{\infty h}$ Character table

$D_{\infty h}$	E	$2C_\infty^\varphi$	...	$\infty\sigma_v$	i	$2S_\infty^\varphi$	...	∞C_2	
Σ_g^+	1	1	...	1	1	1	...	1	x^2+y^2, z^2
Σ_g^-	1	1	...	−1	1	1	...	−1	R_z
Π_g	2	$2\cos\varphi$	...	0	2	$-2\cos\varphi$	...	0	$(R_x, R_y), (xz, yz)$
Δ_g	2	$2\cos 2\varphi$	...	0	2	$2\cos 2\varphi$	...	0	(x^2-y^2, xy)
Σ_u^+	1	1	...	1	−1	−1	...	−1	z
Σ_u^-	1	1	...	−1	−1	−1	...	1	
Π_u	2	$2\cos\varphi$	...	0	−2	$2\cos\varphi$	...	0	(x, y)
Δ_u	2	$2\cos 2\varphi$	...	0	−2	$-2\cos 2\varphi$	...	0	

except π there are two such operations corresponding to rotation clockwise or anticlockwise about this axis. There is an infinite number of C_2 axes perpendicular to the principal axis passing through the central atom, and the corresponding operations are indicated ∞C_2. It is the presence of these 2-fold axes that identifies the group as a D group. The remaining operations are i (the centre of symmetry being at nucleus A), improper rotations about the principal axis ($2S_\infty^\varphi$), and an infinite number of vertical mirror planes containing the principal axis ($\infty\sigma_v$).

We have already commented on the fact that capital Greek letters are used to label the symmetry species in contrast to the Roman letters used for the finite groups. The non-degenerate species are labelled Σ and distinguished by their character under inversion (g or u for +1 and −1 respectively) and by a superfix + or − which is the sign of the character under σ_v. All other species are doubly degenerate. Their character under C_∞^φ is $2\cos n\varphi$ and they are labelled Π, Δ, Φ etc. accordingly as $n = 1, 2, 3$, etc. This labelling is consistent with that used for the molecular orbitals of diatomic molecules (equation 5.26) although we note that it is common practise to use lower case letters for molecular orbitals rather than the upper case letters of the character table.

As a basis for the molecular orbitals of AH_2 molecules we take the $2s$ and three $2p$ orbitals of atom A and the $1s$ orbital of hydrogen ($1s$, $1s'$). We are treating the $1s$ orbital of A as a non-bonding inner shell. It is convenient to choose axes for the molecule which are consistent with the symmetries of the transformation functions shown on the right-hand side of the character table.

We see that z belongs to Σ_u^+ and is unchanged by the σ_v operation. It follows that we should choose z to lie along the internuclear axis and x and y must be perpendicular to this. Clearly x and y are mixed together by the operation C_∞^φ and hence they must belong to a doubly degenerate species, and we see that this is Π_u. The transformation properties of the $2p$ orbitals of atom A are the same as the three unit vectors along the Cartesian axes, so that we can identify $2p_z$ as a Σ_u^+ species whilst $2p_x$ and $2p_y$ together belong to Π_u.

The $2s$ orbital of A is unchanged by any operation of the group and must therefore belong to the totally symmetric species Σ_g^+.

Finally we come to the two hydrogen orbitals ($1s$ and $1s'$). Taken individually the orbitals do not belong to any one symmetry species because the effect of some operations is not to leave the orbital alone or to change its sign, but to change it into the other $1s$ orbital. For example

$$i\ 1s = 1s', \tag{7.32}$$

and

$$i\ 1s' = 1s. \tag{7.33}$$

In this respect the two $1s$ orbitals are like the $2p_x$ and $2p_y$ orbitals in that they must be considered together. There is a difference however in that it is possible to find combinations of $1s$ and $1s'$ which do belong to separate non-degenerate species.

By adding or subtracting equations (7.32) and (7.33) we deduce that

$$i\,(1s + 1s') = (1s + 1s') \tag{7.34}$$

and

$$i\,(1s - 1s') = (1s' - 1s) = -(1s - 1s'). \tag{7.35}$$

In other words, the character for the combination $(1s + 1s')$ is $+1$ and for $(1s - 1s')$ it is -1. An examination of the effect of the other operations on these two combinations shows that $(1s + 1s')$ belongs to the species Σ_g^+ and $(1s - 1s')$ to Σ_u^+. We can refer to these two combinations as *symmetry adapted basis functions*.

Table 7.10. The symmetry species of the basis functions for the molecular orbitals of linear HAH

Σ_g^+	Σ_u^+	Π_u
$2s$	$2p_z$	$2p_x$, $2p_y$
$1s + 1s'$	$1s - 1s'$	

We are therefore in a position to catalogue all our basis functions in terms of the symmetry species to which they belong and this is done in Table 7.10.

We can now use the result of the preceeding section that only basis functions of the same symmetry will combine together to form molecular orbitals. The molecular orbitals of HAH that can be formed from our basis are therefore of three kinds. Firstly, the Π_u orbital consists of degenerate $2p$ orbitals of atom A and as there are no hydrogen orbitals of this symmetry they are non-bonding. Secondly, there are Σ_g^+ orbitals which are mixtures of the basis functions $2s$ and $1s + 1s'$.

These molecular orbitals will be two in number and will have the general form

$$\psi_g = c_1\, 2s + c_2\, (1s + 1s'). \tag{7.36}$$

Group theory can give us no information about the values of the coefficients c_i and c_2 nor about the energies of the orbitals. From the information in Chapter 6, however, we deduce that one of these orbitals, is bonding, and for this the two coefficients have the same sign; the other is antibonding, and the coeffients have opposite signs. By similar argument, there will be two Σ_u^+ molecular orbitals of the form

$$\psi_u = c_3 2p_z + c_4 (1s - 1s') \tag{7.37}$$

one of which will be bonding and the other antibonding.

We can set up an interaction diagram between these basis functions to give a molecular energy scheme as in Figure 7.5. Because the $2s$ orbital of A has a lower energy than the $2p$ orbitals, the bonding orbital Σ_g^+ will have a lower energy than the antibonding Σ_u^+, regardless of the relative energies of the basis atomic orbitals. Figure 7.5 shows the actual calculated energies of the molecular orbitals for a linear configuration of the water molecule.

We turn now to a bent AH_2 system which belongs to the symmetry group C_{2v} whose character table is Table 7.4. Bent AH_2 has a lower symmetry than the linear molecule because the order of the group is smaller. We therefore expect that the basis functions fall into a smaller number of symmetry species, that is, more of them will contribute to the same molecular orbital.

It is usual practice to choose the Cartesian coordinates for a C_{2v} molecule such that z is the 2-fold symmetry axis. However, in this exercise we are trying to establish a connection between the molecular orbitals of the linear and bent molecules and it is convenient to keep common space fixed axes as in Figure 7.5, so that x is the C_2 axis and xz is the plane containing the nuclei. With this non-standard choice of axes the functions given on the right-hand side of the character table cannot be used directly to establish the symmetry species of the basis orbitals.

The symmetry species of the atomic orbitals of A are determined by writing down the characters for each of the symmetry operations: 1 indicates that the operation leaves the orbital unchanged and -1 that it changes its sign. The result of applying this rule is shown in Table 7.11 (remember that σ_v is reflection in the plane perpendicular to the plane of the molecule, in this case yz).

As in the case of the linear molecule the two hydrogen orbitals do not individually belong to one of the symmetry species because the operations C_2 and σ_v' change one basis function into the other. However, the combinations $1s + 1s'$ and $1s - 1s'$ are clearly symmetry adapted and have the characters $(1,1,1,1)$ and $(1,-1,1,-1)$ respectively. We can therefore classify our basis functions as shown in Table 7.12.

A comparison of Tables 7.10 and 7.12 shows that the only difference between the linear and bent structure is that $2p_x$, which was non-bonding for linear structures, interacts with $2s$ and with $(1s + 1s')$ for bent structures. Thus the A_1

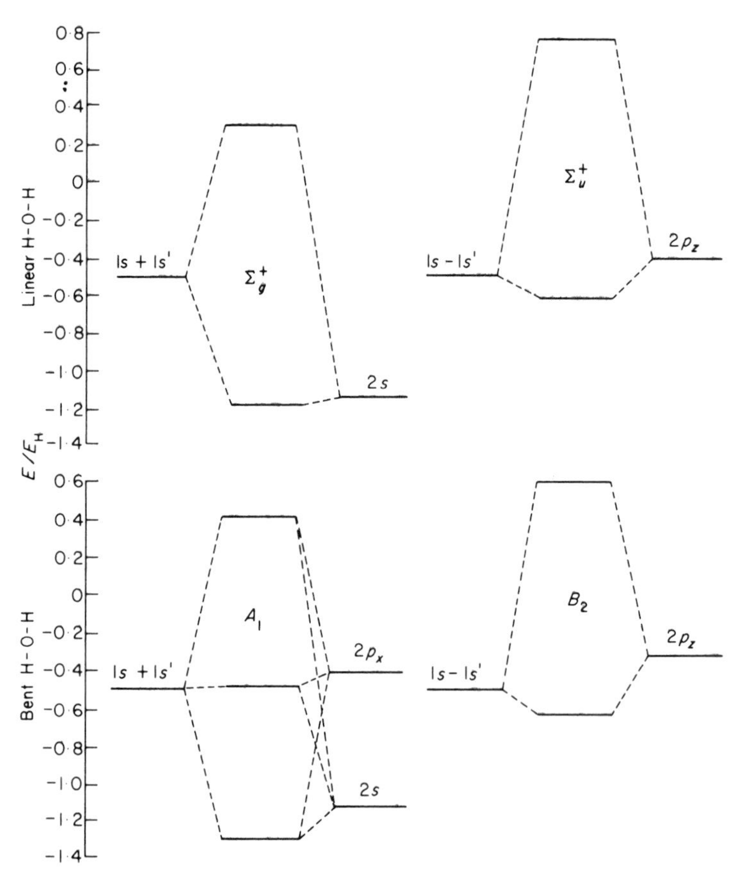

Figure 7.5 Orbital interaction scheme for linear and bent configurations of the molecule HOH. The non-bonding orbitals ($2p_x$, $2p_y$ for linear and $2p_y$ for bent) are not shown.

orbitals will be a mixture of three basis functions, and the energies of these orbitals will be obtained by solving a 3 x 3 secular determinant such as (6.55). In this case it is more difficult to make a qualitative judgement about the relative order of the orbitals. Figure 7.5 shows the interaction scheme for H_2O in its equilibrium geometry as a typical case.

Although one cannot predict without calculation the relative order of the molecular orbitals of bent AH_2 molecules by the above arguments we can go further than symmetry allows by considering the behaviour of the orbitals of such molecules as we gradually bend the molecule from the linear configuration. Such an analysis was first used to great effect by Walsh in the 1950s to predict the equilibrium geometries of simple polyatomic molecules in various electronic states, and the subject of the next section is the Walsh diagram.

Table 7.11. The effect of the C_{2v} symmetry operations on the basis functions for bent AH_2 using the same system of axes as for linear AH_2

	E	C_2	σ_v	σ_v'
$2s$	1	1	1	1
$2p_x$	1	1	1	1
$2p_y$	1	−1	1	−1
$2p_z$	1	−1	−1	1

Table 7.12. The symmetry species of the basis functions for the molecular orbitals of bent AH_2

A_1	B_2	B_1
$2s$	$2p_z$	$2p_y$
$2p_x$	$1s - 1s'$	
$1s + 1s'$		

7.5. Walsh diagrams

We have already seen that the relative order of energy of the atomic orbitals and the electron occupation of these orbitals is the key to our understanding of the periodic classification of the elements. In Section 6.2 we showed how the stability and properties of homonuclear diatomic molecules could be explained by the same principles.The question we raise now is whether it is useful to extend this approach to polyatomic molecules. Can we make any broad rationalization of the properties of such molecules without quantitative calculations on them individually? This was first attempted on any large scale by Walsh† and the systems he studied were small polyatomic molecules with some symmetry: the AH_2 moleules considered in the last section will serve to illustrate his approach. It must be emphasized that some elements of symmetry are essential for this approach to succeed because it is only then that we can reach some conclusions about the form of the molecular orbitals and energies without quantitative calculations.

We start with the assumption that there is some pattern of the molecular orbitals for AH_2 systems which is common to all atoms A, or at least to atoms which are close in the Periodic Table. For example, in the AH_2 series we have some information about the structure of the ground and some excited electronic states

†A. D. Walsh, *J. Chem. Soc.*, 2260–2331 (1953).

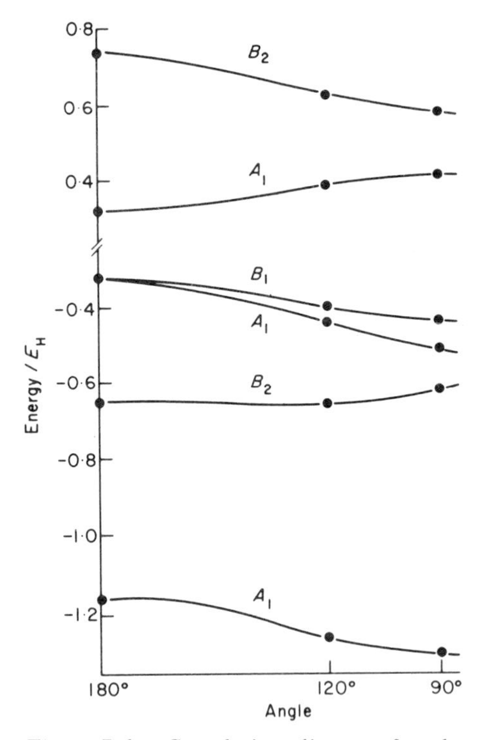

Figure 7.6 Correlation diagram for the molecular orbitals of an HAH molecule as a function of its bond angle. The dots are calculations on the ground state of H_2O by Pitzer and Merrifield.

for atoms A belonging to groups 3, 4, 5, and 6 of the Periodic Table,† and we would like to rationalize the data on all of these. The most important question to be asked is perhaps whether the molecule is linear or bent.

A Walsh diagram shows the variations in the molecular orbital energies of a molecule with varying bond angle in the molecule. We will later examine similar diagrams in which bond lengths are varied in order to rationalize some features of chemical reactivity. The term *correlation diagram* covers any figure which shows how energies of orbitals or states of a molecule vary continuously with some internal coordinate of the molecule.

Figure 7.6 shows the correlation diagram for the molecular orbitals of an AH_2 molecule where the variable is the bond angle. The diagram has been interpolated from calculations of Pitzer and Merrifield‡ on H_2O at 90°, 120°, and 180°. We shall now see how the energy changes can be explained.

† G. Herzberg, *Molecular Spectra and Molecular Structure*, Volume III, *Electronic Structure of Polyatomic Molecules,* Van Nostrand, New York, 1966.

‡ R. M. Pitzer and D. P. Merrifield, *J. Chem. Phys.*, **52**, 4782 (1970).

The lowest energy orbital (A_1) is stabilized on bending (from 180°) for two reasons. Firstly, the wavefunction of the orbital becomes more flexible by virtue of the admixture of $2p_x$ and by the variational prinicple this is likely to stabilize the orbital. Secondly, the orbital has a component ($1s + 1s'$) which is bonding between the two hydrogen atoms and any deformation which brings the hydrogens atoms together should stabilize the orbital.

The lowest B_2 orbital has a component ($1s - 1s'$) and this is H–H antibonding. In addition, for the bent molecule the hydrogen orbitals are not end-on to the two lobes of the $2p_z$ orbital so that overlap will be smaller in the linear configuration and there will be a smaller interaction. These factors explain why the B_2 orbital increases in energy on bending.

The two components of the Π_u orbital of the linear molecule become A_1 and B_1 on bending and the degeneracy is lifted. The A_1 component is lowered in energy and it can be argued that this is mainly due to the contribution from $2s$ which is the lowest energy of the basis orbitals. The B_1 component is a pure $2p$ orbital and we have described it as non-bonding. It is seen that its energy does change a little with bond angle and this is probably due to a decrease in the energy of repulsion between electrons in B_1 and those in other orbitals.

The two highest orbitals, A_1 and B_2, are not occupied by electrons in the ground state of H_2O. They change in energy in the opposite direction to orbitals of the same symmetry which are occupied. Thus the A_1 occupied orbitals are stabilized on bending but the A_1 unoccupied is destabilized. This is in accord with the qualitative conclusion reached in Section 6.3 (see equations 6.69 and 6.70) that the energies of bonding and antibonding orbitals are related, the more bonding the one the more antibonding the other.

In the lowest states of all the molecules BH_2, CH_2, NH_2, and H_2O the A_1 and B_2 orbitals are all fully occupied. The A_1 orbital favours a bent structure and the B_2 favours a linear structure, but less strongly. The other electrons are (one to four in number in this series) all in orbitals which favour a bent structure. The observed bond angles for the ground states are 131°, 130°, 103°, and 105° for BH_2, CH_2, NH_2, and H_2O respectively and we note the closing up of this angle as more electrons are added to these orbitals. The ground state of CH_2 has unpaired electrons showing that there must be one electron in each of the A_1 and B_2 orbitals that correlate with Π_u of the linear molecule. The big change in angle occurs on going from CH_2 to NH_2, and this shows that it is the electrons in the A_1 orbital which strongly favour a bent structure and those in the B_1 orbital are rather neutral in their effect.

Walsh diagrams such as that of Figure 7.6 can also be used to explain the changes in geometry on ionization or on electron excitation of a molecule. For example, the photoelectron spectrum of H_2O shows that when an electron is removed from the highest occupied A_1 orbital (which strongly favours a bent structure) the resulting state of the ion is linear. Likewise the first excited state of BH_2, which involves promotion of an electron from A_1 to B_1, is linear.

Walsh diagrams are qualitatively useful but not infallible. For example, we would predict from Figure 7.6 that BeH_2, a four-valence-electron molecule, would be bent if we took the relative energies of the lowest A_1 and B_2 orbitals of H_2O as a guide to the situation in BeH_2. However, calculations on BeH_2† show that, on bending, the B_2 orbital increases in energy substantially more than the A_1 decreases so that overall the linear structure is preferred. In other words, the quantitative features of a correlation diagram depend on the specific atoms involved and it is only the qualitative features that can be useful if we take a single diagram to apply to a family of molecules.

†S. D. Peyerimhoff, R. J. Brenker, and L. C. Allen, *J. Chem. Phys.*, **45**, 734 (1966).

Chapter 8

Molecular Orbitals and the Electron-pair Bond

8.1. Electron equivalence and antisymmetry

In Chapter 3 we examined the form of the wavefunctions of atomic orbitals and in Chapter 5 we turned our attention to molecular orbitals. An orbital, whether atomic or molecular, is a one-electron function but we have seen in the earlier chapters that it is a good basis for a description of many-electron systems. In this section we deal briefly with the question of how the wavefunctions of many-electron systems are constructed from orbitals. We require this interlude in order to be able to discuss the nature of equivalent orbitals and localized electron pairs in molecular orbital theory.

In Chapter 4 we described how the postulates of electron spin and the Pauli exclusion principle were needed to explain atomic spectroscopy and the Periodic Table. Our starting point for many-electron wavefunctions must therefore be functions that represent both the spin and the space properties of an electron. The simplest approach to this was developed by Pauli who identified the two possible spin states of the electron with notional spin wavefunctions usually given the symbols α and β. We can formally think of a spin coordinate or variable (s) for this wavefunction so that the total wavefunction for a single electron is a product of a space and spin part. From each space orbital ϕ_r we can construct two space-spin functions, namely

$$\phi_r(x, y, z)\,\alpha(s) \quad \text{and} \quad \phi_r(x, y, z)\,\beta(s), \tag{8.1}$$

which are defined in a 4-dimensional space. These functions are called *spin – orbitals.*

Let us now examine the wavefunction for the helium atom. According to the aufbau principle we have two electrons in a $1s$ atomic orbital with opposite spins. If we label the coordinates of the two electrons by suffixes 1 and 2, then we can write a total wavefunction to describe this situation as

$$\Psi = \phi_{1s}(x_1, y_1, z_1)\alpha(s_1)\phi_{1s}(x_2, y_2, z_2)\beta(s_2), \tag{8.2}$$

which we will abbreviate to

$$\Psi = 1s(1)\alpha(1)1s(2)\beta(2). \tag{8.3}$$

We note that the aufbau principle makes no specific reference to the effect that these two electrons have one on the other. Adopting the probability interpretation of the wavefunction given in Chapter 2, we can deduce, from (8.3), that the probability of finding electron 1 at the position x_1, y_1, z_1 with a spin α and electron 2 at x_2, y_2, z_2 with spin β is

$$1s(1)^2 \; 1s(2)^2 . \tag{8.4}$$

This is just the product of the separate probabilities for each electron in this orbital to be at these points: in statistical terminology the positions of the electrons are *uncorrelated*.† However, some correlation must exist between the positions of electrons because of the repulsion between them. It follows that wavefunctions like (8.3) cannot be exact solutions of the many-electron Schrödinger equation. Nevertheless, they are a good starting point from which to build up better wavefunctions.

The obvious deficiency of the function (8.3) is that it implies that we can distinguish between the two electrons. If we are simply seeking a wavefunction to represent two electrons with opposite spins in a 1s orbital then we could equally well have written

$$\Psi' = 1s(1)\beta(1)1s(2)\alpha(2). \tag{8.5}$$

As Ψ and Ψ' are equally acceptable they must have equal weight in any wavefunction based upon them. We are therefore led to consider two possible combinations which satisfy this criterion, namely

$$1s(1)\alpha(1)\; 1s(2)\beta(2) \pm 1s(1)\beta(1)\; 1s(2)\alpha(2). \tag{8.6}$$

There is nothing in the Schrödinger equation to say which, if either, of these combinations is possible. However, the evidence from spectroscopy is unambiguous and shows that *only the combination with the negative sign is allowed*. This, and indeed all wavefunctions for electrons, must be *antisymmetric* to the exchange of the labels of any two electrons. That is, the wavefunctions must change sign on this operation. Electrons are said to be *Fermi particles*, or fermions.

No state has ever been detected in spectroscopy which violates the antisymmetry rule. This very important restriction on wavefunctions can be treated formally as follows. Let $\mathscr{P}_{12}$ be the *permutation operator* which interchanges electrons 1 and 2. When it acts on the combination given in equation (8.6) which has the negative sign we obtain the result

$$\begin{aligned} &\mathscr{P}_{12}\{1s(1)\alpha(1)1s(2)\beta(2) - 1s(1)\beta(1)1s(2)\alpha(2)\} \\ &\quad = \{1s(2)\alpha(2)1s(1)\beta(1) - 1s(2)\beta(2)1s(1)\alpha(1)\} \\ &\quad = -\{1s(1)\alpha(1)1s(2)\beta(2) - 1s(1)\beta(1)1s(2)\alpha(2)\}. \end{aligned} \tag{8.7}$$

In general n electrons may be allocated to n spin–orbitals in $n!$ ways but only

†Tossing coins is a typical uncorrelated action. The probability of getting heads twice is the product of the probabilities of getting heads at each toss.

one linear combination of these $n!$ terms satisfies the antisymmetry rule. This unique combination may be written in a compact form as a determinant. For example, the two-electron function used in (8.7) may be written

$$\begin{vmatrix} 1s(1)\alpha(1) & 1s(1)\beta(1) \\ 1s(2)\alpha(2) & 1s(2)\beta(2) \end{vmatrix} \equiv 1s(1)\alpha(1)1s(2)\beta(2) - 1s(1)\beta(1)1s(2)\alpha(2), \quad (8.8)$$

and for n electrons in the spin–orbitals $\psi_a, \psi_b \ldots \psi_k$ we have the determinant

$$\begin{vmatrix} \psi_a(1)\psi_b(1) \ldots \psi_k(1) \\ \psi_a(2)\psi_b(2) \ldots \psi_k(2) \\ \text{-----------} \\ \psi_a(n)\psi_b(n) \ldots \psi_k(n) \end{vmatrix} \quad (8.9)$$

The antisymmetry property is satisfied by reason of the property of a determinant that if we exchange two rows (equivalent to exchanging the labels of two electrons) we change its sign.

The Pauli exclusion principle is a manifestation of the antisymmetry rule. If there were two electrons in an atom with the same four quantum numbers then two electrons would have the same space and spin wavefunctions. In other words two spin-orbitals would be the same. It would follow that two columns of the determinant would be identical, and it is well known property of determinants that in this case, on expansion, the determinant will be zero (all the individual terms from the expansion will cancel exactly). In other words, we cannot construct a non-zero wavefunction that satisfies the antisymmetry rule if we assign more than one electron to any spin – orbital.

Determinants like (8.9) are the most convenient functions from which to construct many-electron wavefunctions. As they individually satisfy the antisymmetry condition, any linear combination of determinants is also antisymmetric. The rules for evaluating the Hamiltonian and overlap integrals with such determinantal wavefunctions were first formulated by J. C. Slater and they are therefore commonly referred to as *Slater determinants*.

8.2. Equivalent orbitals and localized orbitals

Molecular orbitals are one-electron functions which are, in general, delocalized over several nuclear centres in a molecule. This point has been brought out in two respects in earlier chapters. Firstly, we have related molecular orbitals to the process of ionization of a molecule. According to Koopmans' theorem the orbital energy is the negative of the ionization potential, and, further, the distribution of positive charge in the ion is given by the wavefunctions of the molecular orbital from which the electron has been removed. Secondly, we have seen that molecular orbitals have a spatial distribution which belongs to one of the symmetry species of the appropriate symmetry group. For example, in the case of H_2O we deduced bonding molecular orbitals which were distributed over both OH bonds, one symmetric to rotation about the 2-fold axis and one antisymmetric.

As a result of the delocalization of molecular orbitals over many (perhaps all) nuclear centres in a molecule there may be no obvious relationship between the molecular orbitals of chemically related molecules. The orbitals of methane for example have no obvious relationship to the orbitals of ethane or cyclohexane. However, there is plenty of chemical and physical evidence to show that there is a strong similarity between the CH bonds in these compounds. For hydrocarbons the length of a CH bond is always approximately 1.06 Å, its stretching force constant is always approximately 500 Nm^{-1} and the energy required to break the bond is approximately 400 $\mathrm{kJ\ mol}^{-1}$. Likewise the length, force constant, and bond energy of CC single bonds are always approximately constant, although in a highly strained molecule like cyclopropane there are deviations from the norm.

At first sight there appears to be no simple explanation of near-constant bond properties in molecular orbital theory, and this was probably one reason why the alternative theory of the chemical bond, known as valence bond theory, had such strong support from chemists during the first twenty years of quantum chemistry. However, in 1949 Lennard-Jones showed that transferable bond properties could be understood from molecular orbital theory, the key being the determinantal representation of many-electron wavefunctions.

An important property of determinants is that the value of the function we get on expansion is not changed if we add columns of the determinant one to another. Taking a 2 x 2 determinant, for example, it is easy to confirm by expansion that the following identity holds for any value of λ.

$$\begin{vmatrix} a & b \\ c & d \end{vmatrix} = \begin{vmatrix} a & (\lambda a + b) \\ c & (\lambda c + d) \end{vmatrix} = ad - bc. \tag{8.10}$$

Let us now examine a determinantal wavefunction for two electrons, one in spin-orbital ψ_a and one in spin-orbital ψ_b and carry out the following column adding operations·

$$\begin{vmatrix} \psi_a(1) \psi_b(1) \\ \psi_a(2) \psi_b(2) \end{vmatrix} \xrightarrow[\text{col. 2 + col. 1}]{\text{replace col. 1 by}} \begin{vmatrix} [\psi_a(1) + \psi_b(1)] & \psi_b(1) \\ [\psi_a(2) + \psi_b(2)] & \psi_b(2) \end{vmatrix}$$

$$\Big\downarrow \text{ replace col. 2 by (col. 2} - \tfrac{1}{2}\text{col. 1)}$$

$$\begin{vmatrix} [\psi_a(1) + \psi_b(1)] & [\psi_b(1) - \psi_a(1)]/2 \\ [\psi_a(2) + \psi_b(2)] & [\psi_b(2) - \psi_a(2)]/2 \end{vmatrix}$$

$$-\frac{1}{2}\begin{vmatrix} [\psi_a(1) + \psi_b(1)] & [\psi_a(1) - \psi_b(1)] \\ [\psi_a(2) + \psi_b(2)] & [\psi_a(2) - \psi_b(2)] \end{vmatrix} \xleftarrow{\text{take out factor } -\frac{1}{2}} \tag{8.11}$$

Apart from a multiplying constant, in this case $-½$, identical wavefunctions can be constructed by having electrons in ψ_a and ψ_b or by having electrons in $(\psi_a + \psi_b)$ and $(\psi_a - \psi_b)$. We emphasize that this is only a property of antisymmetric functions and is not a property of simple product wavefunctions, thus:

$$\psi_a(1)\psi_b(2) \neq [\psi_a(1) + \psi_b(1)]\,[\psi_a(2) - \psi_b(2)]\,. \tag{8.12}$$

There is an infinity of transformations of the type shown in (8.10) and hence the functions $\psi_a + \psi_b$ and $\psi_a - \psi_b$ are in no way special and do not necessarily have any physical significance. Insofar as they are one-electron functions they can be called spin-orbitals, but unlike ψ_a and ψ_b they have no relationship to the ionization process nor, in general, are their spatial components (the orbitals) the solutions of any one-electron Schrödinger equation. There are, however, some orbitals produced by transformations of this type that have physical significance and these are called *equivalent* orbitals and *localized* orbitals.

The term equivalent orbital can only be used when the molecule possesses some elements of symmetry. Equivalent orbitals are functions that are identical except for their spatial disposition in the molecule. We shall see, for example, that it is possible to form four equivalent orbitals for CH_4 each of which is associated with one of the CH bonds. The symmetry operations of the molecule transform one equivalent orbital into itself or some other member of the set. Thus equivalent orbitals, unlike molecular orbitals, do not belong to one symmetry species. Like the atomic orbitals from which they are formed they will span several symmetry species: they satisfy equations like (7.31) rather than (7.12).

Localized orbitals are functions which are confined as far as possible to different regions of the molecule. There are several recipes for quantifying this condition, one of the most common being to minimize the total electron repulsion between electrons in *different* localized orbitals. In other words, we require electrons in different localized orbitals to be as far from each other as possible.

When a molecule has symmetry the localized orbitals become equivalent orbitals or, *vice versa*, equivalent orbitals are found to satisfy the localization criterion. Although the localization criterion is more general than the condition of symmetry equivalence, it is more difficult to apply and we will therefore illustrate the above points by considering a molecule having symmetry, namely CH_4.

We first apply the simplifying techniques of symmetry group theory described in the last chapter to deduce the form of the molecular orbitals of CH_4. The character table for the tetrahedral group T_d to which CH_4 belongs is given in Table 8.1. As a basis for the valence molecular orbitals we take the $2s$ and three $2p$ orbitals of the carbon atom and the four hydrogen $1s$ orbitals (h_1, h_2, h_3, h_4). The carbon $2s$

Table 8.1. Character table for the tetrahedral group T_d

T_d	E	$8C_3$	$3C_2$	$6S_4$	$6\sigma_d$	
A_1	1	1	1	1	1	$(x^2 + y^2 + z^2)$
A_2	1	1	1	−1	−1	
E	2	−1	2	0	0	$(x^2 - y^2), (2z^2 - x^2 - y^2)$
T_1	3	0	−1	1	−1	
T_2	3	0	−1	−1	1	$(x, y, z), (xy, xz, yz)$

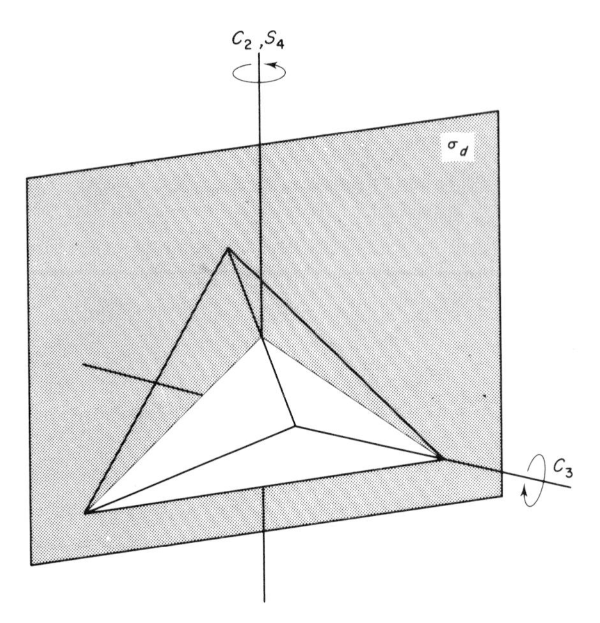

Figure 8.1 Symmetry elements for the tetrahedral group T_d. Only one element of each type is shown.

orbital is clearly unchanged by any symmetry operation of the group and must belong to the species A_1. We see from the character table that the functions (x, y, z) transform together as the T_2 species which is triply degenerate and we conclude that the $2p_x$, $2p_y$, and $2p_z$ orbitals likewise transform as T_2.

The hydrogen orbitals individually do not belong to any one symmetry species and we must form combinations which do. This was the procedure adopted in the last chapter when we examined the molecular orbitals of AH_2. The symmetry operations of the T_d group are a little complicated, but the axes and planes of symmetry are illustrated in Figure 8.1. The S_4 operations involve a rotation about the C_2 axes by $2\pi/4$ followed by reflection in a plane perpendicular to the axis. Each operation of the group sends one of the hydrogen basis functions into one of the set of four. Let us consider, for example, the effect of one of the eight C_3 operations (an anti-clockwise rotation by $2\pi/3$ about the 3-fold axis passing through h_1) on each of our basis functions. The result is as follows (see Figure 8.2)

$$C_3\, h_1 \to h_1,$$
$$C_3\, h_2 \to h_3,$$
$$C_3\, h_3 \to h_4,$$
$$C_4\, h_4 \to h_2. \tag{8.13}$$

We see from this that

$$C_3(h_1 + h_2 + h_3 + h_4) \to (h_1 + h_2 + h_3 + h_4). \tag{8.14}$$

It is not difficult to confirm that each of the symmetry operations leaves the combination $(h_1 + h_2 + h_3 + h_4)$ unchanged so that this combination must belong to the totally symmetric species A_1.

There are three other independent combinations of the hydrogen orbitals which have to be found.† We can anticipate that they must belong, like the $2p$ orbitals, to the T_2 species. The reasons for this are firstly that they are three in number and secondly that there is clearly some overlap (and therefore interaction) between the $2p$ orbitals and the hydrogen orbitals.

A simple way to find the required combinations is to consider the hydrogen orbitals in two pairs (h_1, h_2) and (h_3, h_4). Each (C)H_2 fragment has C_{2v} symmetry and, following the same procedure used for AH_2 molecules in Section 7.3, we deduce that the appropriate symmetry adapted basis functions would be

$$h_1 + h_2, h_1 - h_2, h_3 + h_4, h_3 - h_4. \tag{8.15}$$

We now note that the pair (h_1, h_2) is related to (h_3, h_4) by an S_4 operation of the T_d group so we can guess that it is necessary to take sums and differences of the functions (8.15) to obtain symmetry adapted functions for the T_d group as a whole. The required combinations are therefore

$$\begin{aligned}
\phi_1 &= (h_1 + h_2) + (h_3 + h_4),\\
\phi_2 &= (h_1 - h_2) + (h_3 - h_4),\\
\phi_3 &= (h_1 + h_2) - (h_3 + h_4),\\
\phi_4 &= (h_1 - h_2) - (h_3 - h_4).
\end{aligned} \tag{8.16}$$

The first of these is the A_1 combination already discussed and the other three belong to the species T_2. It can be seen that with coordinate axes defined as in Figure 8.2 the combination $h_1 - h_2 + h_3 - h_4$ has the same general spatial properties as $2p_x$, $h_1 + h_2 - h_3 - h_4$ resembles $2p_y$ and $h_1 - h_2 - h_3 + h_4$ resembles $2p_z$. There is therefore a non-zero overlap between these pairs of functions but no overlap outside these pairs; for example, $2p_x$ does not overlap with $h_1 + h_2 - h_3 - h_4$.

The molecular orbitals of CH_4 which have been calculated from these basis functions are as follows:‡

$$\begin{aligned}
\psi_1 (A_1) &= 0.58\ 2s + 0.19\ (h_1 + h_2 + h_3 + h_4),\\
\psi_2 (T_2) &= 0.55\ 2p_x + 0.32\ (h_1 - h_2 + h_3 - h_4),\\
\psi_3 (T_2) &= 0.55\ 2p_y + 0.32\ (h_1 + h_2 - h_3 - h_4),\\
\psi_4 (T_2) &= 0.55\ 2p_z + 0.32\ (h_1 - h_2 - h_3 + h_4).
\end{aligned} \tag{8.17}$$

From these four molecular orbitals it is possible to form four linearly

†There are some theorems of group theory which help in this problem, but they have not been dealt with in this book because of their reliance on some knowledge of matrix algebra.

‡R. M. Pitzer, *J. Chem. Phys.*, **46**, 4871 (1971). The $2s$ orbital is orthogonalized to the $1s$ orbital and the small contribution from the $1s$ orbital in ψ_1 has been neglected.

independent combinations which are equivalent to one another except for their orientation in space. The general form of the wavefunctions of these equivalent orbitals is

$$\theta = c_1\psi_1 + c_2\psi_2 + c_3\psi_3 + c_4\psi_4. \tag{8.18}$$

Note that the four hydrogen orbitals h_1, h_2, h_3, h_4 have the property of equivalence that we are seeking: they are equivalent except for their positions in the molecule. We also can see that on inverting the transformation contained in (8.16) we revert to these hydrogen orbitals except for a multiplying constant, thus:

$$\begin{aligned}
\phi_1 + \phi_2 + \phi_3 + \phi_4 &= 4h_1,\\
\phi_1 - \phi_2 + \phi_3 - \phi_4 &= 4h_2,\\
\phi_1 + \phi_2 - \phi_3 - \phi_4 &= 4h_3,\\
\phi_1 - \phi_2 - \phi_3 + \phi_4 &= 4h_4.
\end{aligned} \tag{8.19}$$

This therefore suggests that we form equivalent orbitals by taking the molecular orbitals (8.17) in the same combinations as the ϕ_i in (8.19). If this is done and we introduce a factor of ½ so that the equivalent orbitals are normalized (the molecular orbitals in equation 8.17 are normalized and orthogonal) we obtain the following functions:

$$\begin{aligned}
\theta_1 &= \frac{1}{2}(\psi_1 + \psi_2 + \psi_3 + \psi_4)\\
&= 0.29\ 2s + 0.28\ (2p_x + 2p_y + 2p_z) + 0.58\ h_1 - 0.07\ (h_2 + h_3 + h_4),\\
\theta_2 &= \frac{1}{2}(\psi_1 - \psi_2 + \psi_3 - \psi_4)\\
&= 0.29\ 2s + 0.28\ (-2p_x + 2p_y - 2p_z) + 0.58\ h_2 - 0.07\ (h_1 + h_3 + h_4),\\
\theta_3 &= \frac{1}{2}(\psi_1 + \psi_2 - \psi_3 - \psi_4)\\
&= 0.29\ 2s + 0.28\ (2p_x - 2p_y - 2p_z) + 0.58\ h_3 - 0.07\ (h_1 + h_2 + h_4),\\
\theta_4 &= \frac{1}{2}(\psi_1 - \psi_2 - \psi_3 + \psi_4)\\
&= 0.29\ 2s + 0.28\ (-2p_x - 2p_y + 2p_z) + 0.58\ h_4 - 0.07\ (h_1 + h_2 + h_3).
\end{aligned} \tag{8.20}$$

These four functions are clearly equivalent in all respects except for their orientation in space. They each have the same amount of $2s$ and $2p$ density for example. θ_1 has a large coefficient of h_1 and equal positive weightings from the three $2p$ orbitals and it therefore has its maximum amplitude along the direction of the Ch_1 bond. There are small negative contributions from the other three hydrogen orbitals. Similarly θ_2, θ_3, and θ_4 have their maximum amplitudes along the Ch_2, Ch_3, and Ch_4 bonds respectively. Because of these small negative coefficients the

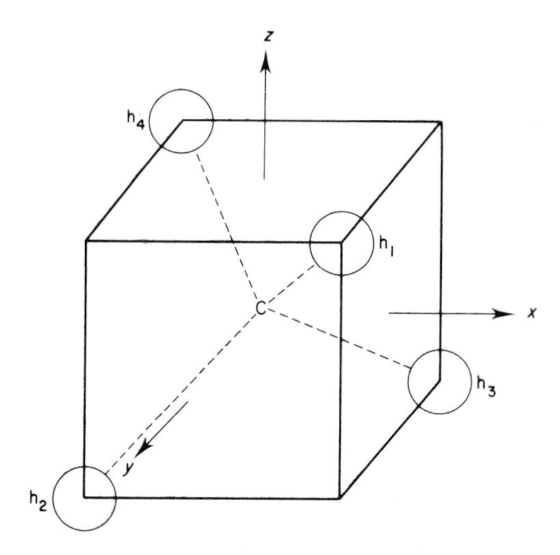

Figure 8.2 The relationship of the hydrogen orbitals, h_1, h_2, h_3, h_4 to the coordinate axes, and hence to the $2p_x$, $2p_y$, and $2p_z$ orbitals of the central carbon atom.

functions cannot be said to be entirely localized in one CH bond region, but the degree of localization is clearly appreciable. These negative coefficients are necessary to satisfy the condition that the functions are mutually orthogonal, or linearly independent.

The functions (8.20) satisfy the localization criterion that the total repulsion energy of electrons in different equivalent orbitals is minimized, although this is not something we can prove in this book. For molecules having less symmetry than CH_4 the condition of equivalence is not generally sufficient to define optimum localized orbitals.

For example, the occupied valence molecular orbitals of H_2O are (from Figure 7.6) of symmetry A_1, B_2, A_1 and B_1 in order of increasing energy. If we form the combinations

$$\psi(A_1) \pm \psi(B_2), \tag{8.21}$$

by the transformation (8.11) then we obtain two equivalent orbitals which are symmetric to reflection in the plane of the molecule (σ_v') (see Table 7.4, Figure 7.1) because both A_1 and B_2 orbitals are symmetric to this operation, but which have mixed symmetries under C_2 or σ_v. These two equivalent orbitals are roughly localized in the OH bonds. Similarly, if we form combinations

$$\psi(A_1) \pm \psi(B_1), \tag{8.22}$$

we obtain equivalent orbitals which are symmetric under σ_v but of mixed symmetries for C_2 and σ_v'. These are equivalent functions that are directed on opposite sides of the plane of the nuclei and are called *lone-pair* orbitals. However, we have not yet defined the functions $\psi(A_1)$ that appear in (8.21) and (8.22) because there are two A_1 molecular orbitals to be considered. We must therefore add a localization condition to define the wavefunctions of the two sets of equivalent orbitals unambiguously. When this is done it results in lone-pair orbitals (8.22) which have very little contribution from the hydrogen $1s$ orbitals, and it is for this reason that the adjective lone-pair is used.

If one carries out a calculation on a hydrocarbon such as ethane using the localization criterion then, amongst the set of orbitals so constructed, there will be functions with a close resemblance to those of (8.20). These will be CH localized orbitals. It is because of this similarity – but we emphasize not any equality – that one can expect bond transferable properties in such molecules. The justification would be that as there is only one localized orbital in any one CH bond then the properties of this bond (length, force constant, etc.) should be reflected in the wavefunction of this orbital. We have not proved rigorously that bond transferable properties follow from this but the interpretation of such properties within the molecular orbital framework is certainly as satisfactory as their interpretation within the alternative valence-bond framework.

Finally, we come to the question of whether it is always possible to transform molecular orbitals to localized orbitals. The answer is that it is not, and the molecules for which it is not possible are chemically very interesting. As an example we consider benzene, the molecular orbitals of which will be examined in the next chapter. The valence molecular orbitals of benzene fall into two symmetry types, those that are unchanged on reflection in the plane of the ring, which are labelled σ orbitals, and those that change sign on this operation, which are labelled π orbitals. Calculations show that there are twelve bonding σ orbitals and three bonding π orbitals and in the ground state of benzene all are occupied by two electrons.

It is possible to make an equivalent orbital transformation of the twelve σ orbitals to give twelve functions, six of which will be found to be localized in the CH bonds and six in the CC bonds. We can say that the σ molecular orbitals are *bond localizable*. However, from the three π orbitals it is only possible to construct three equivalent orbitals and clearly these cannot be localized in the CC bonds as they are insufficient in number. At best the π orbitals can be localized over two bonds or three atoms. Thus the π molecular orbitals are not bond localizable. Chemists commonly refer to σ molecular orbitals as being localized but, as we have seen, all molecular orbitals are delocalized. The correct terminology is that σ orbitals are localizable in bonds but in a molecule like benzene the π orbitals are not.

8.3. Hybridization

In the early quantum mechanical development of valence theory which was strongly influenced by the Lewis theory of the electron-pair bond, a concept was introduced known as hybridization. This, as we shall see in Chapter 13, is an essential concept of valence bond theory. In molecular orbital theory its position is somewhat weaker, but in the development of this book it is here that we first meet the concept.

In Lewis theory the covalent bond was considered to arise from the sharing of electrons by atoms so that each atom completes a stable octet of electrons. The transposition of this idea to quantum mechanics was to associate the covalent bond with the overlap of two atomic orbitals each containing single electrons and the pairing of the spins (one α and one β) of these two electrons.

In a molecule like CH_4 with four equivalent covalent bonds, there is an obvious difficulty with this atomic orbital description because, although there are four valence orbitals on the carbon to overlap with the hydrogen orbitals ($2s$, $2p_x$, $2p_y$, $2p_z$), these four orbitals are not all of the same type. The concept of hybridization was introduced to get around this difficulty: the orbitals which one invokes to form the covalent bonds are not necessarily atomic orbitals, but may be hybrid (or mixed) atomic orbitals.

From the four carbon valence atomic orbitals it is possible to form four equivalent hybrid orbitals. These have the wavefunctions

$$\begin{aligned}\psi_1 &= \tfrac{1}{2}(2s + 2p_x + 2p_y + 2p_z),\\ \psi_2 &= \tfrac{1}{2}(2s - 2p_x + 2p_y - 2p_z).\\ \psi_3 &= \tfrac{1}{2}(2s + 2p_x - 2p_y - 2p_z),\\ \psi_4 &= \tfrac{1}{2}(2s - 2p_x - 2p_y + 2p_z).\end{aligned} \tag{8.23}$$

We note that each of these hybrid orbitals has the same ratio of s density to p density; the density being defined by the squares of the coefficients of the atomic orbitals in the wavefunctions. Thus ψ_1 has $\tfrac{1}{4}s$ density and $\tfrac{1}{4}$ weighting of each of the three p orbitals, that is, a net $\tfrac{3}{4}\,p$ density. We refer to such a hybrid as sp^3, the connotation being that sp^n is a hybrid where the ratio of p to s density is n:1.

A contour diagram of one sp^3 hybrid is shown in Figure 8.3. It is seen that the hybrid has a strong directional character and the direction of largest amplitude is that of the positive lobe of the resultant p orbital in its wavefunction. For example, the combination

$$2p_x + 2p_y + 2p_z, \tag{8.24}$$

is another $2p$ orbital, with the same shape as its components, whose positive lobe is directed towards the point (1,1,1) in Cartesian coordinate space. Thus ψ_1 of (8.23) is an sp^3 hybrid whose maximum amplitude is in this direction. Analogous arguments apply to each of the four hybrid orbitals (8.23), and we therefore conclude that the four hybrids are directed towards the four corners of a tetrahedron.

The connection between hybrid atomic orbitals and molecular orbital theory can now be seen by comparing the functions (8.23) and (8.20). To a good approximation the equivalent orbitals (8.20) are a linear combination of one sp^3 hybrid of the carbon and the hydrogen orbital to which it is directed. The fact that the s:p ratio is not exactly 1:3 arises in molecular orbital theory because the s and p

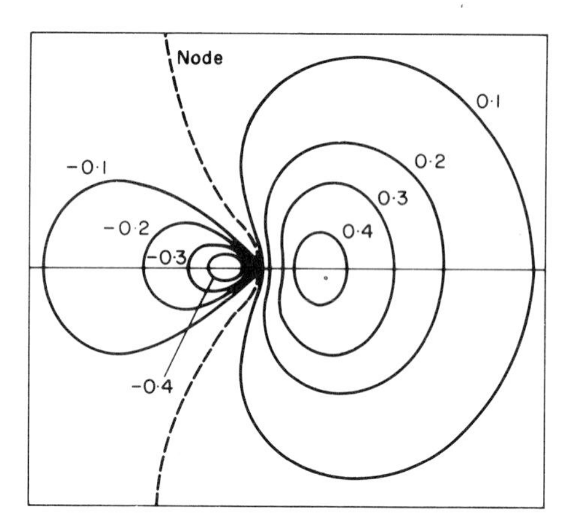

Figure 8.3 Contours of a carbon sp^3 hybrid orbital.

orbitals are not equally used in the bonding molecular orbitals. It is seen from the molecular orbital wavefunctions (8.17) that the $2s$ orbital has a slightly greater weighting in the bonding orbitals than the $2p$. The converse would be true for the antibonding orbitals.

In *ab-initio* molecular orbital theory it makes no difference to the final outcome whether one takes as a basis, hybrid orbitals or atomic orbitals because hybrid orbitals are themselves linear combinations of atomic orbitals. In some approximate

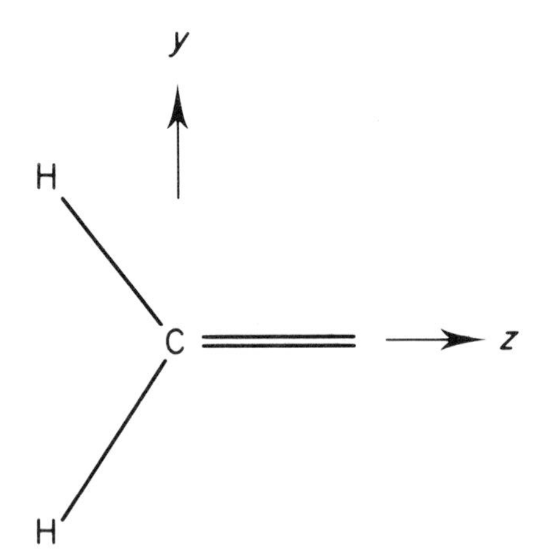

Figure 8.4 Axes used for the carbon sp^2 hybrid orbitals defined by equation (8.26), and which are a suitable basis for describing bonding in alkenes or benzenoid compounds. The x is vertical to the plane of the paper.

or empirical molecular orbital theory there can be a difference between the two bases if some of the small overlap or Hamiltonian integrals are neglected. Thus if one takes as a basis for CH_4, carbon sp^3 hybrid orbitals and $h_1 - h_4$ as defined in Figure 8.2, and one then neglects all terms which couple a hybrid with the hydrogen orbitals to which it is not directed, ψ_1 of (8.23) with h_2, h_3, and h_4, for example, then the resulting wavefunctions will represent bond localized orbitals which will be approximations to the exact localized equivalent orbitals. Such functions are much easier to calculate than the exact functions and they are very useful for giving a qualitative description of the bonding in large molecules.

The most important feature of hybrid orbitals is that they have a maximum amplitude in one direction of space and they are therefore ideally suited for a discussion of the bonding in those molecules for which bond angles are an important characteristic. For example, the paraffinic hydrocarbons have bond angles about the carbon atom which are disposed tetrahedrally (angles 109°) and the sp^3 hybrids described above are a natural basis in which to discuss the bonding in such molecules.

Benzenoid or ethylenic hydrocarbons are planar molecules with bond angles of, or near to, 120°, and sets of hybrids which subtend such an angle at the carbon are the sp^2 hybrids. These hybrids have $\frac{1}{3}$ s density and $\frac{2}{3}$ p density and hence their wavefunctions have the general form

$$\psi = \sqrt{\frac{1}{3}} \cdot 2s + \sqrt{\frac{2}{3}} \cdot 2p \tag{8.25}$$

with the maximum amplitude of the hybrid being in the direction of the positive lobe of the p orbital. Precise wavefunctions of these hybrids in terms of axis-fixed p orbitals can readily be constructed. For example, for the axes defined in Figure 8.4 the wavefunctions of the sp^2 hybrids are as follows:

$$\begin{aligned}
\psi_1 &= (2s + \sqrt{2} \cdot 2p_z)/\sqrt{3} \\
\psi_2 &= \left[2s + \sqrt{2}\left(\frac{\sqrt{3}}{2} \cdot 2p_y - \frac{1}{2} \cdot 2p_z\right)\right]/\sqrt{3} \\
\psi_3 &= \left[2s - \sqrt{2}\left(\frac{\sqrt{3}}{2} \cdot 2p_y + \frac{1}{2} \cdot 2p_z\right)\right]/\sqrt{3}.
\end{aligned} \tag{8.26}$$

Note that the $2p_x$ orbital, which has a node in the plane of the molecule, makes no contribution to these hybrids.

For linear molecules (e.g. acetylene) one commonly uses as a basis sp hybrids whose wavefunctions (given the z axis as the molecular internuclear axis) are

$$\begin{aligned}
\psi_1 &= (2s + 2p_z)/\sqrt{2}, \\
\psi_2 &= (2s - 2p_z)/\sqrt{2}.
\end{aligned} \tag{8.27}$$

However, our emphasis here on what one might call the integral sp^n hybrids, which are referred to so often by chemists, should not be taken to imply that non-integral hybrids cannot be defined. These will be useful for situations in which the bond angles are not exactly 180, 120, or 109°. Thus a general wave function for an sp^n hybrid, n being integral or not, is

$$\psi = (2s + \sqrt{n}.2p)(1 + n)^{-\frac{1}{2}}. \tag{8.28}$$

Even for a molecule like acetylene the use of integral hybrids to discuss the bonding has no quantitative basis for there is no reason to suppose that the 'best' hybrid to construct the CH bond has the same s:p ratio as that used to construct the CC triple bond. Molecular orbital calculations show that the carbon $2p_z$ orbitals make a larger contribution than the $2s$ to the CH bonds but a smaller contribution to the CC bond in this molecule.

Our discussion of hybrid orbitals has been limited to carbon compounds as this is where the concept has found greatest use in chemistry. The generalization to other atoms and to other orbitals requires no extension of the concept. For a set of one s and three p orbitals it is only possible to obtain four linearly independent functions and hence a maximum of four hybrid orbitals. If an atom has more than four directed bonds, as for example does the sulphur in SF_6, then a hybrid basis must be constructed which uses d orbitals in addition to s and p. It is possible to obtain six independent and equivalent hybrids which are directed to the corners of an octahedron by using s, p_x, p_y, p_z, d_{z^2}, and $d_{x^2-y^2}$. These are called sp^3d^2 hybrids.

Standard combinations are available for other regular figures such as the trigonal bipyramid (sp^3d or spd^3). However, the interest in this area has declined in recent years as molecular orbital theory has taken over from valence bond theory as the predominant approach to the theory of the chemical bond, and it is becoming increasingly rare outside carbon chemistry for the chemical bond to be described in a hybridization framework.

An exception to this statement is found in the qualitative approach to molecular geometry which we describe in the next section, this was first suggested by Sidgwick and Powell in 1940 and later extended considerably by Nyholm and Gillespie.

8.4. Electron pair repulsion: the Sidgwick–Powell approach to molecular geometry

In the equivalent orbital description of CH_4 the eight valence electrons have spatial wavefunctions which are θ_1 to θ_4 of (8.20) each spatial function being associated with electrons of α and β spins. We can loosely speak of two electrons being in each equivalent orbital, and these are referred to as bond-localized electron pairs.

The electronic energy of a molecule can be divided into four contributions: the kinetic energy of the electrons, the electron–nuclear attraction, the nuclear–nuclear repulsion, and the electron–electron repulsion. Some *ab-initio* calculations on CH_4 have been carried out which allow us to deduce the values for these quantities, and these are given in Table 8.2. It is seen that the electron–nuclear

Table 8.2. *Ab-initio* calculations on CH_4 by Snyder and Basch (*Molecular Wave Functions and Properties*, Wiley-Interscience, New York, 1972)

Kinetic energy	40.172 4
e–n attraction	−119.767 0
n–n repulsion	13.391 8
e–e repulsion	26.020 5
Total	$-40.182\ 3E_H$

attraction energy is the largest contribution, but all are important and much larger than the CH bond dissociation energy, $0.16E_H$.

In any model in which we allocate electrons to specific orbitals, either molecular orbitals or equivalent orbitals, we can further subdivide these energies into parts. In particular we can calculate the electron repulsion energy between electrons in the same orbital and that between electrons in different orbitals. It is not surprising that the latter is smaller for bond localized equivalent orbitals than it is for molecular orbitals, and as we have noted the minimization of this term is the most common procedure for calculating localized equivalent orbitals when symmetry criteria are not available.

For CH_4 the repulsion between two pairs of electrons which are in different bond-localized equivalent orbitals is approximately $1.62E_H$, so that the total pair–pair repulsion energy between the valence electrons is approximately $6 \times 1.62 = 9.72E_H$.

Suppose that we were to calculate the energy of CH_4 for another nuclear configuration with the same basis of atomic orbitals. The total energy would be less negative and each of the contributions given in Table 8.2, would have different values. Moreover, the relative amount of inter-pair and intra-pair electron repulsion would change. The question we ask is whether any one of these many contributions to the energy can be recognized as having a dominant role in determining the most stable nuclear configuration for CH_4. Sidgwick and Powell in 1940 suggested that there was one such contribution and that was the repulsion between valence shell electron pairs. This idea was extended by Nyholm and Gillespie into a set of rules of which the most important may be stated as follows:

> 'Pairs of electrons in the valence shell preferentially adopt that arrangement which maximizes their separation'

The suggestion of Sidgwick and Powell has some support from *ab-initio* calculations as can be seen from the example of NH_3. This molecule is pyramidal (bond angle 107°) and is isoelectronic with CH_4. Its eight valence electrons can also be described in terms of four localized pairs. The difference from CH_4 is that only three of these are bond-localized pairs and the fourth is a lone pair of electrons.

The relationship can best be pictured by removing a proton from CH_4 and replacing it by an extra positive charge on the central atom. If this is done without making a major change in the electronic structure then the four bond-localized pairs of CH_4 become three bond-localized pairs of CH_3 plus the lone pair which occupies roughly the fourth tetrahedral position in space.

Equivalent orbitals have been calculated by Kaldor† from the molecular orbitals of NH_3 using the minimum pair repulsion criterion. This has been done for the pyramidal configuration, which we know from spectroscopic evidence to be the most stable configuration of NH_3, and also for a planar configuration. Kaldor found that the electron pair repulsion increased on passing from the pyramidal to the planar configuration by an amount ranging from $0.24E_H$ to $0.44E_H$ depending on the molecular orbitals used to perform the calculation. Most of this increase came from an increased repulsion between the lone pair on the one hand and the three bond pairs on the other.

Although the increase found in this calculation may appear small, it is very much larger than the increase in total energy on passing from the pyramidal to the planar structure which is only $0.01E_H$ or 25 kJ mol^{-1}. Thus the change in pair repulsion energy is much larger than the change in total energy and although it is obviously an important contribution it must be largely offset by changes in other terms. Whether we are right to follow the Sidgwick–Powell hypothesis that it is the only term that needs to be considered is a more controversial issue.

There is no doubt that the Sidgwick–Powell hypothesis, as developed by Nyholm and Gillespie is a simple and powerful rationale for the shape of many molecules. In using it the first step is to determine the number of valence electron pairs associated with an atom. This is obtained from the classical valence structure by counting one pair for each single bond and adding the number of lone-pair electrons. We have already seen that CH_4 and NH_3 have four electron pairs. The isoelectronic species H_2O would be another molecule with this number. All of these molecules have structures which are based upon a tetrahedral arrangement of electron pairs at the central atom, hence to a first approximation their bond angles should be 109°.

The minimum energy spatial arrangements for two to six electron pairs are given in Table 8.3. The only one of these which is not obvious is the trigonal bipyramid. The other likely arrangement for five pairs is the square pyramid and this, in fact, is only slightly less favourable (see Figure 8.5).

Examples of molecules with 6 electron pairs at the central atom are the species SF_6, IF_5, and XeF_4. In SF_6 the six valence electrons of sulphur, together with one from each of the fluorine atoms, will form six SF single bonds. In IF_5 only five of the seven valence electrons of iodine are used to form single bonds, the other two make up a lone pair. In XeF_4, four of the eight valence electrons of Xe are used for bonding and the other four make two lone pairs.

The first rule we have given above for electron-pair repulsion does not allow us to determine the geometry of XeF_4. It can be seen from Figure 8.6 that two spatial arrangements of the fluorine atoms are consistent with an octahedral arrangment of

†U. Kaldor, *J. Chem. Phys.*, **46**, 1981 (1967).

Table 8.3. Minimum energy arrangements for electron pairs

Number of pairs	Arrangement
2	linear
3	equilateral triangle
4	tetrahedron
5	trigonal bipyramid
6	octahedron

the six electron pairs. A second rule must therefore be invoked and this is that electron pair repulsion decreases in the following order:

lone pair–lone pair > lone pair–bond pair > bond pair–bond pair

(lp–lp > lp–bp > bp–bp).

There is therefore a tendency for lone pairs to keep as far away as possible and in XeF_4 they will occupy *trans* positions. XeF_4 should therefore have a square planar structure, and this has been confirmed experimentally.

The order of electron-pair repulsion described by the second rule provides an explanation for the differences in bond angles in the iso-electronic series CH_4, NH_3, and H_2O. These angles are 109.5°, 107.3°, and 104.5° respectively. In NH_3, for example, a distortion which closes up the HNH bond angles will reduce lp–bp repulsion at the expense of an increase in bp–bp repulsion. A similar distortion in H_2O would also reduce lp–lp repulsion. We can also understand why in the six-electron pair species BrF_5, the bromine atom is below the plane of the four basal fluorine atoms, as shown in Figure 8.7.

Even with the second rule of electron pair repulsion the shapes of 5-pair molecules are not unambiguous. In ClF_3, for example, three of the seven chlorine valence electrons are used for Cl—F bonding leaving two lone pairs. If the only

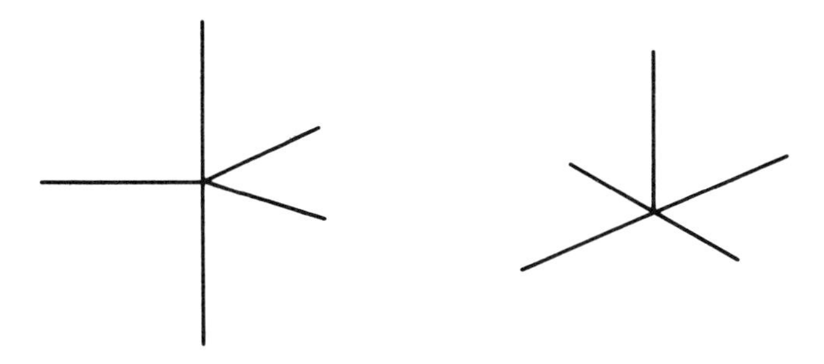

Figure 8.5 Trigonal bipyramid and square planar arrangements for five electron pairs. The trigonal bipyramid has a slightly lower pair repulsion energy.

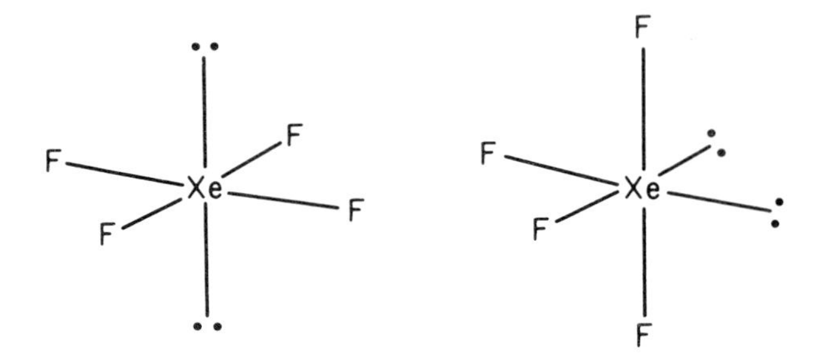

Figure 8.6 Possible structures for XeF_4. The square planar arrangement conforms to experiment.

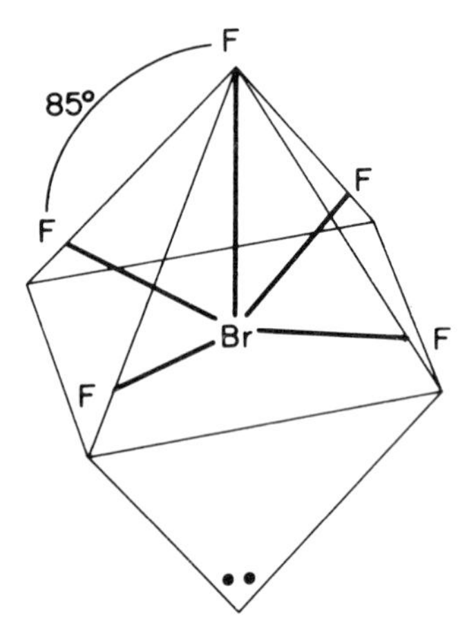

Figure 8.7 The structure of BrF_5 showing the reduction in FBrF angle produced by lone pair –bond pair repulsion.

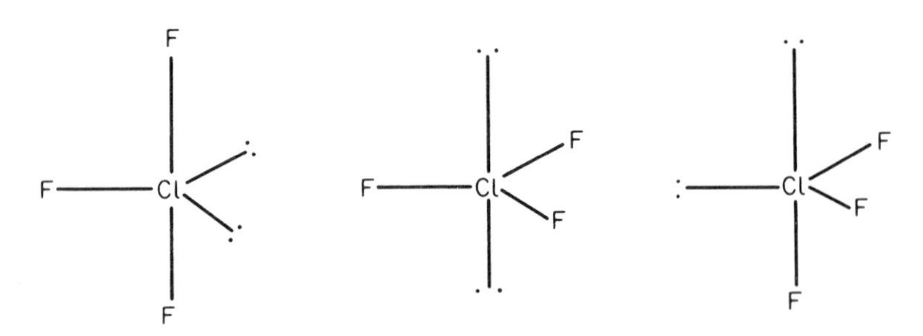

Figure 8.8 Possible structures for ClF_3. The T-shaped structure (left) conforms to experiment.

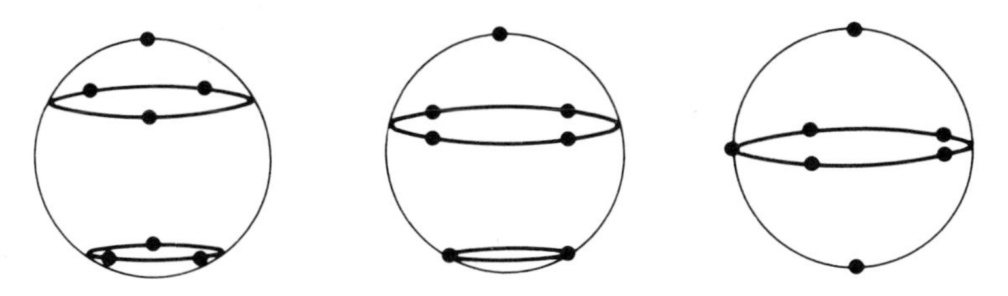

Figure 8.9 Possible arrangements for seven electron pairs. Examples of all three are known.

criterion was to minimize lp–lp repulsion then the two lone pairs should occupy the polar positions in a trigonal bipyramid and ClF_3 would be a planar equilateral triangular molecule (symmetry D_{3h}). However, in all 5-pair species the lone pairs occupy equatorial positions as shown in Figure 8.8. Thus ClF_3 is approximately a T-shaped molecule. We conclude that the six lp–bp repulsions have a bigger influence than the one lp–lp repulsion in this case.

For more than six electron pairs there are many less data on which to test the Nyholm–Gillespie rules. As for the 5 pair case there is more than one spatial arrangement of the electron pairs that have approximately the same energy. For seven electron pairs examples are known for each of the geometries shown in Figure 8.9.

Chapter 9

The Independent-electron Model

9.1. Independent-electron Hamiltonians

In Chapter 6 we saw that the essential step in carrying out calculations by the LCAO MO model is the evaluation of the Hamiltonian and overlap integrals which arise in the secular equations (6.74). In our discussion of both atomic orbitals (Section 3.4) and molecular orbitals (Section 6.1) we have emphasized that the Hamiltonian contains a term which allows for the average repulsion of an electron by all other electrons in the molecule. In our discussion of the molecular orbitals of both LiH and HF (Section 6.4) we took account of this repulsion in one case by an *ab-initio* calculation and in the other by making an empirical adjustment to the orbital energies.

MO theory without electron repulsion is computationally rather simple because the Hamiltonian then consists only of kinetic energy, electron–nuclear attraction and nuclear–nuclear repulsion terms and Hamiltonian integrals can be calculated with relative ease. The Hamiltonian (6.24) for H_2^+ is of this type, and explicit expressions for the relevant integrals in this case have been given in equations (6.25)–(6.27). It is only when we include electron repulsion terms that the Hamiltonian integrals have to be calculated by an iterative procedure, as in the SCF method described in Section 3.4, and the computational work is very difficult. It was because of these computational difficulties that, before the advent of large computers, most calculations with MO theory were based on a model in which electron repulsion was not specifically included in the Hamiltonian. Such models are referred to as independent-electron models.

Use of an independent-electron model does not necessarily mean that electron repulsion is neglected, but rather that the contributions from this term to the Hamiltonian integrals are assumed to be independent of the final electron distribution in the molecule. In other words we may calculate the Hamiltonian integrals or choose empirical values for these without regard to the final outcome of the calculations, and there is no iterative SCF procedure in an independent-electron calculation.

It is clear that an independent-electron model will only be satisfactory if we can estimate in advance the final electron distribution in the molecule. The model was never very satisfactory for transition metal complexes because the electron distribution in such molecules is not easily predictable from experimental data and varies considerably from one molecule to another. On the other hand there is good evidence for many organic hydrocarbons that the individual atoms do not have large net charges and, as we shall see, application of an independent-electron model leads to final results which are consistent with experiment.

Most independent-electron calculations are empirical rather than *ab-initio*, that is, experimental data are used to determine the Hamiltonian integrals. As a successful empirical theory requires a greater output of predicted results than there is input of adjustable parameters, a large class of molecules for which a common set of parameters may be used will provide the most appropriate application of such a theory. Organic hydrocarbons are just such a class because of the low polarity of their bonds. In contrast we could not expect to have a common value for the Hamiltonian integral

$$\int \phi_{1s} \mathscr{H} \phi_{1s} \, dv, \tag{9.1}$$

where ϕ_{1s} is the hydrogen 1*s* orbital, in the compounds LiH, H_2, and HF, because we know that the charge on the hydrogen is very different in the three molecules.

The starting point for all MO calculations is the LCAO expansion.

$$\psi = \sum_n c_n \phi_n. \tag{9.2}$$

We have seen that the substitution of this expansion into the Schrödinger equation leads to (equation 6.73)

$$\sum_n c_n (\mathscr{H} - E)\phi_n = 0. \tag{9.3}$$

and if we multiply this by each of the basis functions in turn and integrate we obtain the secular equation, one for each value of m (equation 6.74)

$$\sum_n c_n (H_{mn} - ES_{mn}) = 0, \tag{9.4}$$

where S_{mn} and H_{mn} are defined by equations (6.48) and (6.51) respectively. The energies of the orbitals are obtained by equating the secular determinant to zero, that is

$$| H_{mn} - ES_{mn} | = 0, \tag{9.5}$$

and each value of E that is a solution to this equation can be substituted back into (9.3) to obtain the coefficients c_n, apart from a normalizing factor.

$S_{mn} (m \neq n)$ is called the overlap integral and we have seen in Chapter 6 that this has an important role in the theory of valence. In view of this it is perhaps surprizing that the independent-electron theory developed by Hückel which has been very successful in interpreting the properties of organic molecules, is based upon the assumption (or approximation) that all overlap integrals are zero. In other words the secular determinant is assumed to have the form

$$| H_{mn} - E\delta_{mn} | = 0, \tag{9.6}$$

where δ_{mn} is the Kronecker delta which was introduced in equation (6.35) to define the orthonormality condition.

The zero-overlap assumption can be justified by arguing that we can always take combinations of the ϕ to form an orthogonal basis set, the members of which are in 1:1 correspondence with the original basis ϕ. Since the integrals are to be empirically determined the use of the zero-overlap assumption means that our integrals really relate to the orthogonal basis, for which zero overlap is exact.

The integrals that occur in equation (9.5) are of two types, those that occur on the diagonal of the determinant, H_{mm}, and those off the diagonal, $H_{mn}(m \neq n)$. The diagonal terms are generally referred to as the Coulomb integrals and are given the symbol α_{mn}, or just α if there is only one type of orbital. The off-diagonal integrals are called the resonance integrals and are given the symbol β_{mn}, or just β if there is only one type. These terms are historical in origin and are not particularly appropriate for modern usage, but since they are widely used we will keep to the convention in this book. The precise way in which the Coulomb and resonance integrals are treated as empirical parameters in independent electron theories depends on the type of molecule considered. The simplest of all these theories is that developed by Hückel which is especially applicable to planar unsaturated hydrocarbons. All the atomic orbitals are taken to be similar and of identical energy. In addition to neglecting the overlap integral we also neglect the resonance integral H_{mn} unless the two orbitals ϕ_m and ϕ_n belong to atoms directly bonded together $(m \rightarrow n)$ in which case we give it the constant value β. The Coulomb integral H_{mm} is written as α.

We can illustrate the method by considering a homo-nuclear diatomic molecule such as H_2. If we label the two atoms a and b, the secular equations (9.4) with the zero-overlap assumption become

$$\text{for } m = \text{a}, \; c_\text{a}(\alpha - E) + c_\text{b}\beta = 0, \tag{9.7}$$

$$\text{for } m = \text{b}, \; c_\text{b}\beta + c_\text{b}(\alpha - E) = 0, \tag{9.8}$$

$$\text{where } \alpha = H_\text{aa} = H_\text{bb}, \tag{9.9}$$

$$\beta = H_\text{ab} = H_\text{ba}, \tag{9.10}$$

$$S_\text{ab} = S_\text{ba} = 0. \tag{9.11}$$

Equations (9.7) and (9.8) have a non-trivial solution only if the determinant of the multipliers is zero

$$\begin{vmatrix} \alpha - E & \beta \\ \beta & \alpha - E \end{vmatrix} = 0, \tag{9.12}$$

which on expansion gives a quadratic equation with the solutions

$$E = \alpha \pm \beta. \tag{9.13}$$

Hence from (9.6) we have two molecular orbitals with energy $E = \alpha + \beta$ and $E = \alpha - \beta$. We can determine the value of the coefficients by substituting

$E = \alpha \pm \beta$ back into the secular equations (9.7), and we obtain

$$c_a = \pm c_b. \tag{9.14}$$

To determine their absolute value we need to apply the normalization condition.

The un-normalized wavefunctions are

$$\psi = c_a(\phi_a \pm \phi_b), \tag{9.15}$$

and we require

$$\int \psi^2 \, dv = c_a^2 \int (\phi_a^2 \pm 2\phi_a\phi_b + \phi_b^2) \, dv = 1. \tag{9.16}$$

Because of the zero-overlap assumption (9.11) this becomes in both cases

$$2c_a^2 = 1; \quad c_a = 1/\sqrt{2}. \tag{9.17}$$

It is a general result in the zero-overlap model that for normalization the sum of the squares of the coefficients must be unity:

$$\sum_n c_n^2 = 1. \tag{9.18}$$

Combining the results of expression (9.13) and (9.16) we have the following two solutions:

$$\begin{aligned} \psi_2 &= \sqrt{\tfrac{1}{2}}\,(\phi_a - \phi_b), \quad E_2 = \alpha - \beta, \\ \psi_1 &= \sqrt{\tfrac{1}{2}}\,(\phi_a + \phi_b), \quad E_1 = \alpha + \beta. \end{aligned} \tag{9.19}$$

The wavefunctions can be compared with equations (6.18) and (6.19) and we see that $\psi_1 = \psi_g$, $\psi_2 = \psi_u$ when we take $S_{ab} = 0$. Similarly the energies are identical with those in equation (6.23), if we make the Hükel simplifications (9.9)–(9.11). By analogy with the *ab-initio* result we conclude that β must be a negative quantity so that ψ_1 (ψ_g) is the bonding and ψ_2 (ψ_u) the antibonding orbital. Sketches of these orbitals are given in Figure 6.1.

The most important feature of the above calculation is that we have obtained the wavefunctions and algebraic expressions for the energies without any need to calculate the integrals α and β. The theory is therefore ideally structured to be treated as an empirical model because the parameters can be chosen by fitting experimental data *after* the calculations have been performed.

9.2 Hückel π-electron theory

In this section we consider the molecular orbitals of planar unsaturated hydrocarbons such as ethylene and benzene. These molecules are planar and the rules of symmetry group theory described in Chapter 7 can be used to divide the molecular orbitals into two classes, those that are symmetric to reflection in the plane of the molecule and those that are antisymmetric.

If we choose axes for ethylene as in Figure 9.1 then the hydrogen 1*s* orbitals and

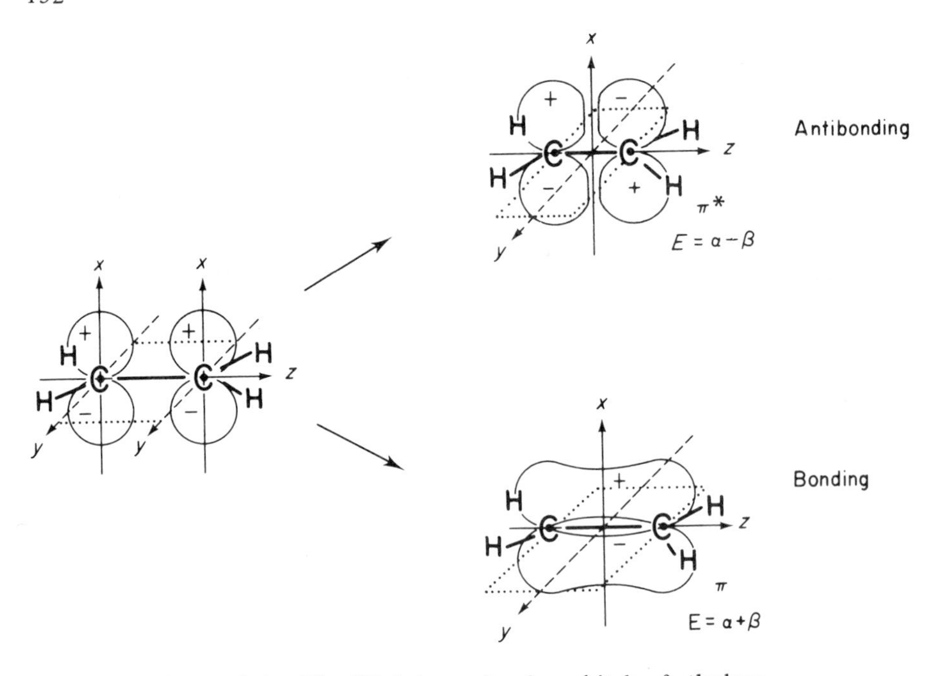

Figure 9.1 The Hückel π molecular orbitals of ethylene.

the carbon $2s$, $2p_y$, and $2p_z$ orbitals will all be unchanged by reflection in the yz plane. These can be combined to give molecular orbitals which also have this property and these are traditionally referred to as σ orbitals because their spatial characteristics are similar to the σ orbitals of diatomic molecules shown in Figure 6.4. The set of occupied σ molecular orbitals can be transformed to bond-localized equivalent orbitals according to the procedure described in Section 8.2 and in this case they can be approximated as linear combinations of hydrogen $1s$ and carbon sp^2 hybrids whose wavefunctions are given by expression (8.22).

For all planar unsaturated hydrocarbons there is a set of bonding σ orbitals of this type, which can be localized, and moreover they have similar characteristics to the molecular orbitals of saturated hydrocarbons such as ethane. They are strongly bonding and their associated ionization potentials are above 11 eV. The very tightly bound nature of σ electrons makes saturated hydrocarbons transparent to ultra-violet radiation of longer wavelength than 2 000 Å because the excitation of electrons from bonding to antibonding orbitals requires higher energy than provided by this radiation. The lack of chemical reactivity of saturated hydrocarbons towards ordinary polar reagents is another reflection of the tightly bound nature of the electrons.

The only valence atomic orbitals of ethylene which are anti-symmetric to reflection in the molecular (yz) plane are the $2p_x$ orbitals of each carbon atom. These will combine to give molecular orbitals which are also anti-symmetric to reflection and these are called π orbitals; again because their spatial characteristics are similar to those of diatomic molecule π orbitals as shown in Figure 6.4. However, the π orbitals of diatomic molecules occur in degenerate pairs $\pi(x)$ and $\pi(y)$ whereas for planar molecules such as ethylene there is no equivalence of the x and y axes. The π orbitals of ethylene are only of $\pi(x)$ type so that the use of the symbol π, which was originally an angular momentum symbol for diatomic orbitals, is not entirely appropriate for molecules like ethylene but its use is now common practice for unsaturated molecules in general.

The ionization potentials of π electrons are smaller than those of σ electrons and in large aromatic hydrocarbons may be as low as 6 eV. It is the low excitation energy of electrons from π-bonding to π^*-anti-bonding molecular orbitals that is responsible for the absorption in the visible and near ultra-violet of such compounds. Moreover it is the comparatively weak bonding of the π electrons of unsaturated hydrocarbons that is responsible for their greater chemical reactivity compared with saturated hydrocarbons. Many of the interesting physical and chemical properties of unsaturated hydrocarbons are due to the presence of the π electrons, and Hückel π-electron theory is based on the assumption that in explaining the differences between one such molecule and another (e.g. ethylene and benzene) we can for the most part ignore the σ orbitals. We are therefore concerned only with molecular orbitals derived from the $2p_x$ atomic orbitals of each unsaturated carbon atom.

The Hückel treatment of the π orbitals of ethylene follows exactly the same pattern as that we have given in Section 9.1 for H_2. The wavefunctions and energies are those of equation (9.19) except that ϕ represents a carbon $2p$ orbital rather than a hydrogen $1s$ orbital. The relationship between the atomic and molecular orbitals is illustrated in Figure 9.1. Such figures are to show the positions of the nodes and the relative signs of the wave function; they are not intended to depict 'shape' accurately.

The value of the Hückel approach to conjugated molecules is illustrated by the example of butadiene (Figure 9.2). The general Hückel secular equation is

$$c_m(\alpha - E) + \sum_{n \to m} c_n \beta = 0, \tag{9.20}$$

where $\sum_{n \to m}$ indicates a sum over all centres n bonded to m. Dividing these equations by β and introducing the parameter

$$x = \frac{\alpha - E}{\beta} \tag{9.21}$$

allows us to write the general equation as

$$xc_m + \sum_{n \to m} c_n = 0. \tag{9.22}$$

If the four $2p_x$ atomic orbitals in butadiene are labelled ϕ_a, ϕ_b, ϕ_c, and ϕ_d as in Figure 9.2 the four secular equations for butadiene are

$$m = \mathrm{a} \qquad xc_a + c_b \qquad = 0, \tag{9.23}$$

$$m = \mathrm{b} \qquad c_a + xc_b + c_c = 0 \tag{9.24}$$

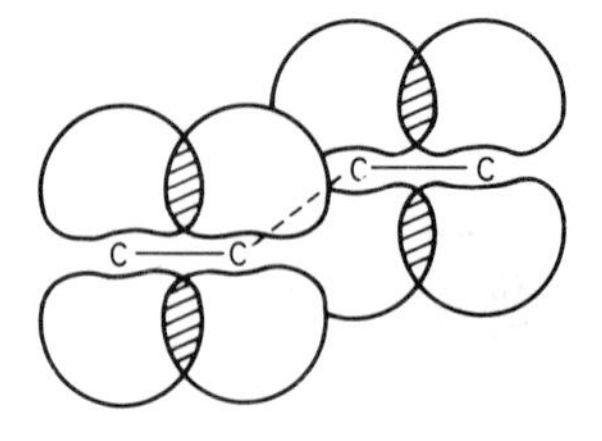

Figure 9.2 The overlap of carbon $2p_x$ orbitals in butadiene.

$$m = \mathrm{c} \qquad c_\mathrm{b} + xc_\mathrm{c} + c_\mathrm{d} = 0, \tag{9.25}$$

$$m = \mathrm{d} \qquad c_\mathrm{c} + xc_\mathrm{d} = 0, \tag{9.26}$$

and the secular determinant is

$$\begin{vmatrix} x & 1 & 0 & 0 \\ 1 & x & 1 & 0 \\ 0 & 1 & x & 1 \\ 0 & 0 & 1 & x \end{vmatrix} = 0. \tag{9.27}$$

Expansion of this determinant leads to the polynomal

$$x^4 - 3x^2 + 1 = 0, \tag{9.28}$$

whose roots are

$$x = \pm \left(\frac{3 \pm \sqrt{5}}{2} \right)^{1/2} = \pm 0.62, \pm 1.62. \tag{9.29}$$

Taking the solution $x = -1.62$ and substituting this back in the secular equations we have

from (9.23): $\quad c_\mathrm{b} = 1.62c_\mathrm{a}$,

from (9.24): $\quad c_\mathrm{c} = -c_\mathrm{a} + 1.62c_\mathrm{b} = 1.62c_\mathrm{a}$,

from (9.26): $\quad c_\mathrm{c} = 1.62c_\mathrm{d}$; $\quad$ i.e. $\quad c_\mathrm{a} = c_\mathrm{d}$.

The normalization condition in the zero overlap model is, from (9.18)

$$c_\mathrm{a}^2 + c_\mathrm{b}^2 + c_\mathrm{c}^2 + c_\mathrm{d}^2 = 1, \tag{9.30}$$

whence

$$c_\mathrm{a} = \pm 0.37. \tag{9.31}$$

Taking the positive solution (the overall sign of the wavefunction is immaterial) gives the wavefunction

$$\psi_1 = 0.37\phi_\mathrm{a} + 0.60\phi_\mathrm{b} + 0.60\phi_\mathrm{c} + 0.37\phi_\mathrm{d} \tag{9.32}$$

whose energy is (from 9.21),

$$E_1 = \alpha + 1.62\beta. \tag{9.33}$$

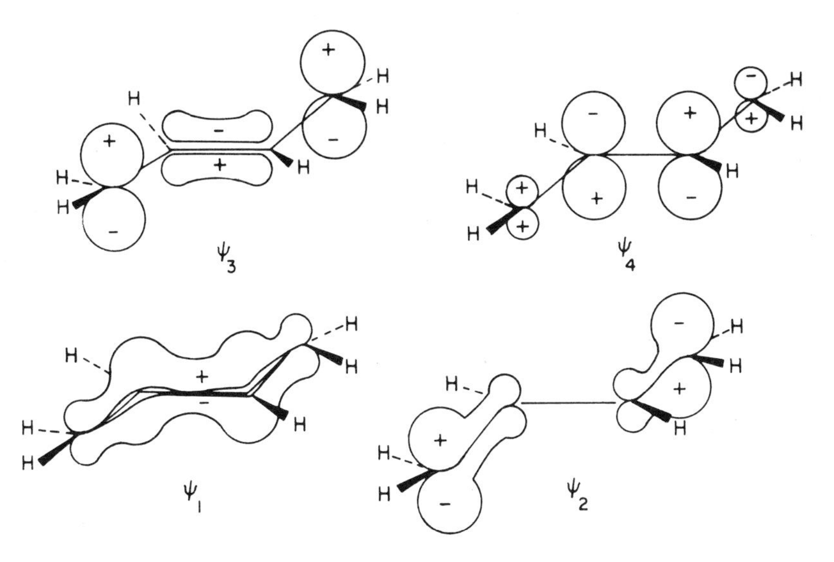

Figure 9.3 Hückel π orbitals of butadiene.

In a similar way we take the remaining three solutions and obtain three further molecular orbitals

$$\psi_2 = 0.60\phi_a + 0.37\phi_b - 0.37\phi_c - 0.60\phi_d; \quad E_2 = \alpha + 0.62\beta \tag{9.34a}$$

$$\psi_3 = 0.60\phi_a - 0.37\phi_b - 0.37\phi_c + 0.60\phi_d; \quad E_3 = \alpha - 0.62\beta \tag{9.34b}$$

$$\psi_4 = 0.37\phi_a - 0.60\phi_b + 0.60\phi_c - 0.37\phi_d; \quad E_4 = \alpha - 1.62\beta. \tag{9.34c}$$

The four orbitals are shown diagramatically in Figure 9.3.

We have seen in Section 9.1 that β is a negative quantity, so that the order of increasing energy is ψ_1 to ψ_4. ψ_1 and ψ_2 have energies less than α and will be bonding orbitals, ψ_3 and ψ_4 have energies greater than α and will be antibonding. Note that the energy of the orbitals increases with the number of sign changes of the orbital along the carbon chain.

In the neutral butadiene molecule each carbon atom contributes one electron (from the $2p_x$ atomic orbital) to the π molecular orbitals. In the ground state these electrons will go two each, with antiparallel spins, into the two bonding orbitals. The sum of the orbital energies of these four electrons is (from equations 9.33 and 9.34.)

$$2(\alpha + 1.62\beta) + 2(\alpha + 0.62\beta) = 4\alpha + 4.48\beta. \tag{9.35}$$

and this is the total π-electron energy if we ignore the effect of one electron on another (as in the independent – electron model). For comparison, the energy of four π electrons in two isolated ethylene orbitals is $4\alpha + 4\beta$, so the difference, 0.48β, which is a negative quantity, represents additional stabilization due to the further delocalization of the π electrons over four carbon centres instead of over two separate pairs.

Aromatic compounds, of which benzene is the archetype, are the most important class of molecules in which delocalization of π electrons controls chemical reactivity. Benzene can be treated in the same way as butadiene. There will be a set of σ molecular orbitals formed from the six sets of carbon sp^2 hybrids and the six hydrogen 1s orbitals, these orbitals are symmetric to reflection in the plane of the molecule. The π orbitals are then derived from six $2p_x$ atomic orbitals of the carbon atoms, and are anti-symmetric to reflection in the molecular plane as illustrated in Figure 9.4.

Figure 9.4 The carbon $2p_x$ orbitals in benzene which form the Hückel π orbitals.

The high symmetry of benzene makes it possible to derive the form of the π molecular orbitals without solving directly the secular equations. However, to emphasize the direct approach we show here the secular determinant which has the following form:

$$\begin{vmatrix} x & 1 & 0 & 0 & 0 & 1 \\ 1 & x & 1 & 0 & 0 & 0 \\ 0 & 1 & x & 1 & 0 & 0 \\ 0 & 0 & 1 & x & 1 & 0 \\ 0 & 0 & 0 & 1 & x & 1 \\ 1 & 0 & 0 & 0 & 1 & x \end{vmatrix} = 0.$$

This determinant on expansion gives the polynomial

$$x^6 + 6x^4 + 9x^2 - 4 = 0, \tag{9.36}$$

which can be factorized to yield six solutions four of which constitute two identical pairs:

$$x = -2, x = -1, x = -1, x = +1, x = +1, x = 2. \tag{9.37}$$

The two pairs of solutions ($x = -1$) and ($x = +1$) show that there are two pairs of degenerate orbitals. This leads to difficulties in determining the coefficients as there is no unique set of coefficients for degenerate orbitals as we have shown in expressions (7.20) to (7.24). A convenient set of coefficients are given in the following equations (9.38) – (9.43) (as before $\psi_1 \rightarrow \psi_6$ represents increasing energy):

$$\psi_1 = \sqrt{\frac{1}{6}}(\phi_a + \phi_b + \phi_c + \phi_d + \phi_e + \phi_f), E_1 = \alpha + 2\beta, \quad E_1 = \alpha + 2\beta, \tag{9.38}$$

$$\psi_2 = ½(\phi_a + \phi_b - \phi_d - \phi_e), \quad E_2 = \alpha + \beta, \quad E_2 = \alpha + \beta, \tag{9.39}$$

$$\psi_3 = \sqrt{\frac{1}{12}}(\phi_a - \phi_b - 2\phi_c - \phi_d + \phi_e + 2\phi_f), \quad E_3 = \alpha + \beta, \tag{9.40}$$

$$\psi_4 = ½(\phi_a - \phi_b + \phi_d - \phi_c), \quad E_4 = \alpha - \beta, \tag{9.41}$$

$$\psi_5 = \sqrt{\frac{1}{12}}(\phi_a + \phi_b - 2\phi_c + \phi_d + \phi_e - 2\phi_f), \quad E_5 = \alpha - \beta, \tag{9.42}$$

$$\psi_6 = \sqrt{\frac{1}{6}}(\phi_a - \phi_b + \phi_c - \phi_d + \phi_e - \phi_f), \quad E_6 = \alpha - 2\beta. \tag{9.43}$$

The total π-electron energy for the ground state of benzene (in the independent-electron model) is given by the sum of the energies of six electrons, 2 each in ψ_1, ψ_2, and ψ_3.

$$2(E_1 + E_2 + E_3) = 6\alpha + 8\beta. \tag{9.44}$$

This can be compared with the energy for three localized π bonds $(6\alpha + 6\beta)$ and shows the very large delocalization energy (2β) in benzene. This stabilization accounts for the relative chemical inertness of benzene. For example it does not react with bromine solution or potassium permanganate at room temperature, whereas ethylene reacts very rapidly with both.

9.3 Charge densities and bond orders

Mechanistic organic chemistry employs differences in charge densities a great deal to provide a theoretical interpretation of many types of reaction. An electron in an orbital $\psi_r = \sum_m c_{rm}\phi_m$ has a density distribution $\sum_m c_{rm}^2 \phi_m^2$ (neglecting all overlap density terms such as $\phi_m\phi_n$). We can thus define the *π-electron charge density* q_m at atom m as follows:

$$q_m = \sum_r n_r c_{rm}^2, \tag{9.45}$$

where c_{rm} is the coefficient of the basis orbital ϕ_m in the molecular orbital ψ_r; the sum is over all the π orbitals with n_r electrons in each orbital ($n_r = 0, 1, 2$). For the ground state of butadiene $n_r = 2$ for the two occupied bonding orbitals and $n_r = 0$ for the antibonding orbitals, thus the charge densities at atoms a and b are

$$q_a = 2c_{1a}^2 + 2c_{2a}^2 = 2(0.37)^2 + 2(0.60)^2 = 1.00, \tag{9.46}$$

$$q_b = 2c_{1b}^2 + 2c_{2b}^2 = 2(0.60)^2 + 2(0.37)^2 = 1.00. \tag{9.47}$$

By symmetry we can see that $q_a = q_d$ and $q_b = q_c$. Thus the π-electron density is unity at each atom in butadiene. Examination of the orbitals (9.38) – (9.40) will show that the same is true for all the carbon atoms in benzene, for example

$$q_a = 2\left(\sqrt{\frac{1}{6}}\right)^2 + 2\left(\sqrt{\frac{1}{12}}\right)^2 + 2(½)^2 = 1. \tag{9.48}$$

To interpret many phenomena in organic chemistry it is desirable to have an estimate of the amount of double-bond character in the bond joining two atoms. By analogy with the charge density at an atom m, we can define a π-electron bond *order* between atoms m and n as follows:

$$p_{mn} = \sum_r n_r c_{rm} c_{rn}. \tag{9.49}$$

For ethylene the two electrons in the bonding π orbital, whose wavefunction is (9.20), give a bond order of

$$2\sqrt{½}\sqrt{½} = 1. \tag{9.50}$$

For butadiene we have

$$p_{ab} = 2c_{1a}c_{1b} + 2c_{2a}c_{2b} = 0.89, \tag{9.51}$$

and

$$p_{bc} = 2c_{2b}c_{1c} + 2c_{2b}c_{2c} = 0.45. \tag{9.52}$$

By symmetry $p_{ab} = p_{bc}$ so that according to Hückel theory the two outer bonds in butadiene have much more double-bond character than the central bond but less than that in an isolated ethylene molecule. The conventional representation of butadiene as a single valence structure ($CH_2{=}CH{-}CH{=}CH_2$) is in close accord with the Hückel picture but fails to show that there is some double-bond character in the central bond.

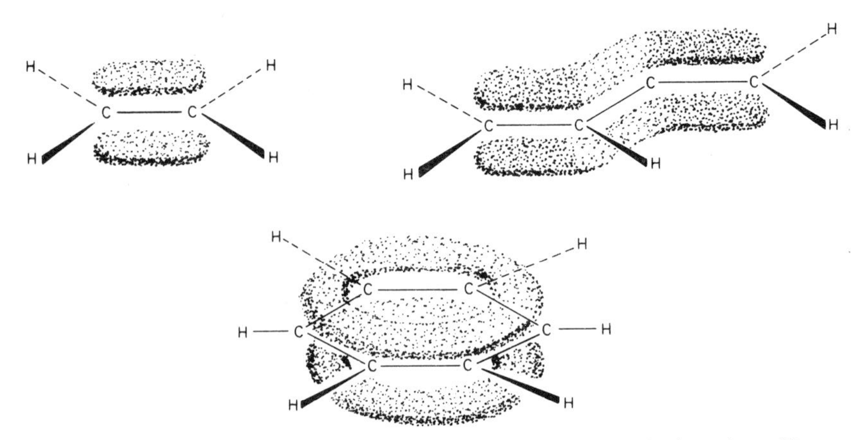

Figure 9.5 A common representation of π bonding in ethylene, butadiene, and benzene.

If we carry out a similar calculation for benzene we obtain a π-electron bond order of 0.67, which is greater than half that for an isolated ethylenic bond. Simply to regard benzene as an equivalent mixture of the two Kekulé structures would ascribe half a π bond to each carbon–carbon link, which is less than the amount predicted by Hückel theory.

Figure 9.5 shows a representation of the π-electron density in ethylene, butadiene and benzene, the shading being darkest where the electron density is greatest.

9.4. Alternant hydrocarbons

There is an important property of the Hückel π orbitals we have determined for ethylene, butadiene, and benzene which we have not discussed. This is that the orbitals occur in pairs with their energies equal to $(\alpha \pm x\beta)$ and such a pair have the same numerical values for the LCAO coefficients but with the sign changing for every alternate atom on going from the bonding to the antibonding member of the pair. Hydrocarbons whose π orbitals have this property are called *alternant* hydrocarbons. They are conjugated molecules whose carbon atoms can be divided into two sets, conventionally labelled starred and unstarred, such that no two atoms of the same set are bonded together. Typical examples are shown in Figure 9.6.

Figure 9.6 Typical alternant hydrocarbons.

The only conjugated hydrocarbons which are not alternant are those with rings containing an odd number of atoms, such as the molecules shown in Figure 9.7.

We have earlier shown (equation 9.22) that the Hückel secular equations take the form

$$xc_m + \sum_{n \to m} c_n = 0, \tag{9.53}$$

where the summation is over atoms n which are bonded to m. If the compound is alternant, then if m is starred n must be unstarred; furthermore if x_r, c_{rm}, and c_{rn} define one solution (i.e. satisfy equation 9.53) then if follows that $-x_r$, c_{rm}, and $-c_{rn}$ must define another. Thus if ψ_r is a bonding orbital of energy $\alpha + x\beta$, there

CH_2

Figure 9.7 Fulvene and azulene, typical non-alternant hydrocarbons.

must be an antibonding orbital of energy $\alpha - x\beta$. If the wavefunction of the bonding orbital ψ_r is

$$\psi_r = \sum_m^{*} c_{rm}\phi_m + \sum_n^{\circ} c_{rn}\phi_n, \tag{9.54}$$

where the summations are over the starred and unstarred atoms respectively, then we can write the wavefunction of the paired antibonding orbital, $\psi_{r'}$ say, as follows:

$$\psi_{r'} = \sum_m^{*} c_{rm}\phi_m - \sum_n^{\circ} c_{rn}\phi_n. \tag{9.55}$$

To convert the bonding orbital into the antibonding orbital we simply change the sign of the coefficients of the unstarred atoms.

We have already shown that the charge density is unity for every atom in both butadiene and benzene. This is true for all neutral alternant hydrocarbons. Because the set of Hückel orbitals is normalized and orthogonal we can invert the set of equations

$$\psi_r = \sum_m c_{rm}\phi_m, \tag{9.56}$$

and express the atomic orbitals in terms of the molecular orbitals as follows:

$$\phi_m = \sum_r c_{rm}\psi_r. \tag{9.57}$$

The normalization of equations (9.57) requires

$$\sum_r c_{rm}^2 = 1. \tag{9.58}$$

Now because of the pairing of the orbitals in an alternant system we have

$$\sum_{\text{bonding}} c_{rm}^2 = \sum_{\text{antibonding}} c_{rm}^2, \tag{9.59}$$

hence by combining (9.58) and (9.59) and using the definition of charge density (9.45) we find for a neutral alternant hydrocarbon

$$2 \sum_{\text{bonding}} c_{rm}^2 \equiv q_m = 1. \tag{9.60}$$

This conclusion is verified by experiment; alternant hydrocarbons have no measurable dipole, in sharp contrast to non-alternant systems like fulvene and azulene.

Neutral alternant hydrocarbons must contain an even number of carbon atoms, but fully conjugated radicals or ions can contain an odd number of atoms. The pairing properties of alternant hydrocarbons are possessed by odd as well as by even alternant systems. It follows that every odd alternant hydrocarbon must have one orbital for which $x = 0$ and hence $E = \alpha$. Since in terms of Hückel theory α is the energy of an isolated atomic orbital, the orbital whose energy is α, is said to be a non-bonding orbital, and is usually given the abbreviation NBO.

Many important conclusions can be drawn from the coefficients of non-bonding orbitals and they have the useful property that they can be determined without solving the whole of the secular equations for the molecule. We note that the general Hückel equation given by (9.22) becomes in the case $x = 0$

$$\sum_{n \to m} c_n = 0, \tag{9.61}$$

that is, the sum of the coefficients around any one atom is zero. Since for an alternant hydrocarbon only starred atoms are bonded to unstarred atoms, it follows that the smaller set (conventionally taken to be the unstarred atoms) have zero coefficients. If we take a polymethine chain as an example and put the coefficients of the NBO for the first carbon atom equal to a, it follows that the coefficient for atom 3 must be $-a$ and that for atom 5 must be $+a$, and so on:

$$\begin{matrix} a & 0 & -a & & a(-1)^{k-2} & 0 & a(-1)^{k-1} \\ \mathrm{C} & \!\!-\mathrm{C}\!\!- & \mathrm{C} & \cdots\cdots\cdots & -\mathrm{C}\text{———} & \mathrm{C}\text{———} & \mathrm{C} \end{matrix} \; .$$

If there are $2k-1$ atoms in the chain, the coefficient of the terminal carbon atom must be $a(-1)^{k-1}$. From the normalization condition we have

$$ka^2 = 1, \tag{9.62}$$

hence the numerical value of a is given by

$$a = k^{-\frac{1}{2}}. \tag{9.63}$$

We shall see in Chapter 14 that this simple way of determining the coefficients of non-bonding orbitals of odd alternant molecules can be extremely useful in predicting the charge distribution in odd alternant cations and anions.

9.5. The correlation of molecular properties by Hückel π-electron theory

Hückel π-electron theory, in spite of its approximations and assumptions, is able to account, at least qualitatively, for many physical and chemical properties which can be attributed to the π electrons of unsaturated hydrocarbons. As the first example we take the electronic spectra of some linear polyenes. The lowest energy absorption is due to the promotion of an electron from the highest occupied molecular orbital (HOMO) to the lowest unoccupied molecular orbital (LUMO). In a polyene both of these will be π orbitals, and we should expect a correlation between the frequency of the absorption band and the difference in energy between the relevant π orbitals.

The HOMO and LUMO energies of the first few linear polyenes are given in Table 9.1. According to the Planck–Einstein relationship (equation 3.1), ΔE is related to the wavelength of the corresponding absorption band by

$$\Delta E = h\gamma = hc/\lambda, \tag{9.64}$$

Table 9.1. Energies of the highest occupied and lowest unoccupied Hückel orbitals of some linear polyenes

	E_{HOMO}	E_{LUMO}	ΔE
Ethylene	$\alpha + \beta$	$\alpha - \beta$	2β
Butadiene	$\alpha + 0.62\beta$	$\alpha - 0.62\beta$	1.24β
Hexa-1,3,5-triene	$\alpha + 0.45\beta$	$\alpha - 0.45\beta$	0.90β
Octa-1,3,5,7-tetraene	$\alpha + 0.35\beta$	$\alpha - 0.35\beta$	0.70β

hence a plot of the reciprocal of the wavelength of the first absorption band for these linear polyenes against the calculated energy difference, ΔE, should be a straight line and from the slope of this a value can be obtained for β. The correlation between theory and experiment is shown in Figure 9.8.

The value of β calculated from the slope of the line in Figure 9.8 is -248 kJ mol^{-1}. The correlation between theory and experiment is very good but we must be cautious about attaching much significance to the numerical value of β particularly as the line does not pass through the origin as it should in this simple form of Hückel theory. We can obtain a similar plot for the u.v. spectra of aromatic hydrocarbons and, perhaps more surprisingly, another for the spectra of unsaturated aldehydes. In this latter case the oxygen is more electronegative than carbon and we would not expect the Hückel approximations to hold. In both cases the straight lines obtained are good, but the slopes are different giving values of β of -260 kJ mol^{-1} and -296 kJ mol^{-1} respectively; and neither of these lines passes through the origin. Probably the biggest source of error is the assumption that all bonds are of equal length, whereas in the linear polyenes they are known to be alternately long and short.

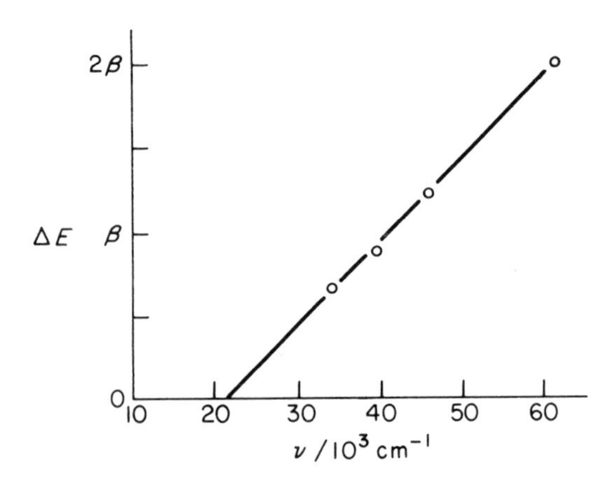

Figure 9.8 Plot of $\nu(=1/\lambda)$ for the first u.v. absorption band against the calculated energy difference between highest occupied and lowest unoccupied molecular orbitals.

The technique of photo electron spectroscopy has been described in Chapter 5. By Koopmans' theorem the ionization potential is the negative of the orbital energy. Thus we would expect a correlation between peaks in the photoelectron spectra of aromatic molecules and the energies of the Hückel orbitals. In practice a good correlation is found (see Figure 9.9), although at the higher energy range the spectra become very difficult to interpret due to overlap of the π and σ levels. The slope gives a value for β of -284 kJ mol^{-1}, reasonably close to the values from the u.v. spectra.

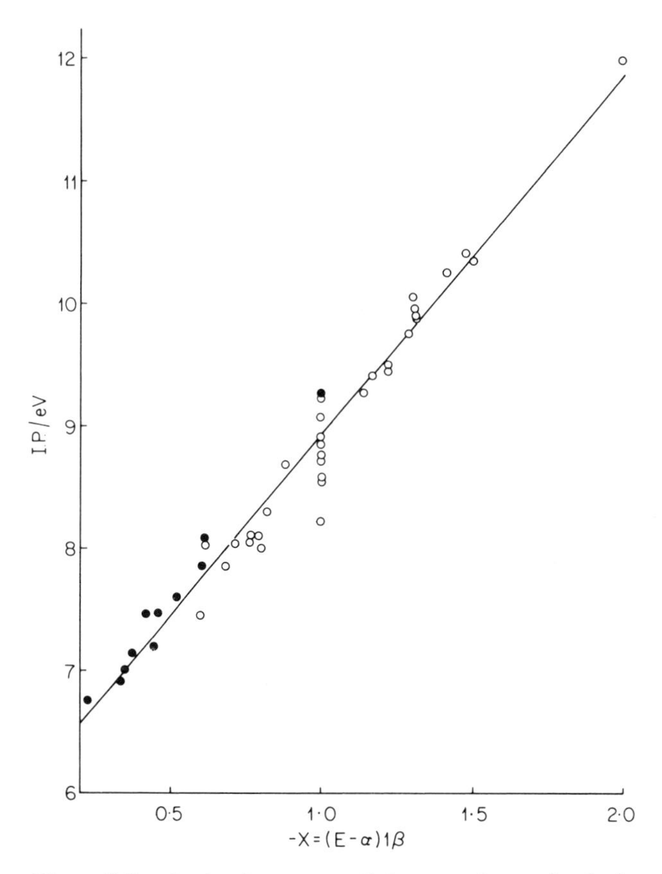

Figure 9.9 Ionization potentials, as determined by photoelectron spectroscopy, plotted against Hückel orbital energies for various unsaturated hydrocarbons. Data are for the following molecules: benzene, naphthalene, pyrene, coronene anthracene, phenanthrene, pentacene, perylene, chrysene, 1,2-benzanthracene, 1,2-benzpyrene, benzo[g,h,i]perylene and ovalene. Closed circles are first I.P.s and open circles are higher I.P.s for which the assignment is less certain. R. Boschi, J. N. Murrell, and W. Schmidt, *Faraday Disc.*, **54**, 116 (1972).

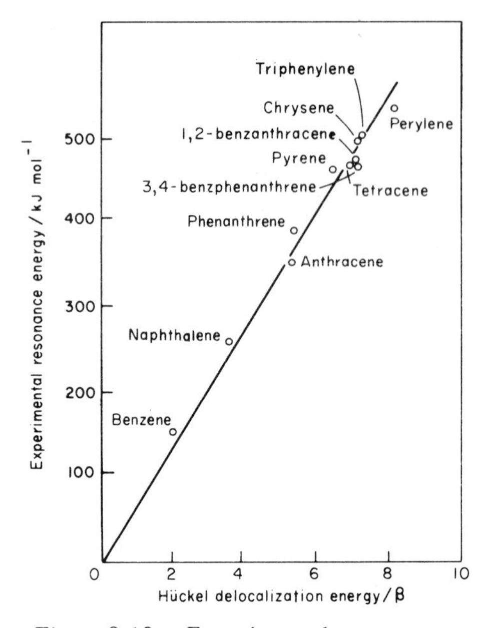

Figure 9.10 Experimental resonance energy, as measured by heat of combustion experiments, plotted against calculated Hückel delocalization energy for various unsaturated hydrocarbons.

Another property we can calculate from Hückel π-electron theory is the delocalization energy, which is the difference between the total π-electron energy and a value which in the sum of 2β for each formal double bond (in a Kekulé structure). We can compare this calculated delocalization energy with the difference between the observed heat of combustion and that predicted on the basis of additivity of bond energies for all the bonds in a single Kekulé structure using the best accepted values [D(C—H) = 226, D(C—C) = 206 and D(C=C) = 507 kJ mol^{-1}]. If we plot the 'observed' delocalization energies against the calculated values (in terms of β) for a series of aromatic compounds we again get an extremely satisfactory correlation (Figure 9.10). From the slope of the curve we again obtain a numerical value for β but this, -71 kJ mol^{-1}, is very different to that obtained from spectroscopic data.

The reason that different values of β are obtained from these different analyses are that energies should properly be evaluated using the complete Hamiltonian for the molecule and not the one-electron Hamiltonian of Hückel theory. The neglect of electron interaction can sometimes be compensated by taking an empirical value of β, but we must not expect Hückel theory to provide absolute as distinct from relative data.

In Section 9.3 we have defined a quantity, the π-electron bond order (equation 9.49), and have shown that for ethylene the π-bond order is unity and that for

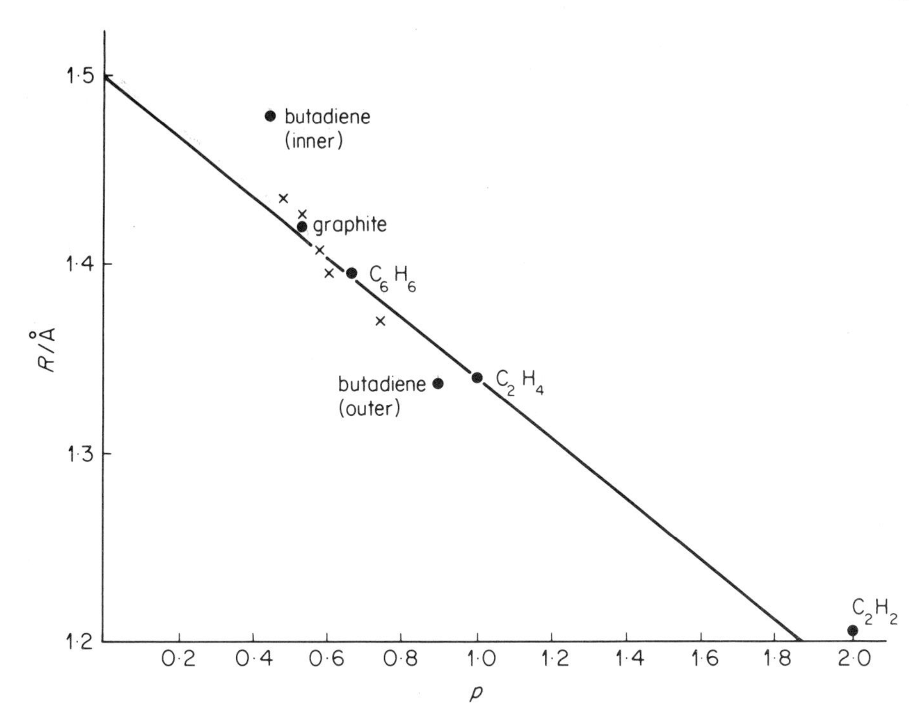

Figure 9.11 Relationship between π bond order and C–C bond length for some alternant hydrocarbons. Data for anthracene are indicated by x, other points as marked.

butadiene the outer bonds have a bond order of 0.89, while the centre bond has a bond order of 0.45. Experimentally it is known that single bonds are longer than double bonds and that triple bonds are shorter still. The π-electron bond order for benzene is 0.67 and the CC bond length for benzene is intermediate between that of ethane and ethylene. It is therefore of interest to examine the correlation between the π-bond order and the bond lengths of alternant hydrocarbons; this is done in Figure 9.11. The correlation is not as good as for the data shown in Figures 9.8 to 9.10. This may in part be because there is experimental uncertainty about the bond lengths in some cases. Points for the key compounds ethylene, benzene, and graphite, whose bond orders are determined by symmetry, lie quite close to the line.

$$r/\text{Å} = 1.50 - 0.16p. \tag{9.65}$$

and this relationship has been widely used for predicting band lengths.

We have seen that we can calculate the π-electron density of each atom by Hückel theory and, in complete agreement with the theory, alternant hydrocarbons are non-polar and possess no dipole moment. Non-alternant hydrocarbons are predicted to be polar and this is in accord with the experiment. Since we can

Table 9.2. Comparison of calculated and observed dipole moments

	Dipole moment/Debye Units	
Compound	observed	calculated by Hückel theory
Fulvene	1.2	4.7
Azulene	1.0	6.9

calculate the charge at each atom we can calculate the total dipole moment for the whole molecule. Table 9.2 shows the calculated and observed dipole moments of two typical-alternant hydrocarbons.

Though Hückel theory does predict that non-alternant hydrocarbons should be polar, it considerably over-estimates the magnitude of the dipole moments of fulvene and azulene. However, one of the assumptions we made in developing Hückel theory was that individual atoms did not carry appreciable net charges so that α could be taken as a constant, independent of the position of the atom in the molecule. This will clearly be a poor approximation if different atoms carry different charges so it is not surprizing that the theory fails when it is applied to polar molecules.

There are two extreme views about the value of Hückel π-electron theory. The first is to dismiss it as a theory based on false assumptions, which is unreliable in its predictions. The other extreme view is to treat the results obtained from Hückel calculations as always having predictive value, and this view often leads to extensions of the approach which are quite unjustified. We saw at the beginning of this chapter that to be successful Hückel theory requires the electron distribution to be fairly regular throughout the molecule and the individual atoms to have no appreciable net charge. Furthermore we emphasized that it was an empirical theory in which experimental data were used to determine the Hamiltonian integrals. We have now shown that for two large classes of compounds, the conjugated polyenes and the aromatic hydrocarbons, the theory can give extremely good correlations with experimental data. We have seen that the value of the Hückel parameter β varies depending on the nature of the property we are studying and this is exactly what we should expect from our derivation. Similarly the theory is less reliable when applied to polar non-alternant molecules and this again is in accord with expectation. Although the Hückel π orbitals have only a qualitative relationship to the true molecular orbitals they provide a valuable model, and we shall see in Chapter 14 that this model can be applied with great success to predicting chemical reactivity.

9.6. Introduction of other atoms into Hückel theory

In introducing Hückel theory we assumed that all the atomic orbitals in an LCAO expansion were similar and had the same energy (Section 9.1). Hückel theory then only required two empirical parameters, the Coulomb integral α and the resonance integral β. We have seen that the Hückel model is extremely successful in correlating experimental data for alternant hydrocarbons, but so far we have only discussed molecules consisting of carbon and hydrogen atoms (we have in fact ignored the latter). If we now wish to try and extend Hückel theory to other atoms in the same conjugated systems, for example $C_6H_5N{=}NC_6H_5$ and $CH_2{=}CH{-}CH{=}O$, then we require values for both the Coulomb and resonance integrals of these other atoms. Changes in α and β are usually expressed in terms of values appropriate to the atoms and bonds of benzene, which we can denote α_C and β_{CC}. Thus for an atom X we have

$$\alpha_X = \alpha_C + h_X\beta_{CC}, \tag{9.66}$$

and

$$\beta_{CX} = k_{CX}\beta_{CC}. \tag{9.67}$$

There is no theoretically sound way of determining values of h_X and k_{CX}; they are better treated purely as empirical parameters which can be adjusted to fit experiment. In qualitative terms the more electronegative is atom X the larger will be h_X and k.

We can regard formaldehyde as ethylene in which one CH_2 group has been replaced by oxygen. To illustrate what happens to the molecular orbitals on such a replacement we will take $h_O = 2$ and $k_{CO} = \sqrt{2}$. Note that we would require different values for h_O and k_{CO} if the oxygen were bound to the conjugated system by a single bond as in methylvinylether or phenol. The Hückel secular equations for formaldehyde are

$$c_C(\alpha_C - E) + c_O k_{CO}\beta_{CC} = 0, \tag{9.68}$$

$$c_C k_{CO}\beta_{CO} + c_O(\alpha_C + h_O\beta_{CC} - E) = 0. \tag{9.69}$$

Putting $x = (\alpha_C - E)/\beta_{CC}$ (equation 9.21), and using the values of h_O and k_{CO} given above we have:

$$\begin{vmatrix} x & \sqrt{2} \\ \sqrt{2} & x+2 \end{vmatrix} = 0. \tag{9.70}$$

The two Hückel orbitals of formaldehyde obtained from these equations are

$$\psi_1 = 0.44\phi_C + 0.85\phi_O, \quad E_1 = \alpha_C + 2.7\beta_{CC}, \tag{9.72a}$$

$$\psi_2 = 0.85\phi_C - 0.44\phi_O, \quad E_2 = \alpha_C - 0.7\beta_{CC}. \tag{9.72b}$$

In the ground state the two π electrons go into ψ_1 which we note is close in energy to the oxygen atomic orbital. We can also see that for this orbital the electron density is much higher around the oxygen atom than it is around the carbon. All these are results which are consistent with our experience of the chemical and physical properties of the carbonyl bond.

We shall return to this picture in Chapter 14, but for the present we must note that by introducing atoms in addition to carbon we are necessarily introducing polarity and we have seen that simple Hückel theory fails to fit experiment satisfactorily when applied to polar molecules. In other words the above treatment of heteroatomic molecules is unlikely to give more than a qualitative picture and even when the parameters h_X and k_{CX} are chosen to fit experiment the usefulness of the treatment will be severely limited.

9.7. Extended Hückel models

In applying Hückel theory to polyenes we consider only the π orbitals and ignore completely the σ orbitals and surrounding hydrogen atoms. Obviously any attempt to introduce more than one type of atomic orbital into the basis of π orbitals will greatly magnify the complications of an independent-electron treatment. In 1963 the development of all-electron theories of this type for hydrocarbons was reported by Hoffmann and by Pople and Santry.

The essential features of the Hoffmann treatment (that of Pople and Santry is not too different) are that the atomic orbital basis consists of all the valence atomic orbitals of the component atoms. For hydrocarbons this consists of hydrogen $1s$ orbitals and carbon $2s$ and $2p$ orbitals. The Coulomb integrals α are given fixed values which have been assigned using spectroscopic data. The accepted values (in atomic units) are:

$$\alpha(1s_H) = -0.5, \quad \alpha(2s_C) = -0.878, \quad \alpha(2p_C) = -0.419. \tag{9.73}$$

The resonance integrals are calculated from

$$\beta_{ab} = \frac{k}{2}(\alpha_a + \alpha_b)S_{ab}, \tag{9.74}$$

where k is an adjustable parameter taken, in the Hoffmann approach, to have the value 1.75. An expression such as this was first suggested by Mulliken and was used in an independent-electron theory of transition metal complexes in the 1950's by Wolfsberg and Helmholz.

The overlap integrals are calculated and included in the secular equations (9.4) so that all S_{ab} and β_{ab}, both between orbitals on neighbouring atoms and between orbitals on non-neighbouring atoms, are included in the calculation. The calculation can be carried out for a fixed geometry or the geometry can be varied to determine the molecular conformation with the minimum energy. This is a very important feature of the method and it has, for example, been found to give a value for the rotational barrier in ethane in good agreement with experiment. The method has also proved qualitatively successful in predicting the shape of potential surfaces in chemical reactions and it has correctly predicted the extent to which σ orbitals can be localized in bonds, the localizability being a quantity that can be estimated by photoelectron spectroscopy and by nuclear magnetic resonance spectroscopy.

However, *Extended Hückel Theory* has many limitations and proves particularly unsatisfactory with polar molecules. To deal with molecules with polar bonds it is necessary to abandon the independent-electron treatment, and to allow specifically for electron–electron repulsion. This is done automatically for all self-consistent-field methods and in particular in any *ab-initio* calculation. There have been many attempts to introduce some degree of iteration into Hückel-type calculations to allow for electron repulsion but such procedures lack the simplicity of Hückel theory without acquiring the reliability of the *ab-initio* approach.

9.8. Generalized orbitals and overlap populations

In Hückel π-electron theory the concepts of charge density and bond order, which were defined in Section 9.3, have been found to correlate very well with experimental data. These definitions have to be extended if they are to be generally useful in molecular orbital theory for two reasons. Firstly because overlap integrals are not always assumed to be zero and secondly because more than one type of atomic orbital on each atom is used in the basis. The extension is needed on both counts for Hoffmann's extended Hückel theory, for example.

An electron in a molecular orbital

$$\psi_r = \sum_m c_{rm}\phi_m \tag{9.75}$$

has a density

$$\psi_r^2 = \sum_m \sum_n c_{rm}c_{rn}\phi_m\phi_n \tag{9.76}$$

and the total electron density function for the molecule is

$$\rho = \sum_r n_r\psi_r^2 = \sum_r \sum_m \sum_n n_r c_{rn}\phi_m\phi_n \tag{9.77}$$

n_r being the number of electrons in orbital ψ_r.

For normalized wavefunctions the integral over space of ψ_r^2 is unity and the integral of ρ is therefore the number of electrons in the molecule (N). We can therefore write

$$N = \sum_r \sum_m \sum_n n_r c_{rm}c_{rn}S_{mn} = \sum_r \sum_m n_r c_{rm}^2 + \sum_r \sum_m \sum_{n\neq m} n_r c_{rm}c_{rn}S_{mn}. \tag{9.78}$$

In a zero-overlap theory the second term on the right hand side of this expression is zero and we can therefore define

$$q_m = \sum_r n_r c_{rm}^2 \tag{9.79}$$

as the electron population of orbital ϕ_m and this is consistent with the requirement that the sum of all orbital populations is equal to N,

$$N = \sum_m q_m. \tag{9.80}$$

The definition (9.79) is identical with the Hückel π-electron charge on an atom (9.45) if there is only one orbital ϕ_m on each atom.

To obtain a quantity which we can relate to the bond order of Hückel theory (9.49), and which we might relate to the strength or length of a bond, we look to the second term on the right hand side of (9.78). For each pair of atomic orbitals we can define an *overlap population* by

$$a_{mn} = 2\sum_r n_r c_{rm} c_{rn} S_{mn}, \tag{9.81}$$

the factor 2 arising from the two ways in which the product $c_{rm}\, c_{rn}$ arises from the double summation in (9.78). As defined by (9.79) q may be considered as half the diagonal part of a, thus

$$q_m = a_{mm}/2. \tag{9.82}$$

The sum of the orbital and overlap populations is then equal to N, for on inserting (9.79) and (9.81) into (9.78) we have

$$N = \sum_m q_m + \sum_{mn} a_{mn}, \tag{9.83}$$

the second summation being over pairs mn $(m \neq n)$ regardless of order. That is, mn is not distinct from nm.

There will, in general, be several overlap populations associated with a bond and to obtain a quantity which is comparable to the Hückel π-bond order, but which measures the contributions from all orbitals, we will sum the quantities a_{mn} for all m on one atom and all n on the other,

$$A_{KL} = \sum_{m(K)} \sum_{n(L)} a_{mn}. \tag{9.84}$$

This is called the *gross overlap population*. Because of the inclusion of the overlap integrals S_{mn} in the definition of a_{mn} (9.81) pairs of orbitals with a large overlap integral will make the largest contribution to the gross overlap population.

From (9.83) it is seen that it is the sum of the orbital and overlap populations which is equal to the total number of electrons in the molecule. If we wish to partition the total number of electrons between the atoms then it is necessary to have some recipe whereby the overlap populations can be divided up amongst the atoms. The simplest such recipe is one suggested by Mulliken, which is to divide a_{mn} equally between the orbitals m and n. This gives a quantity called the *Mulliken orbital population*

$$q'_m = q_m + \frac{1}{2} \sum_{n \neq m} a_{mn}. \tag{9.85}$$

Such terms can be summed for all orbitals on an atom to give the total atom population, and subtracting this from the nuclear charge gives the net atom charge

$$Q_K = Z_K - \sum_{m(K)} q'_m\ . \tag{9.86}$$

As an illustration of these terms we analyse some molecular orbital calculations of acetylene. Table 9.3 shows the bonding molecular orbitals calculated from a

Table 9.3. Molecular orbitals of acetylene and their population analysis. The basis orbitals $\phi_1 \ldots \phi_6$ are for the unprimed atoms in H–C≡C′–H′ and $\phi_1' \ldots \phi_6'$ are corresponding orbitals for the primed atoms. Rounding errors make some of the accumulated totals in error in the last significant figure. Numbers omitted from the table are zero by symmetry or are zero to the accuracy quoted

	LCAO coefficients						
Symmetry	ϕ_1 $1s_C$	ϕ_2 $1s_H$	ϕ_3 $2s_C$	ϕ_4 $2p_{zC}$	ϕ_5 $2p_{xC}$	ϕ_6 $2p_{yC}$	Energy/E_H
$1\sigma_g$	0.702	−0.004	0.017	0.000	–	–	−11.034
$1\sigma_u$	0.703	−0.004	0.031	−0.006	–	–	−11.032
$2\sigma_g$	0.185	−0.120	−0.487	0.162	–	–	−0.963
$2\sigma_u$	0.119	−0.316	−0.353	−0.284	–	–	−0.716
$3\sigma_g$	0.003	0.321	0.081	0.461	–	–	−0.617
$\pi_u(x)$	–	–	–	–	0.616	–	−0.360
$\pi_u(y)$	–	–	–	–	–	0.616	−0.360
q_m	2.070	0.434	0.740	0.638	0.760	0.760	
q_m'	1.995	0.823	1.126	1.057	1.000	1.000	$\sum_m q_m' = 14$

m	n	S_{mn}	a_{mn}	m	n	S_{mn}	a_{mn}
1	2	0.06	−0.016	2	4	0.490	0.427
1	3	0.23	−0.091	2	4′(42′)	0.103	−0.016
1	3′(31′)	0.06	−0.014	3	3′	0.489	0.232
1	4′(41′)	0.10	−0.029	3	4′(43′)	0.462	0.262
2	3	0.50	0.391	4	4′	−0.304	0.192
2	3′(32′)	0.08	−0.008	5	5′(66′)	0.316	0.481

minimal basis of atomic orbitals. These are SCF orbitals but the analysis would follow the same pattern if they were calculated by an independent-electron model. The orbitals are identified by the lower case equivalent of the symmetry species of D_{6h} (see Table 7.9). Only the coefficients of the unprimed orbitals are listed in the table, the coefficients of the primed orbitals are determined by symmetry. Note in particular from the table that the quantities q_m have no direct physical significance in this calculation which is not based on orthogonal orbitals, but that q_m' are a realistic measure of the orbital populations.

The A and Q elements are calculated from the quantities given in Table 9.3 in the following way:

$$A_{HC} = a_{21} + a_{23} + a_{24} = 0.802$$

$$A_{CC} = \sum_{m(\neq 2)} \sum_{n(\neq 2')} a_{mn'} = 1.824$$

$$Q_H = 1 - q_{2'} = 0.177$$

$$Q_C = 6 - q_1' - q_3' - q_4' - q_5' - q_6' = -0.177 \tag{9.87}$$

Table 9.4. Results of a Mulliken population analysis from SCF MO calculations (L. C. Snyder and H. Basch, *Molecular Wave Functions and Properties* Wiley, 1972)

:	A_{HC}	A_{CC}	Q_H	Q_C	ΔH/kJ mol^{-1}	
					pred.	obs.
C_2H_2	0.75	1.98	0.26	−0.26	226.7	227.2
C_2H_4	0.80	1.16	0.17	−0.35	51.6	52.3
C_2H_6	0.77	0.55	0.16	−0.48	−84.4	−84.5

It is interesting to compare these quantities with comparable values for ethylene and ethane. Table 9.3 shows the results of SCF MO calculations using a larger basis of atomic orbitals than the calculation just quoted for acetylene. The difference in the values for acetylene given in this table and given in (9.87) shows that it is important when making such comparisons to have comparable basis sets for all molecules.

The points to be noted from Table 9.4 are firstly, the charge on the hydrogen atoms decreases from acetylene to ethane, which is consistent with the decrease in acidity in this series. Secondly, the ratio of A_{CC} values, which is 3.6: 2.1: 1, is very roughly the same as the ratio of the number of formal CC bonds in the classical valence structures. A more extensive analysis shows that the expression

$$\Delta H_f/\text{kJ mol}^{-1} = 141A_{CC} - 35n_{HC}A_{HC}. \tag{9.88}$$

where n_{CH} is the number of CH bonds in the molecule, predicts heats of formation in excellent agreement with experiment, suggesting that the A factors are a good basis for a bond-additive scheme of heats of formation for such molecules. Expression (9.88) was derived by a 2-parameter fit to the observed ΔH_f of Table 9.4 so the success of this expression is measured by its ability to fit the third heat of formation and others. In fact it fits exactly the heat of formation of allene ($H_2C{=}C{=}CH_2$) but is not so successful for the small ring compounds cyclopropane and cyclopropene so that not too much significance should be placed on this type of analysis.

Chapter 10
Band Theory of Polymers and Solids

10.1. General considerations

It should be clear from the contents of the previous chapters that the mathematical problems encountered in calculating molecular wavefunctions increase as the size of the molecule increases. For example, the size of the basis set of atomic orbitals in the LCAO model of molecular orbitals must increase as the number of atoms increases and this means that more Hamiltonian and overlap integrals must be calculated and the secular equations (6.74) will increase in dimension. By this reasoning, to calculate the energies and wavefunctions of polymers or solids might seem an insuperable problem. The fact that it is not is a result of symmetry, in this case *translational* symmetry.

A crystal consists of a regular array of atoms or molecules. If in a molecular crystal, for instance, we arbitrarily select a reference molecule then all other molecules in the crystal will normally be symmetry-related to it, either by a point group operation analogous to those we met in Chapter 7, by a translational operation, or by some combination of the two. Alternatively, we can think of a crystal as being built up of a basic building block, a unit cell, which is such that the entire crystal can be regarded as built up of these unit cells, the unit cells being related to each other by pure translations only. A useful way of representing these translations is by a three dimensional net of the type shown in Figure 10.1.

In addition to translational symmetry the crystal may have rotational axes of the type discussed in Chapter 7. However, the nature of the translational net places limitations on the rotational axes possible. Thus, the translational net shown in section in Figure 10.2 is such that these could not be C_3, C_4, C_5, or C_6 axes, for instance, perpendicular to the plane of the net but there could be C_2 axes at the positions shown. As the phrase 'could be' implies, these C_2 axes may not exist. Figure 10.3 shows the lattice of Figure 10.2 in which the detailed structure is such that no C_2 axes occur.

We conclude that the existence of translational symmetry places limitations on the rotational axes that may occur. It can be shown that this requirement restricts the allowed *Rotational* axes to be C_1, C_2, C_3, C_4, or C_6. Note in particular that C_5 and C_n, $n > 6$, are not allowed. This is evidently associated with the fact that it is not possible to cover a plane area with pentagons or heptagons (for example) without leaving gaps. It follows that whilst there is an infinite number of point

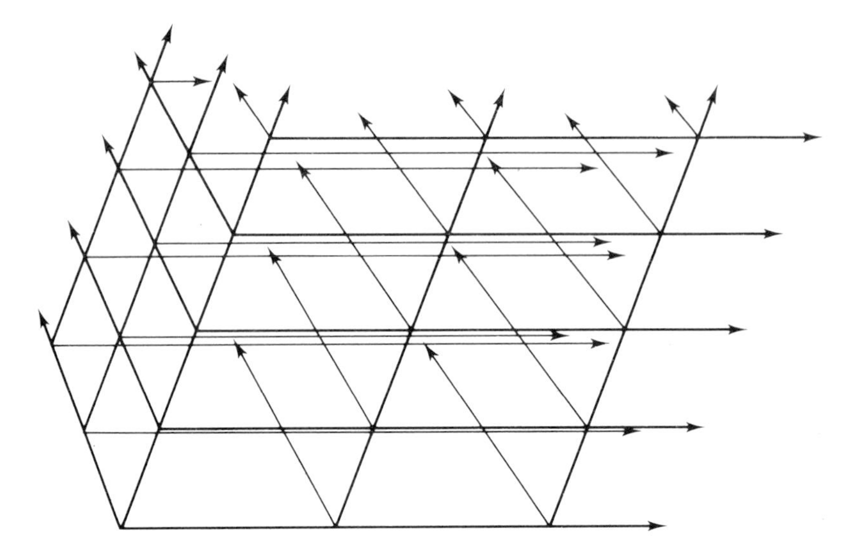

Figure 10.1 A translational net.

groups, only a limited number of these, 32 in all, is allowed for the unit cells of crystals.

The symmetry operations of a crystal are of three kinds, point group operations, translations, and combinations of the two such as a screw rotation (a rotation accompanied by a translation). The set of such operations defines the *space group* of the crystal. The notation we used for point groups in Chapter 7 is called the Schoenflies notation. For space groups, crystallographers usually use a different notation called the Hermann–Maugain or *International* notation. This consists of a string of symbols that define the operations. Thus the symbols 2/m indicates a

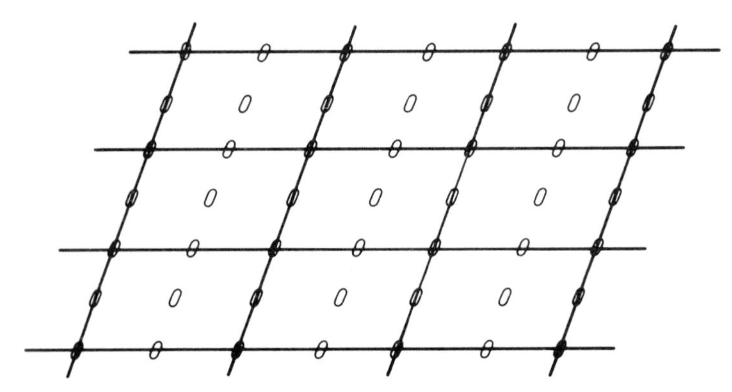

Figure 10.2 Positions of two-fold symmetry (0) in a plane net. Note that if one two-fold axis is chosen then all other two-fold axes duplicate its effect on a point and, in addition, add a translation.

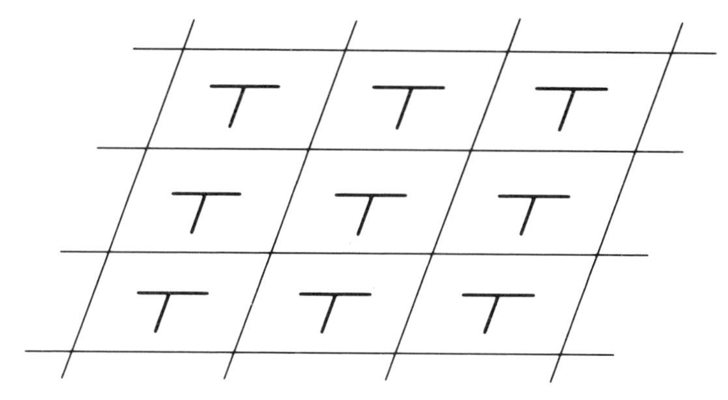

Figure 10.3 The translational net is that of Figure 10.2 but the two-fold axes of that net have been destroyed.

group with a 2-fold rotation axis and a mirror plane perpendicular to this. Only sufficient symbols are given to define the group. Because we will not specifically make use of space groups in this book we will not give a full account of this symbolism and the reader is referred to advanced books on crystallography for this.

A crystal has translational symmetry in three dimensions; a regular polymer can also be considered to have translational symmetry in one dimension or occasionally in two. The surface of a crystal is another two-dimensional array. It is the common approach to all these systems that we consider in this chapter.

In actuality any crystal does not have infinite extension. By treating it as infinite we aim to calculate the properties of the bulk but not of atoms at or near the surface. The precise definition of an infinite array is important when we come to specify the boundary conditions that must be applied to the solutions of the Schrödinger equation. One way to remove the surface of the crystal, in a mathematical sense, is to suppose that if we take an infinite number of translational steps in any direction we eventually arrive back where we started (as if we were walking around the circumference of a large circle). This model leads to the so-called Born–von Karman boundary condition that the wavefunction be unchanged on moving around such a circle.

Not all solids have translational symmetry and those which do not are either amorphous (glasses are of this type) or they may have a microcrystalline structure, the translational symmetry extending only over a relatively small number of unit cells. However, our concern in this chapter is only with the model of an infinite array.

Solids are generally divided into the following four classes:

1. Molecular crystals (e.g. solid naphthalene or solid iodine). The crystal consists of identifiable molecules whose structure is very similar to that of the gas-phase molecule. The molecules are held together by weak forces whose characteristics we shall discuss in Chapter 15. Figure 10.4 shows the position of the molecules in solid naphthalene. There are two molecules in each unit cell.

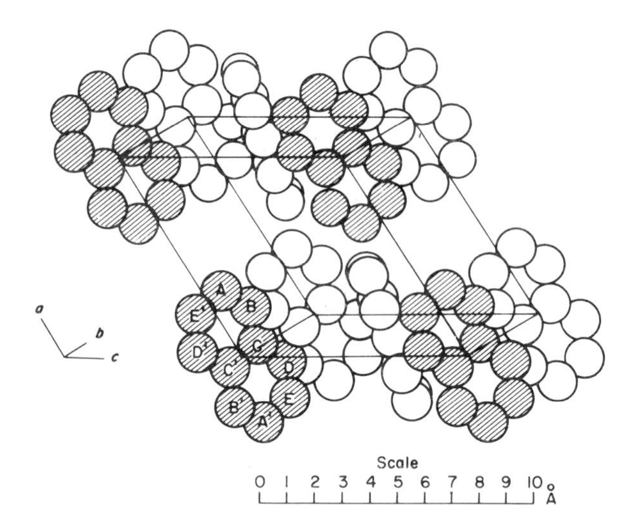

Figure 10.4 The positions of the naphthalene molecules in a unit cell of the crystal [Abrahams *et al.*, *Acta Cryst.*, 2c 238 (1949)].

2. Ionic crystals (e.g. sodium chloride). In these the crystals are built up of a regular array of ions, the anions being surrounded by several cations and *vice versa*. The electrostatic energy of the crystal (per ion) is calculated by summing the electrostatic interaction between one ion and all other ions in the lattice. The results of this may be expressed in the form (SI units)

$$\text{electrostatic energy} = -M\left(\frac{e^2}{4\pi\epsilon_0 R}\right) \tag{10.1}$$

where R is the nearest neighbour distance. M is a constant characteristic of the crystal structure and is call the Madelung constant. Figure 10.5 shows the position of the ions in the cubic crystal of NaCl. Each Na^+ is surrounded by six nearest neighbour Cl^- at a distance R, and there are twelve Na^+ at distance $\sqrt{2}R$, eight Cl^- at $\sqrt{3}R$, six Na^+ at $2R$, and so on. The total potential energy of this one Na^+ in the field of all the neighbours is therefore

$$-\frac{e^2}{4\pi\epsilon_0}\left\{\frac{6}{R}-\frac{12}{\sqrt{2}R}+\frac{8}{\sqrt{3}R}-\frac{6}{2R}+\ldots\right\}. \tag{10.2}$$

There is a similar sum for the potential energy of each Cl^- with all other ions, but if we add all sums as (10.2) for both types of ion we will count every pair interaction twice. Thus the energy per NaCl molecule is just given by the infinite sum (10.2) and when this is evaluated it gives the result $-1.748\ e^2/4\pi\epsilon_0 R$.

As the Madelung constant is invariably greater than unity the electrostatic energy per molecule is greater in the crystal than in an isolated gas-phase

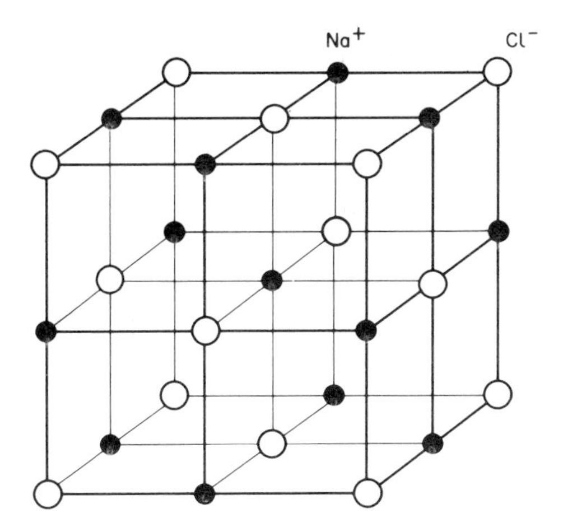

Figure 10.5 Distribution of ions in the NaCl crystal.

molecule. As a result, the nearest neighbour separation in the NaCl crystal is 2.36 Å whereas in an isolated molecule it is 2.81 Å.

There are other contributions to the binding energy of ionic crystals that must be allowed for in accurate calculations but as a first approximation we can assume that the attractive force is purely electrostatic and this is balanced by a repulsion due to partial overlap of the electron densities of nearest neighbours. The nature of such repulsive forces will be discussed in more detail in Chapter 15.

3. Covalent solids (e.g. diamond or silica). In contrast to class 1 solids these must be considered as supermolecules in which atoms are joined to their neighbours by covalent bonds. There are no identifiable molecular sub-units in the crystal.

4. Metals (e.g. sodium, iron). Like classes 2 and 3 these are supermolecules but they cannot be classed as ionic or covalent and have their own characteristic properties. It is the type 3 and 4 solids that we shall discuss in this chapter. Like other classifications we have made in this book examples can be found which are intermediate in character but we hope that these can be understood if we first understand the typical examples of each class. First, however we shall consider a simple but analogous one-dimensional problem.

10.2. The infinite polyene

In Chapter 9 we showed how the Hückel π orbitals of a polyene chain could be calculated. The example we now take of an infinitely long polyene will serve to illustrate some of the most important features of infinite regular systems.

Figure 10.6 shows the Hückel π-molecular orbital energies for such polyenes with up to ten double bonds. A pattern clearly emerges as the number of atoms is

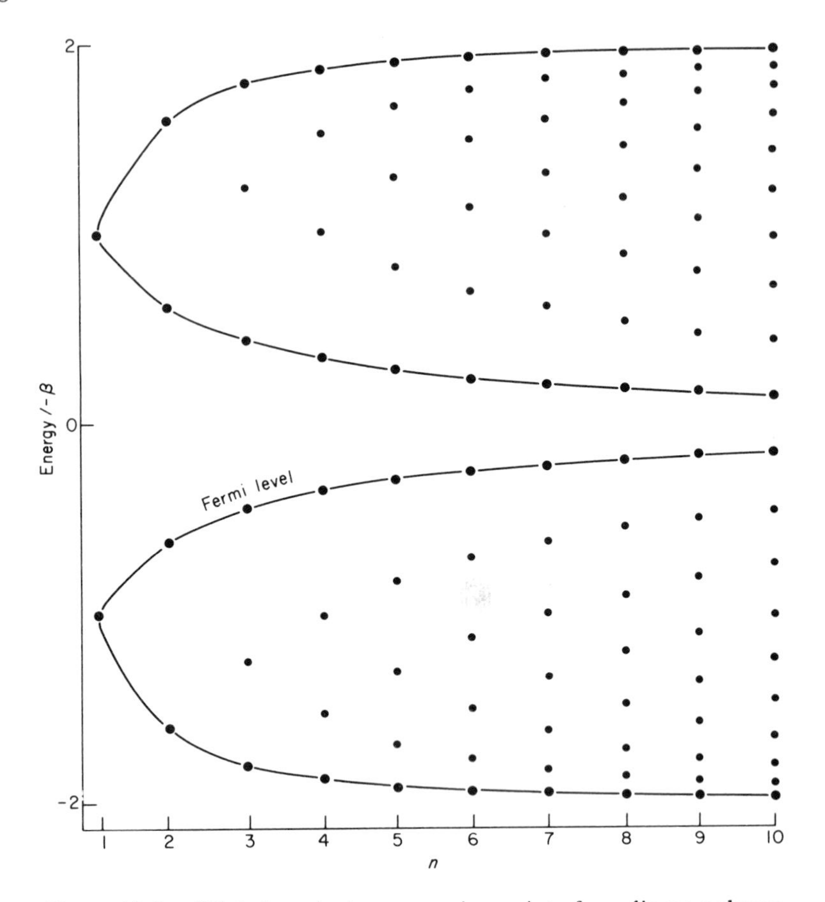

Figure 10.6 Hückel π electron energies points for a linear polyene with n double bonds.

increased. Firstly we note that the positions of the highest and lowest energies appear to approach a finite limit and not the limits $\pm\infty$ that might have been expected. It can be shown that these limits are $\alpha \pm 2\beta$. A second point concerns the distribution of orbital energies between these limits. There are clearly more levels in the proximity of the two limits than there are in the middle.

It is possible to obtain a general expression for the Hückel levels of the linear polyene and this is

$$E_r = \alpha + 2\beta \cos\left(\frac{r\pi}{2n+1}\right); \quad r = 1, 2, \ldots 2n, \tag{10.3}$$

where n is the number of double bonds. The distribution of energy levels therefore follows the distribution of values of $\cos \vartheta$ for equal intervals of ϑ in the range $0 \to \pi$, as shown in Figure 10.7.

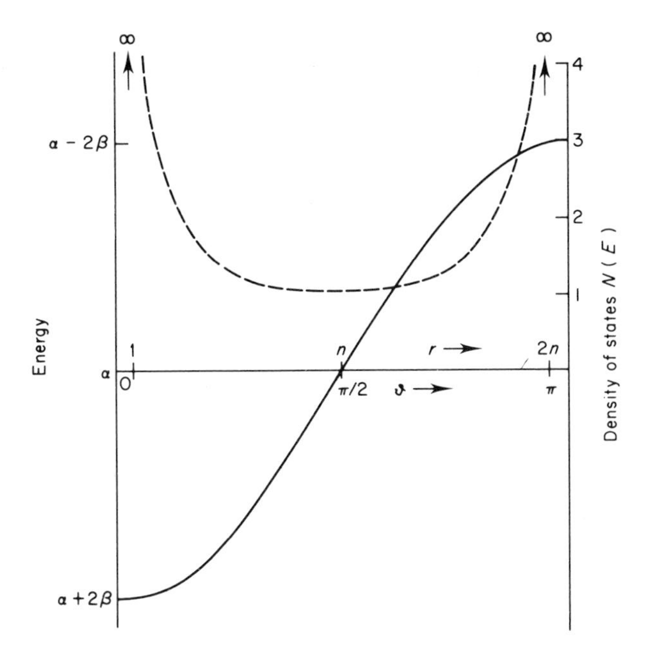

Figure 10.7 Distribution of Hückel energy levels for a polyene with n double bonds, for large values of n. The density of states is the broken line (see equations 10.3 and 10.5).

The energy difference between successive levels is, from (10.3),

$$E_{r+1} - E_r = 2\beta\left[\cos\left(\frac{(r+1)\pi}{2n+1}\right) - \cos\left(\frac{r\pi}{2n+1}\right)\right],$$

$$= -4\beta \sin\left(\frac{(2r+1)\pi}{4n+2}\right) \sin\left(\frac{\pi}{4n+2}\right). \tag{10.4}$$

We can define an energy density of levels $N(E)$ from the reciprocal of (10.4). This is a measure of the number of levels in a fixed energy interval. For a given value of n

$$N(E) \propto \operatorname{cosec}\left(\frac{(2r+1)\pi}{4n+2}\right). \tag{10.5}$$

$N(E)$ is more precisely (and simply) defined as $(\partial r/\partial E)$ and expression (10.5) can be obtained directly by differentiating (10.3).

In the limit $n \to \infty$ the energies defined by (10.3) become a continuous band spreading between $\alpha + 2\beta$ and $\alpha - 2\beta$ and the density of levels is proportional to cosec ϑ as shown in Figure 10.7.

Figure 10.7 shows two of the important features of the energy bands of infinite systems. The first is the energy limits of the top and bottom of the band and the second is the density of levels in the band. For the infinite polyene the density becomes zero at $E = \alpha$ hence we can say that there are two bands for this system, a bonding band spanning energies $\alpha + 2\beta$ to α, and the other, an antibonding band, from α to $\alpha - 2\beta$. In the lowest energy state of the system there are sufficient electrons just to fill the bonding band and the antibonding is empty. The top of the filled band is called the *Fermi level*. Its energy is the negative of the minimum energy required to ionize the system.

An interesting modification of the calculations we have described is to introduce different resonance integrals for alternate bonds. Those bonds which are formally double in the classical valence structure (d) are slightly shorter than those which are formally single (s), a fact we noted for butadiene, and we can allow for this by taking $|\beta_d| > |\beta_s|$. With this modification it is found that a gap appears between the bonding and the antibonding bands. The bonding spreads from $\alpha + \beta_d + \beta_s$ to $\alpha + \beta_d - \beta_s$ and the antibonding from $\alpha - \beta_d + \beta_s$ to $\alpha - \beta_d - \beta_s$. There is a gap of $2|\beta_d - \beta_s|$ between the bands. *Band gaps* are the third general feature we note for infinite systems.

The concept of delocalized orbitals for infinite systems was developed by Bloch in 1928, and this was in fact before the molecular orbital theory had been developed for molecules. However, Bloch did not base his delocalized orbitals

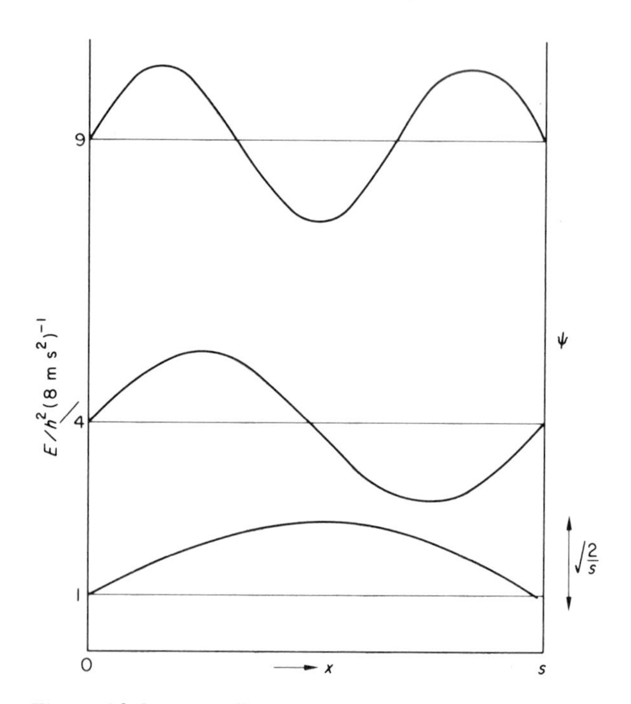

Figure 10.8 The first three energy levels and associated wavefunctions for particle in a one-dimensional box.

(called Bloch orbitals) on the LCAO approximation, as we have done, but thought of them as periodic waves extending through the lattice whose precise form is determined by the periodic potential of the nuclei. This concept is best approached from the model of the completely free electron, that is, an electron which is not subject to any periodic potential. Sommerfeld was the first to apply this model to infinite systems when he published a theory of metallic conduction in 1928.

We consider first the one-dimensional *free-electron model* which we can apply to the linear polyene and which we can compare with the Hückel model just discussed. In this model we ignore the existence of the nuclei, so that an electron is in a uniform potential over the entire length of the molecule, and look in detail at the consequences of this. It has the advantage of being one of the few problems for which the Schrödinger equation can be solved exactly.

We take the potential as constant and of zero magnitude within the molecule but suppose that it rises to an infinite value immediately outside the molecule. With such a potential the electron has zero probability of escaping from the molecule. This situation is shown diagrammatically in Figure 10.8. For obvious reasons this is often referred to as the 'particle in a box' problem, the particle being in our case an electron.

The Schrödinger equation for a particle moving in one dimension in a zero potential is (2.18)

$$\frac{-\hbar^2}{2m}\frac{\partial^2\psi}{\partial x^2} = E\psi, \tag{10.6}$$

and solutions of this have the general form

$$\psi = N\sin\left[\left(\frac{2mE}{\hbar^2}\right)^{1/2}(x+k)\right], \tag{10.7}$$

where N and k are constants of integration.

Because the electron is confined within the limits $0 \leqslant x \leqslant s$ by the infinite potential walls, ψ must be zero outside these limits. The condition that ψ be a continuous function for all values of x leads us to impose the boundary condition that ψ is zero at $x = 0$ and at $x = s$.

If $\psi = 0$ at $x = 0$ then $k = 0$. If $\psi = 0$ at $x = s$ then

$$\psi(s) = N\sin\left(\frac{2mE}{\hbar^2}\right)^{1/2} s = 0, \tag{10.8}$$

whence

$$\left(\frac{2mE}{\hbar^2}\right)^{1/2} = r\pi, \tag{10.9}$$

r being an integer.

Expression (10.9) leads to the quantization condition for the energy

$$E_r = \frac{h^2 r^2}{8ms^2}, \tag{10.10}$$

and inserting this into (10.7) gives the wavefunction

$$\psi_r = N \sin \frac{\pi r x}{s}, \tag{10.11}$$

where we have labelled each wavefunction and energy by the appropriate value of the integer r. These solutions are illustrated in Figure 10.8 from which the analogy with the vibrations of a stretched string is evident.

In applying these particle-in-a-box solutions to the π molecular orbitals of a linear polyene we have to relate the length of the molecule, s, to the number of carbon atoms in the polyene. If the carbon–carbon bond length is d and there are n double bonds, then the distance between the first and last carbon atoms along the zig-zag chain is $(2n-1)d$. However, this is the distance between the first and last nuclei and we can reasonably assume that the zero potential extends half a bond length at either end, so that the effective length of the molecule is $s = 2nd$. This very simple model is actually quite good for predicting the wavelength of the first absorption band of a polyene. This band arises from the excitation of an electron from the highest occupied π orbital of the ground state to the lowest unoccupied orbital (c.f. Table 9.1). In the case butadiene (for example), which has four π electrons, the highest occupied orbital is ψ_2 and the lowest unoccupied is ψ_3. The energy difference between these levels is obtained from (10.10) with $s = 4d$ as

$$E_3 - E_2 = \frac{h^2(3^2 - 2^2)}{8m(4d)^2}. \tag{10.12}$$

Taking d as 1.4 Å we predict that the absorption band is at

$$\lambda = \frac{hc}{E_3 - E_2} = 2\,068 \text{ Å}, \tag{10.13}$$

which is very close to the observed value of 2 200 Å. The free-electron treatment of the finite polyene can be extended to the infinite chain. Provided that the same boundary conditions as in the finite case are chosen, expressions (10.10) and (10.11) are quite general.

The energy density of states can be obtained from (10.10) as

$$N(E) = \frac{2\partial r}{\partial E} = \frac{8ms^2}{h^2 r} = \left(\frac{8ms^2}{h^2 E}\right)^{1/2}. \tag{10.14}$$

It is customary to include the factor of two in this definition because each orbital can take two electrons.

If each atom in the chain contributes one electron to the system, as for the π orbitals of the polyene, then for N atoms the lowest $N/2$ orbitals will be filled. The

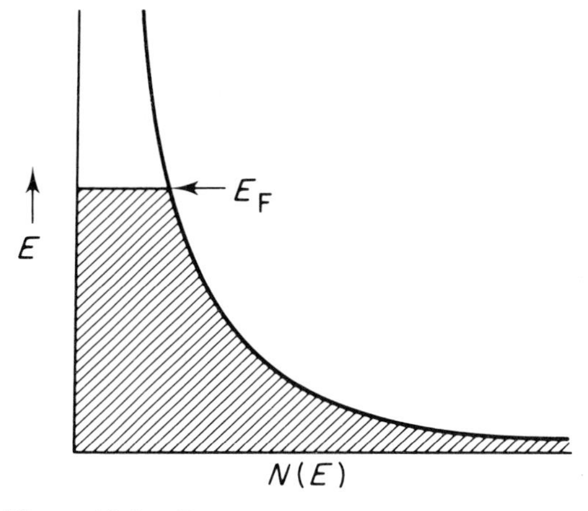

Figure 10.9 Density of states $N(E)$ for the one-dimensional free-electron model. The hatched area indicates the occupied levels at 0 K.

top occupied orbital will have $r = N/2$ and from (10.10), this defines the Fermi level as

$$E_F = \frac{h^2 N^2}{32ms^2}. \tag{10.15}$$

For infinite chains s/N is just the distance apart of neighbouring atoms (d) and

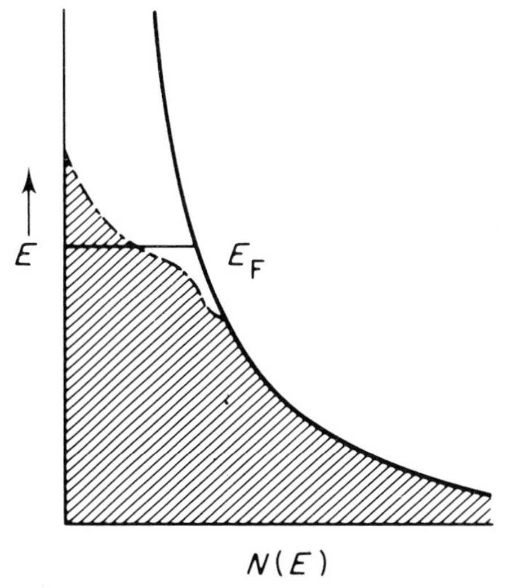

Figure 10.10 Density of states for the one-dimensional free-electron model showing (hatched area) the levels occupied at finite temperature.

hence the Fermi level is

$$E_F = \frac{h^2}{32md^2}. \tag{10.16}$$

Equation (10.14) is shown graphically in Figure 10.9 and the shading indicates the levels filled to the Fermi level. Above this level are empty orbitals. However, this figure is really only applicable at 0 K; above this temperature there will be a Boltzmann population of the energy levels above E_F (and depopulation of those below).

A discussion of the free-electron model at temperatures above 0 K requires a detailed statistical treatment which is beyond the scope of the present text. We shall content ourselves with the statement that at such temperatures the distribution will be of the general form shown in Figure 10.10.

10.3. Metals and covalent solids

The two descriptions of the infinite polyene that have been given in the last section, the LCAO and the free-electron models, can be extended to three dimensions to give a general description of covalent solids and of metals.

The central concept of both models is the energy band, and the way in which electrons fill these bands largely determines the properties of the solids. Covalent solids such as diamond have bands which are either completely filled or are empty, and there is a band gap between the highest filled level and the lowest empty level.

Metals have partially filled bands and it is this that gives them their characteristic properties of high thermal and electrical conductivity. Negligible amounts of energy are needed to excite electrons from filled to empty orbitals, and as these orbitals are delocalized over the entire metal the energy of the electrons can readily be distributed. Likewise light of any wavelength will be absorbed and, suprisingly enough, it is this which gives rise to the phenomenon of metallic reflectivity.

For a realistic description of a three dimensional solid the models used in the last section must be extended in two respects. Firstly in the LCAO model, usually called the *tight-binding approximation* because the electrons are strongly associated with the nuclei through the atomic orbitals, we must allow for more than one atomic orbital and more than one electron for each atom. In the free-electron model we must allow for the potential energy of attraction of the electrons to individual nuclei. With such a potential, which will exhibit the periodicity of the space group of the lattice, the free-electron model becomes the *nearly-free-electron* model.

Let us first see the effect of having more than one atomic orbital for each atom. Solid beryllium will illustrate the points we wish to make. Let us imagine that we have a lattice in which the interatomic spacing is very large (the metal actually has a close-packed hexagonal structure but that is not relevant for our qualitative arguments). The energy bands will be clustered around the atomic orbital energies of the free atoms (1s, 2s, 2p, 3s, 3p, 3d, etc.) and the bands will be very narrow

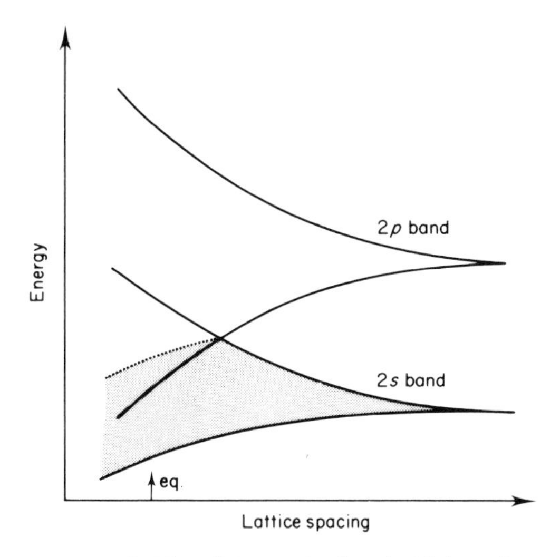

Figure 10.11 Schematic bandwidths of a Be lattice as a function of the lattice spacing. The broken line corresponds to the level of filling. The equilibrium spacing is marked eq.

because the width of the band depends on the interaction Hamiltonian integral between neighbouring orbitals. For example, we saw in the last section that the width of the bands for the linear polyene is proportional to the resonance integral, β.

If we now imagine the lattice spacing to be gradually decreased then the bands will get wider but they will still be spread around the orbital energies of the free atoms. At certain critical lattice spacings two neighbouring bands may become so wide that they overlap one another on the energy scale and in this case we must refer to one composite band rather than two separate bands Figure 10.11 illustrates this situation schematically for solid beryllium. At the lattice spacing in the crystal the bands that originate in the 2*s* and 2*p* levels of the atoms overlap one another; the 1*s* band, however, is still well separated from the 2*s*.

Each beryllium atom contributes four electrons to the solid. Two of these originate from the 1*s* orbital of the atom and these will be sufficient to fully occupy the 1*s* band. The other two originate in the 2*s* orbital and are sufficient to fill the 2*s* band. At large lattice spacings the ground state of the solid would therefore have a filled 1*s* band and filled 2*s* band and there would be a gap between the latter and the empty 2*p* band. In contrast to a metal it would require a substantial energy, equal to the 2*s* – 2*p* band gap, to promote electrons from the occupied to the vacant levels and this energy would have to be provided from some source such as light or an electric battery. Solid beryllium with a large lattice spacing would be an insulator.

However, at the equilibrium distance in solid beryllium the 2*s* and 2*p* bands overlap and together they are capable of holding eight electrons from each atom.

The two electrons that are actually provided by the atoms will therefore only partly fill the combined bands and there is no energy gap between filled and empty levels. Solid beryllium is a typical metal.

The question of whether a solid is a metal or a non-metal therefore depends on three factors. Firstly, the spacing between the orbital energies in the isolated atoms, secondly on the lattice spacing, and thirdly on the number of electrons that are provided by the atoms. Solid lithium, for example, would be a metal even if the 2*s* and 2*p* bands did not overlap because the 2*s* band is only half filled. The transition elements are metals because of the incomplete filling of bands that arise from *nd* orbitals and also because of the overlap of *nd* with $(n + 1)s$ and $(n + 1)p$ bands.

Increasing the pressure on a solid will decrease the lattice spacing and this will increase the band widths. As Bernal first pointed out, under a sufficiently high pressure all solids are expected to display metallic conduction because even for those which are normally insulators there will, at high pressures, be an overlap between the highest energy filled band and the lowest energy empty band. Hydrogen at 4.2 K has been found to show metallic conduction under a few megabars of pressure.† Diamond and silica have also been converted to a metallic form.

In discussing the free-electron model of solids it is convenient to start from expressions (10.10) and (10.11) and to define a quantity k by

$$k = \frac{\pi r}{s}. \tag{10.17}$$

Note that this has the dimensions of reciprocal length. On substituting (10.17) into (10.11) the wavefunction is given in the form

$$\psi_k = N \sin kx, \tag{10.18}$$

and the wavelength of this function is $2\pi/k$.

Inserting (10.17) into (10.10) gives the energy as

$$E_k = \frac{h^2 k^2}{8\pi^2 m} = \frac{\hbar^2 k^2}{2m}. \tag{10.19}$$

Note that in (10.18) and (10.19) we have labelled each solution of the Schrödinger equation by k rather than by r. For the infinite system ($s \to \infty$), k becomes a continuous variable spanning the range $0 \to \infty$ and the energy spectrum of such a system is continuous.

Expression (10.18) satisfies the boundary conditions that the wavefunction be zero at the beginning and at the end of the lattice. For an infinite system such a boundary condition becomes inappropriate and, in the Born–van Karman cyclic boundary condition already mentioned, another independent solution is allowed which is

$$\psi_k' = N \cos kx. \tag{10.19}$$

†Y. Yakalov, *New Scientist*, 71, 478 (1976).

This has the same energy as ψ_k. These two wavefunctions ψ_k and ψ'_k, have the same wavelength but differ in phase by $\pi/2$. Because ψ_k and ψ'_k have the same energy, it follows from equations (7.20) to (7.24) that an alternative description of the system may be made in terms of the set of complex functions

$$\theta_k = e^{ikx}, \tag{10.20}$$

where k is a continuous variable in the range $-\infty < k < +\infty$. This follows from the definition of the trigonometric functions

$$e^{ikx} = \cos kx + i \sin kx,$$
$$e^{-ikx} = \cos kx - i \sin kx. \tag{10.21}$$

We can now see how the periodic potential of the nuclei will perturb the energy levels. For values of $|k|$ close to zero the wavelength of the free-electron waves is long, and such waves will feel only the broad average of the periodic potential and will be changed little in energy. For very large values of $|k|$ the wavelength will be short compared with the lattice spacing and there will be many oscillations of the wave over a length in which the potential changes little: these short waves will also be little influenced by the potential. However, for intermediate values of $|k|$ there will be waves whose length is approximately equal to the lattice spacing, or to a simple multiple (2 say) of the lattice spacing, and these will be strongly influenced by the periodic potential of the lattice.

Consider, for example, a value of $|k|$ such that the wavelength is twice the lattice spacing. We can choose the two solutions for this value of $|k|$ such that the nodes of one and the antinodes of the other coincide with the nuclei as shown in Figure 10.12. Such waves can always be found from some linear combinations of $\cos kx$ and $\sin kx$ or of e^{ikx} and e^{-ikx} which give the phases we require. In the free-electron model these two waves have the same energy, but the periodic potential of the nuclei will obviously have a large stabilizing effect on the wave with antinodes at the nuclei but a much smaller effect on the wave with nodes at the nuclei. In other words, the periodic potential removes the degeneracy of the two states and the splitting manifests itself as a band gap in the energy spectrum. The states at the top of the lower energy band have wavefunctions like θ_k and those at the bottom of the upper band have wavefunctions like θ'_k.

Figure 10.12 Free electron waves which are strongly perturbed by the periodic potential of the lattice.

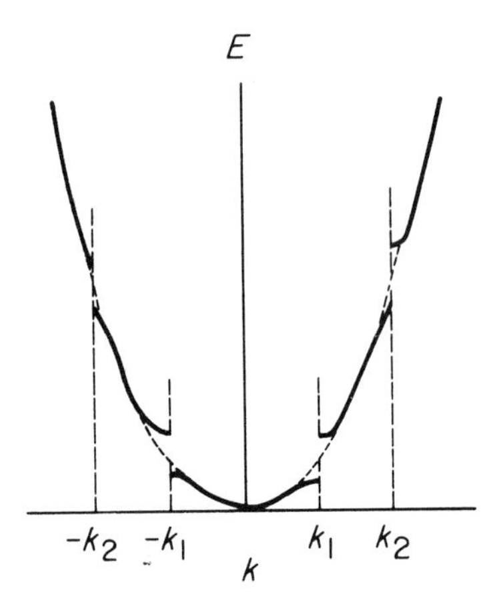

Figure 10.13 The energy levels of the one-dimensional free-electron model (broken curve) and the nearly-free-electron model (solid interrupted curve) as a function of k. In the nearly-free-electron model band gaps occur for $|k| = |k_1|$ and $|k_2|$.

The allowed energy levels as a function of k are shown schematically in Figure 10.13. For certain values of $|k|$, determined by the lattice spacing, there are discontinuities in the energy spectrum for the reasons just given. These gaps divide up the $|k|$ space into zones which are called *Brillouin zones*. The region from $|k| = 0$ to the first break is called the first Brillouin zone; from there up to the second break is called the second zone, and so on. These zones have the dimensions of reciprocal length, because that is the dimension of k.

The nearly-free-electron model in three dimensions follows the pattern just described for one dimension with one important new feature. The free-electron waves in three dimensions can be developed from the generalization of (10.20), namely

$$\theta_{k_x k_y k_z} = e^{ik_x x}\, e^{ik_y y}\, e^{ik_z z} = e^{i(k_x x + k_y y + k_z z)}. \tag{10.22}$$

We can think of k_x, k_y, and k_z as the three components of a vector $\mathbf{k}$

$$\mathbf{k} = (k_x, k_y, k_z). \tag{10.23}$$

Because the lattice spacing differs from one direction to another in the crystal, the energy breaks will occur for different values of k in different directions. These values of k define a shape in k 'reciprocal' space, and in three-dimensional lattices they define the three-dimensional Brillouin zones. Thus k space is divided up into zones whose precise form depends on the crystal structure. Within the zones the

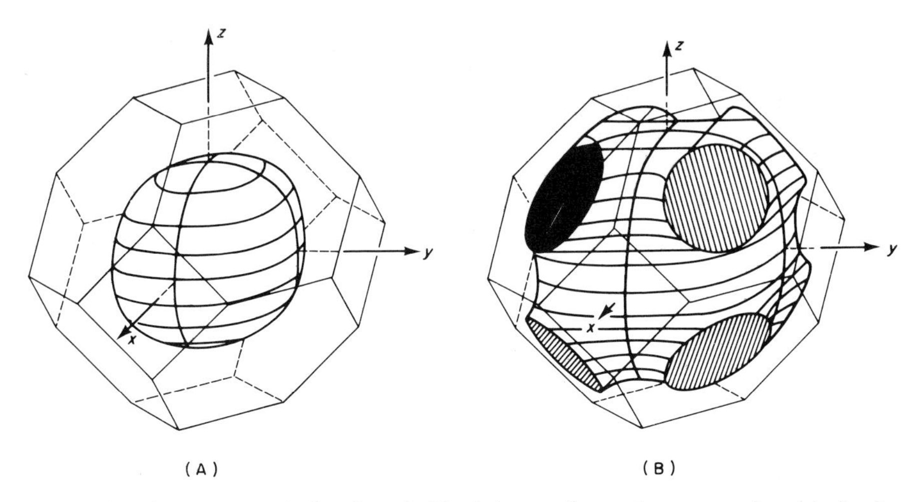

Figure 10.14 Shape of the first Brillouin zone for a face-centred cubic lattice showing the surfaces of constant energy in k space for (A) the zone nearly empty and (B) the zone nearly full.

energy is a continuous function of k but across the boundaries of the zones the energy is discontinuous.

Figure 10.14 shows the shape of a Brillouin zone for a face-centred cubic lattice, a structure exhibited by many metallic elements. It is a regular polyhedron with a mixture of hexagonal and square faces. The surfaces of constant energy in k space at the bottom of the first zone are nearly spherical as shown in figure 10.14A, and this is in accord with the completely-free-electron model, but at the top of the zone they are distorted by the polyhedron as shown in Figure 10.14B. The highest energy points in the zone correspond to the vertices of the polyhedron.

10.4. Semiconductors

We have so far considered solids to be either insulators or conductors. An intermediate category exists which are semiconductors, and in view of their considerable commercial importance it is appropriate to give a brief explanation of their properties.

If the highest energy completely filled band is separated from the lowest energy empty band by a small energy gap then some electrons may be thermally excited from the filled to the empty band. The highest filled band is often referred to as the *valence* band, and the empty band is called the *conduction band* because any electrons that enter it will conduct electricity.

The population of thermally excited electrons will be governed by statistical principles (to be precise by the so-called Fermi–Dirac statistics) but the most important criterion is that there will be a significant number only if the energy gap is of the order of kT (k being the Boltzmann constant) or less.

The higher the temperature the greater the number of conduction electrons so that the conductivity will increase, or the resistivity decrease, with increasing temperature. This contrasts with the situation for metals whose resistivity *increases* with increasing temperature a phenomenon due to the motion of the nuclei. Nuclear motion, which manifests itself as lattice vibrations, tends to destroy the perfect translational symmetry of the lattice and this perturbs the electronic wavefunctions and effectively reduces the length over which they are delocalized.

In addition to the conduction produced by thermally excited electrons in the conduction band there will be conduction from the vacancies produced in the valence band. The conduction band is said to give n-type (n for negative electron) conduction and the valence band p-type (p for positive hole) conduction. Both, of course, arise from the movement of electrons down the electrostatic field.

We can picture n-type conduction as the unrestricted movement of an electron along the lattice from one lattice site to the next because most energy levels in the band are empty. However, in p-type conduction the electrons can only move if there is an unfilled level at the neighbouring site down field, and moving electrons down field into a vacancy is equivalent to moving the positive vacancies upfield. This is illustrated schematically in Figure 10.15.

The excitation of electrons from valence to conduction bands can also be stimulated by light of sufficient frequency to span the band gap. The necessary condition is

$$h\nu > E_C - E_F$$

where E_F is the Fermi level and E_C is the energy of the bottom of the conduction band. Conductivity produced in this way is called photoconductivity.

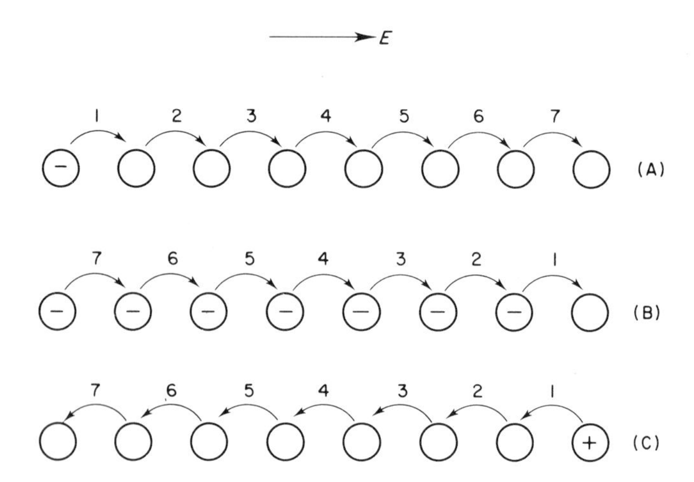

Figure 10.15 Movement of electrons in successive steps (numbered in the order in which they occur) down an electric field in: (A) an almost empty band, (B) an almost filled band. (B) is equivalent to the movement of a positive hole upfield as shown in (C).

Germanium and silicon, with band gaps of approximately 60 and 100 kJ mol^{-1} respectively, are typical semiconductors. The conductivity of silicon at 20°C is approximately 10^{-2} ohm^{-1} cm^{-1}, which is between values typical of an insulator (10^{-12} ohm^{-1} cm^{-1}) and of a metal (10^{5} ohm^{-1} cm^{-1}). These substances are called intrinsic semiconductors because their conductivity is a property of the pure element. Of perhaps greater importance are the so-called impurity semi-conductors which form the basis of the transistor.

Even a small amount of impurity incorporated into a perfect lattice, or defects in a lattice such as a misplaced atom, can modify the structure of the Brillouin zones. Carefully selected impurities can have particularly large effects, and it is these that are commercially important.

The impurities used to 'dope' silicon or germanium are elements which have one more electron in their valence shell, such as phosphorus or arsenic, or elements which have one less electron in their valence shell such as gallium or indium.

If phosphorus is added to the silicon lattice it occupies a site equivalent to the silicon atoms but contributes one extra electron to the system. It is almost as if one were adding Si^- to the lattice. Such extra electrons must go into the conduction band because the valence band in the crystal is full. Thus phosphorus impurities give n-type conduction. A group 3 element like gallium however, will go into the lattice rather like Si^+ and thus create a hole in the valence band so that there will be p-type conduction.

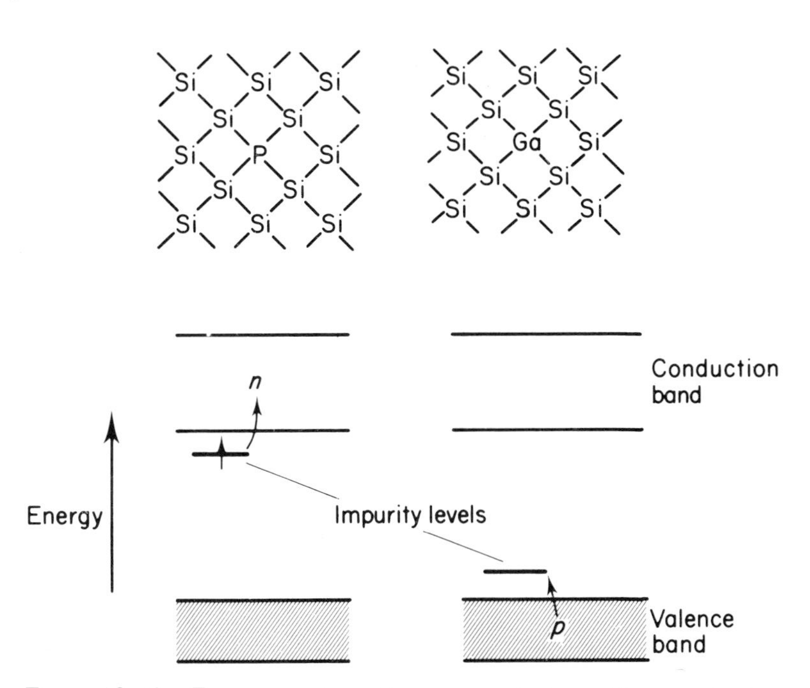

Figure 10.16 Energy in impurity semiconductors. The conduction arises from the excitation of electrons out of or into impurity levels, as shown.

n-type and p-type conduction can be distinguished by what is called the Hall effect. If a magnetic field (**H**) is applied at right angles to the electric field (**E**), then electrons and positive holes both experience a force which is at right angles to **H** and **E**; but these forces are in opposite directions. Thus by measuring the direction of the current at right angles to **H** and **E** one can decide whether conduction is by holes or by electrons.

A more detailed analysis of the band structure of silicon or germanium shows that the energy levels associated with the impurities are not completely absorbed into the Brillouin zone structure of the pure crystal. Rather they represent localized impurity levels which are either just above the Fermi level of the pure crystal for group 3 impurities or just below the bottom of the conduction band of the pure crystal for group 5 impurities. This is shown schematically in Figure 10.16. It is the thermal excitation of electrons into or out of these impurity levels that is responsible for the conduction, hence such impurity semiductors have the same temperature dependence as that of an intrinsic semiconductor.

Chemistry has long been concerned with understanding the properties of isolated atoms, molecules, or ions but there seems currently to be an increase in the study of solid state chemistry. The concepts which we have introduced in this chapter, and which are part of solid state physics, will probably be increasingly important for chemists in the future.

Chapter 11

The Concept of a Perturbation

11.1. Solution of the Schrödinger equation for perturbed systems

In this chapter we show how the wavefunctions and energies for a complicated problem can be determined using as a starting point a simpler, related problem for which the wavefunctions and energies are known. The method is known as perturbation theory and relies for its validity on the assumption that the difference between the Hamiltonians of the complicated and the simpler system, which we call the perturbation to the Hamiltonian, has only a small effect on the energies and wavefunctions.

Perturbation theory has widespread application in physics and chemistry as shown by the following three examples.

A large part of our knowledge of molecular structure and molecular properties comes from observing the effects of electric or magnetic fields on molecular spectra. The spectroscopic changes induced by an electric field (the Stark effect) and by a magnetic field (the Zeeman effect) are usually interpreted through perturbation theory in which the effect of the electric or magnetic field is treated as a perturbation to the Hamiltonian of the field-free molecule.

We have seen in Chapter 7 that the techniques of group theory are a valuable aid in solving the Schrödinger equation for molecules with symmetry. In some case it is useful to relate the solutions for a molecule with low symmetry to those for a high symmetry molecule. For example, both pyridine (C_5H_5N) and toluene ($C_6H_5CH_3$) have low symmetries but they can both be related to benzene, which has the high symmetry, D_{6h}, by considering in the first case the replacement of a CH group by N as a perturbation and in the second case the replacement of H by CH_3 as a perturbation. Such an approach brings out the fact that there are some physical and chemical properties of these three molecules which are closely related: their u.v. absorption spectra, for example.

Probably the most important application of perturbation theory to valence theory is to the energy levels of transition metal complexes. We shall see in the next chapter that the d-electron energies of such atoms (and their ions) are often little changed by placing the atom in an environment where it experiences the electrostatic field of neighbouring atoms or ions. Thus the spectra of transition metal ions in solution or in the solid state may be interpreted on the basis of a perturbation due to these neighbours (the ligand or crystal field) on the spectra of the free ion.

The equations of perturbation theory are based on the assumption that both the wavefunction and the energy of a state are smoothly varying functions of the strength of the perturbation, and that both can be expanded as a power series in a strength parameter. If the Hamiltonian for the system of interest is $\mathscr{H}$ and that for the unperturbed or reference system is $\mathscr{H}^\circ$, then the strength of the perturbation is measured by λ defined by

$$\mathscr{H} - \mathscr{H}^\circ = \lambda \mathscr{H}', \tag{11.1}$$

where $\mathscr{H}'$ contains any algebraic factors that describe the perturbation explicitly.

When the perturbation arises from an applied field then λ can be taken as the strength of the field. For example, an electron in a uniform electric field F_z applied in the z direction has a potential energy (in atomic units) $F_z \,.\, z$ and with this as the perturbation term we can take $\lambda = F_z$ as the strength parameter and $\mathscr{H}' = z$ as the algebraic factor. When the perturbation arises from a unique situation, say the replacement of one atom by another, then λ is a dummy parameter which may have physical reality for only one particular value.

If the expansion of the energy and wavefunction in powers of λ is written as

$$E = E^\circ + \lambda E' + \lambda^2 E'' + \ldots, \tag{11.2}$$

$$\Psi = \Psi^\circ + \lambda \Psi' + \lambda^2 \Psi'' + \ldots, \tag{11.3}$$

then E', Ψ', etc. are defined by the derivatives of E and Ψ at $\lambda = 0$

$$E' = (\mathrm{d}E/\mathrm{d}\lambda)_0, \; \Psi' = (\mathrm{d}\Psi/\mathrm{d}\lambda)_0,$$

$$E'' = \frac{1}{2}(\mathrm{d}^2 E/\mathrm{d}\lambda^2)_0, \; \Psi'' = \frac{1}{2}(\mathrm{d}^2 \Psi/\mathrm{d}\lambda^2)_0, \text{ etc.} \tag{11.4}$$

Substitution of (11.2) and (11.3) into the Schrödinger equation

$$(\mathscr{H} - E)\Psi = 0, \tag{11.5}$$

gives

$$[\mathscr{H}^\circ + \lambda \mathscr{H}' - (E^\circ + \lambda E' + \lambda^2 E'' + \ldots)]\,(\Psi^\circ + \lambda \Psi' + \lambda^2 \Psi'' + \ldots) = 0. \tag{11.6}$$

If this equation is valid for all values of λ then the functions multiplying each power of λ must individually be zero. For example, taking these terms from (11.6) which are independent of λ we have

$$(\mathscr{H}^\circ - E^\circ)\Psi^\circ = 0, \tag{11.7}$$

and we recognize this as the Schrödinger equation for the unperturbed Hamiltonian $\mathscr{H}^\circ$.

The terms in (11.6) which are the first power in λ may be collected together as

$$\lambda[(\mathscr{H}^\circ - E^\circ)\Psi' + (\mathscr{H}' - E')\Psi^\circ], \tag{11.8}$$

and it follows that

$$(\mathscr{H}^{\circ} - E^{\circ})\Psi' + (\mathscr{H}' - E')\Psi^{\circ} = 0. \tag{11.9}$$

This is known as the *first-order perturbation equation*. It is, like the Schrödinger equation itself, a differential equation, which in principle could be solved to give E' and Ψ'. In practice the equation is usually solved by the same approach as that described in Section 6.3 by choosing an expansion set for Ψ' and finding the coefficients in this. The most convenient functions for the expansion are the solutions of (11.7), which are the unperturbed or zeroth-order states.

Let us now concentrate attention on a particular solution of (11.7), Ψ_m°, E_m°, which under the influence of the perturbation evolves to the wavefunction Ψ_m whose energy is E_m. The first-order perturbation equation for this state is, from (11.9),

$$(\mathscr{H}^{\circ} - E_m^{\circ})\Psi'_m + (\mathscr{H}' - E'_m)\Psi_m^{\circ} = 0. \tag{11.10}$$

We now expand the first-order wavefunction in terms of the unperturbed wavefunctions

$$\Psi'_m = \sum_n c'_n \Psi_n^{\circ}, \tag{11.11}$$

and substituting this into (11.10) we obtain

$$\sum_n c'_n (\mathscr{H}^{\circ} - E_m^{\circ})\Psi_n^{\circ} + (\mathscr{H}' - E'_m)\Psi_m^{\circ} = 0. \tag{11.12}$$

Multiplying (11.12) from the left by Ψ_m° (more strictly its complex conjugate) and integrating over the space of the coordinates gives the equation

$$\sum_n c'_n(H_{mn}^{\circ} - E_m^{\circ}S_{mn}) + (H'_{mm} - E'_m S_{mm}) = 0, \tag{11.13}$$

where the same symbolism has been used for the integrals as in (6.48) and (6.50). Because the expansion functions are eigenfunctions of $\mathscr{H}^{\circ}$, and they also can be taken as normalized, we have, according to (6.35) and (6.40)

$$S_{mn} = \delta_{mn}, \tag{11.14}$$

and

$$H_{mn}^{\circ} = E_n^{\circ}\delta_{mn}. \tag{11.15}$$

It therefore follows that only the last term in (11.13) is non-zero and from this we obtain the result

$$E'_m = H'_{mm} \equiv \int \Psi_m^{\circ} \mathscr{H}' \Psi_m^{\circ} \, dv. \tag{11.16}$$

This is called the *first-order energy*. By analogy with (5.9) it can be considered as the average value of the operator $\mathscr{H}'$ for the unperturbed state Ψ_m°.

The first-order wavefunction (11.11) is obtained by multiplying (11.12) by each of the unperturbed basis functions in turn, other than Ψ_m°. For example, multiplying by Ψ_r° and integrating gives

$$\sum_n c_n'(H_{rn}^\circ - E_m^\circ S_{rn}) + (H_{rm}' - E_m' S_{rm}) = 0, \tag{11.17}$$

which from (11.14) and (11.15) reduces to

$$c_r'(E_r^\circ - E_m^\circ) + H_{rm}' = 0, \tag{11.18}$$

whence

$$c_r' = \frac{H_{rm}'}{E_m^\circ - E_r^\circ}. \tag{11.19}$$

The *first-order wavefunction* is therefore

$$\Psi_m' = \sum_n{}' \left(\frac{H_{nm}'}{E_m^\circ - E_n^\circ} \right) \Psi_n^\circ, \tag{11.20}$$

where the prime on the summation sign means that we exclude $n = m$.

It can be seen from (11.20) that the contribution of any basis function Ψ_n° to the first-order wavefunction depends on two factors. The first is the magnitude of the interaction integral

$$H_{mn}' = \int \Psi_n^\circ \mathscr{H}' \, \Psi_m^\circ \, \mathrm{d}v. \tag{11.21}$$

Following the discussion of such integrals in Chapter 7, we know that a necessary condition for this to be non-zero is that the two basis functions belong to the same symmetry species in the symmetry group of the Hamiltonian $\mathscr{H}$.

The second factor is the term in the denominator $E_m^\circ = -D_n^\circ$. From this it follows that states which in the unperturbed system have energies close to that of Ψ_m° will, other things being equal, make a larger contribution to Ψ_m' than those whose energy is far away. Expression (11.20) is clearly invalid if $E_m^\circ = E_n^\circ$ and H_{mn}' is non-zero. Special perturbation techniques must be used for such cases of degeneracy.

The only other perturbation term that we shall use in this book, and it is the only other one of general interest, is the *second-order energy*. This can be obtained from the terms of (11.16) which are of order λ^2. For a particular state m these are

$$(\mathscr{H}^\circ - E_m^\circ)\Psi_m'' + (\mathscr{H}' - E_m')\Psi_m' - E_m''\Psi_m^\circ = 0. \tag{11.22}$$

Expanding Ψ'' in terms of the basis functions as in (11.11)

$$\Psi_m'' = \sum_n c_n'' \Psi_n^\circ, \tag{11.23}$$

and using (11.11), converts (11.22) into

$$\sum_n [c_n''(\mathscr{H}^\circ - E_m^\circ)\Psi_n^\circ + c_n'(\mathscr{H}' - E_m')\Psi_n^\circ] - E_m''\Psi_m^\circ = 0. \tag{11.24}$$

Multiplying this on the left by Ψ_m° and integrating gives, after using (11.14), (11.15), and (11.16),

$$\sum_n c'_n H'_{mn} - E''_m = 0, \tag{11.25}$$

whence on introducing the expression we have already obtained for the coefficients (11.19) we find

$$E''_m = \sum_n{}' \frac{H'_{mn}H'_{nm}}{E^\circ_m - E^\circ_n}. \tag{11.26}$$

The numerators in (11.26) are all positive and equal to $(H'_{mn})^2$ for real wavefunctions or $|H'_{mn}|^2$ for complex wavefunctions. We therefore conclude that states Ψ°_n which are higher in energy than Ψ°_m give a negative contribution to the second-order energy and states which are lower give a positive contribution. We have already obtained expressions of this type for the case of just two interacting states, (6.69) and (6.70), and we see that these expressions are consistent with the more general perturbation expressions (11.16) and (11.26)

Finally we note that (11.26) like (11.20) is not valid for the case of interacting degenerate states. When there is degeneracy in the states of the unperturbed system then one must first find combinations of such states which do not interact with each other (the integrals H'_{mn} are zero) following which the standard expressions of perturbation theory may be applied. The required combinations can always be found by solving the secular equations (6.74) for each set of degenerate states with $\mathcal{H}'$ as the Hamiltonian.

11.2. The energy levels of many-electron atoms

In Chapters 3 and 4 an account has been given of the electronic structure of the ground states of the atoms based upon the concept of atomic orbitals and the building-up principle. Electronic states are identified in the first place by the location of electrons in atomic orbitals. This defines the so-called *electron configuration* of the state and such configurations are restricted by the Pauli exclusion principle which requires that no orbital can contain more than two electrons.

For partially filled shells of p, d, f orbitals, etc. there is more than one way of allocating electrons to the orbitals and of assigning spins to these electrons subject to the exclusion principle. Not all of these allocations have the same energy because of the repulsion between electrons. For example, two electrons in the same p orbital will have a greater repulsion than two electrons in different p orbitals. Thus a single electron configuration can give rise to several atomic states which will have different energies.

The SCF model of atomic structure, which has been described in Section 3.4, allows for electron repulsion only through a spherical average of the electron density, hence the energy of a configuration does not take account of the specific allocation of electrons to orbitals. However, the spherically averaged electron repulsion accounts for the gross features of atomic energy and the difference between this and the precise repulsion for a specific atomic state can be considered as a perturbation. It is because of the success of this perturbation model that we generally assume that the same set of orbitals can be used for all states arising from a given configuration.

There are further effects which influence atomic energy levels that are magnetic in origin. An electron moving around the nucleus can produce a magnetic field and this will interact with the spin magnetic moment of the electron. This is called *spin–orbit coupling.* Different orientations of the electron spin relative to the orbital magnetic field will have different energies. Spin–orbit coupling is another perturbation which may be applied to the atomic Hamiltonian, using the atomic wavefunctions calculated without spin–orbit coupling as the unperturbed wavefunctions. The size of this particular perturbation increases with increasing charge on the nucleus because the electrons move more rapidly, and hence give rise to bigger magnetic fields, the closer they are to the nucleus. For very heavy atoms (e.g. the actinides) spin–orbit coupling can become a larger perturbation than that due to the non-spherically-averaged electron repulsion.

We can illustrate the points made above for any many-electron atom. Table 11.1 shows in the right-hand column the wavenumbers (units of inverse wavelength in cm^{-1}) of the first few states of the carbon atom as deduced from atomic spectroscopy. Such a unit is conventionally used in electronic spectroscopy and may be converted to energies in kJ mol^{-1} by multiplying by 0.011 96.

The lowest energy configuration of the carbon atom is $1s^2 2s^2 2p^2$ and, as we shall see, this allows for 15 distinct allocations of electrons to the three $2p$ orbitals. These group into five different atomic states whose degeneracy is listed in the table. The degeneracy will be removed if the atom is placed in a magnetic field (Zeeman splitting of a state). The first excited configuration is formed by promoting an electron from the $2p$ shell to the $3s$ to give $1s^2 2p 3s$. For this configuration 12

Table 11.1. The perturbation model of atomic states showing the effect of two perturbations (electron repulsion and spin–orbit coupling) on the first two configurations of the carbon atom

Configuration	Wavenumber/ cm^{-1}	Term	Wavenumber/ cm^{-1}	State	Wavenumber/ cm^{-1}	Degeneracy
$1s^2 2s^2 2p^2$	2 140.6	^3P	29.6	$^3P_0$	0	1
				$^3P_1$	16.4	3
				$^3P_2$	43.5	5
		^1D	10 193.7	$^1D_2$	10 193.7	5
		^1S	21 648.4	$^1S_0$	21 648.4	1
$1s^2 2s^2 2p 3s$	60 775.6	^3P	60 373.4	$^3P_0$	60 333.8	1
				$^3P_1$	60 353.0	3
				$^3P_2$	60 393.5	5
		^1P	61 982.2	$^1P_1$	61 982.2	3
		(effect of electron repulsion)		(effect of spin-orbit coupling)		

electron allocations are possible and four distinct atomic states result as shown in the table.

In the left-hand column of Table 11.1 are listed the wavenumbers of the two configurations as deduced from the averages (weighted by the degeneracies) of the atomic states. The difference between them, 58 635 cm^{-1}, is seen to be large compared with the separation of states arising from a given configuration. These energies can be taken as the unperturbed energies (solutions of equation 11.7) in our scheme.

The second column shows the result of applying the perturbation due to the non-spherically-averaged electron repulsion. The ground configuration divides into three so-called *terms* and the excited configuration into two terms. These terms are given a spectroscopic label that shows the total orbital angular momentum and the total spin angular momentum of the electrons.

The orbital angular momentum of a single electron in an atom is proportional to the l quantum number of the orbital it occupies, and the spectroscopic label of the orbital follows the convention (see Section 3.2) that the quantum numbers

$$l = 0, 1, 2, 3, 4, \ldots$$

are associated with the spectroscopic symbols

$$s, p, d, f, g, \ldots \quad \text{respectively.} \tag{11.27}$$

Likewise the total orbital angular momentum of all the electrons is proportional to an integer quantum number L and is indicated by the corresponding upper case symbol. Thus the quantum numbers

$$L = 0, 1, 2, 3, 4, \ldots$$

are associated with the spectroscopic symbol

$$S, P, D, F, G, \ldots \quad \text{respectively.} \tag{11.28}$$

Atomic orbitals except s and all atomic terms except S have non-zero orbital angular momenta.

A single electron has a spin quantum number $s = ½$ and may have either of two components ($m_s = ½$ or $-½$) along the axis, which is any unique direction in space determined by an external field (see Section 4.1). The spins of two electrons in different orbitals may be parallel or antiparallel so that the total spin quantum number may be $S = 0$ or $S = 1$. The spin multiplicities of these two states $(2S + 1)$ are 1 and 3 respectively. The first is called a singlet and the second a triplet spin state. The spin multiplicity (number of spin components) of a state is indicated as a pre-superfix to the angular momentum symbol thus ^3P (read as triplet P) or ^1D (read as singlet D).

It can be seen from Table 11.1 that for both configurations the lowest energy term is the one having the greatest spin multiplicity. There is a general rule to this effect which was first proposed by Hund, and which has few exceptions; none for ground state configurations. A second rule due to Hund is that if there is more than one term from a configuration with the same spin multiplicity then these terms decrease in energy with increasing L value.

The explanation for Hund's rules lies in the detailed form of the wavefunctions of the individual terms, a subject which is beyond the scope of this book. The explanation usually given is that the order is determined by the decrease in electron repulsion as S and L increase. However, this explanation has been refuted by accurate calculations of wavefunctions in some cases, and it has been found that the small differences in the electron–nuclear energy are larger than the differences in electron repulsion.† However, this may be an argument such as 'which came first, the chicken or the egg', because the changes in electron–nuclear energy can be thought to have their origin in the relaxation of the orbitals to mitigate the electron repulsion.

Only terms which have a non-zero total orbital angular momentum and hence a non-zero field from orbital motion of the electrons) and non-zero total spin, can show spin-orbit coupling. In Table 11.1 only the ^3P terms satisfy this requirement. The effect of spin–orbit coupling is to split a term with quantum numbers (L, S) into $2S + 1$ separate states (if $L \geqslant S$) or $2L + 1$ separate states if $S \geqslant L$. Thus for a ^3P *term* $(L = 1, S = 1)$ we have by either condition 3 states in all. These are indicated by a suffix which shows the *total* angular momentum (orbital plus spin) which is represented by a quantum number given the symbol J. J can take the values

$$J = L + S, L + S - 1, \ldots, L - S \quad \text{if } L \geqslant S$$

or

$$J = S + L, S + L - 1, \ldots, S - L \quad \text{if } S \geqslant L.$$

If either S or L is zero only one state arises from each term and the J value of this is equal to $L + S$.

Notice in Table 11.1 that the splitting of the ^3P terms into states is only a few cm^{-1} and this is much smaller than the splitting of configurations into terms. This is true for most electron configurations of all except the heaviest elements.

Let us now give a closer look at the allocation of electrons to orbitals in the configuration $1s^2 2s^2 2p^2$. The pair of electrons in the $1s$ orbital must, by the Pauli exclusion principle, have opposite spins and therefore only one spatial and spin description of these electrons can be given: the same applies to the two electrons in the $2s$ orbital. In contrast, there are 15 ways of arranging the electrons in the $2p$ orbitals subject to the requirement of the Pauli principle: these are listed in Table 11.2. The three $2p$ orbitals have been identified by their m quantum numbers (defined by 3.26 and 3.27) and their spin components have been identified by opposite facing arrows: $\uparrow \equiv (m_s = \tfrac{1}{2})$, $\downarrow \equiv (m_s = -\tfrac{1}{2})$.

As the m and m_s quantum numbers indicate the components of the orbital and spin angular momenta in some specified direction in space (conventionally the z axis) we can simply add the m values of each electron to find the component of the

†R. P. Messmer and F. W. Birss, *J. Phys. Chem.*, **73**, 2085 (1969);
J. P. Colpa, *Molec. Phys.*, **28**, 581 (1974).

Table 11.2. The fifteen possible arrangements of electrons for the carbon $2p^2$ configuration which are allowed by the Pauli exclusion principle

	$m =$ 1	0	−1	$M_L = \Sigma m$	$M_S = \Sigma m_s$
1	↑↓			2	0
2	↑	↑		1	1
3	↑	↓		1	0
4	↓	↑		1	0
5	↓	↓		1	−1
6	↑		↑	0	1
7	↑		↓	0	0
8		↑↓		0	0
9	↓		↑	0	0
10	↓		↓	0	−1
11		↑	↑	−1	1
12		↑	↓	−1	0
13		↓	↑	−1	0
14		↓	↓	−1	−1
15			↑↓	−2	0

total orbital angular momentum (L), and add the m_s values of each electron to find the component of the total spin angular momentum (S) along the z axis, these are given the symbols M_L and M_S respectively.

It can be seen from Table 11.2 that allocation (1) has $M_L = 2$ and as this is the maximum M_L value in the table it must be one component of a D state $(L = 2)$ whose five components have $M_L = 2, 1, 0, -1, -2$. There is a strict analogy here with the 5-fold degeneracy of d orbitals $(l = 2, m = 2, 1, 0, -1, -2)$. As allocation (1) has $M_S = 0$ it must be a singlet spin state $(S = 0, M_S = 0)$.

Allocation (2) in Table 11.2 has $M_L = 1$ and $M_S = 1$. This must be one component of a ^3P *state* $(L = 1, S = 1)$ whose nine components arise from all pairs that can be formed from the sets $M_L = 1, 0, -1$ and $M_S = 1, 0, -1$.

The ^1D and ^3P states account for 14 of the possible allocations of electrons in the configuration (which allocations belong to which state in detail need not concern us and is a more difficult problem). This leaves only one allocation unaccounted for and the only term which has just one component is ^1S $(L = 0, S = 0)$. We therefore deduce, in agreement with Table 11.1, that the ground configuration of carbon gives rise to the terms ^3P, ^1D, and ^1S. Confirmation that the second configuration in Table 11.1 gives the terms ^1P and ^3P can be obtained by similar arguments.

Atomic spectroscopy is an extensive subject whose theory can be taken in sufficient depth to form a book in itself. In this chapter we have given only a brief survey but a sufficient basis to understand the subject of the next chapter which is the ligand field theory of transition metal complexes.

Chapter 12
Ligand Field Theory

12.1. The concept of strong and weak ligand fields

Transition elements exhibit a much wider variety of chemical bond types than any main group element. We saw, for example, in section 6.7 that they can exist in many different formal oxidation states. They can form complex ions with electronegative elements such as the halogens, or with stable molecules such as H_2O and NH_3. They can form compounds with organic radicals such as CH_3 and with stable organic molecules like benzene. They can form bonds with main group metals or with other transition metals. In short, there is almost no end to the variety of compound they can form.

No one model of the chemical bond will be equally successful in explaining the properties of all compounds of the transition elements. Molecular orbital theory is the most flexible chemical bond theory but for the transition elements it suffers on the one hand from being computationally expensive at the *ab-initio* level and on the other very difficult to parameterize at the semi-empirical level. It is only in recent years that molecular orbital calculations have produced any satisfactory explanations of the structure and spectroscopy of transition metal compounds. In contrast to this, an empirical theory known as ligand field theory has proved very successful for interpreting the properties of transition metal compounds of an important, if limited, class.

The term *ligand* is used for the atoms or molecular groups that surround a transition metal in its complexes. Ligand field theory aims to explain the properties of the complex in terms of the perturbation to the energies and wavefunctions of the isolated transition metal atom or ion arising from the presence of the ligands. Ligand field theory is therefore valid when perturbation theory is valid and that is when the changes in the wavefunctions and energies which are induced by the ligands are relatively small.

Ligand field theory originated from the work of van Vleck in the 1930s, on the spectroscopy of transition metal ions in crystals. It was then referred to as crystal field theory. The colour of ruby, for example, arises from the presence of Cr^{3+} ions as an impurity in aluminium oxide. In the crystal each Cr^{3+} is surrounded by six oxide anions (O^{2-}) and it is the field of these oxide anions which perturbs the energy levels of Cr^{3+}. In crystal field theory this perturbation was originally conceived as being purely electrostatic in origin. Today, in ligand-field theory,we

Table 12.1. The number of $3d$ electrons in the di- and tri-positive ions of the first series of transition elements

	Sc	Ti	V	Cr	Mn	Fe	Co	Ni	Cu	Zn
M^{2+}	1	2	3	4	5	6	7	8	9	10
M^{3+}	0	1	2	3	4	5	6	7	8	9

would recognize that part of this perturbation arises from the overlap of orbitals of the ligands and orbitals of the transition metal.

The transition elements differ from the main group elements in having many stable compounds in which there are incomplete shells of electrons. These electrons occupy orbitals which, to a first approximation, are the *d* orbitals of the free atoms. For the first transition series the valence electrons occupy the $3d$ orbitals but the $4s$ and $4p$ orbitals have energies which are very close to that of the $3d$ orbitals and they are undoubtedly important for bond formation. We shall examine their role in more detail later in this chapter. However, as we shall see, many properties, particularly the spectroscopic and magnetic properties of the complexes, can be understood by considering only the changes in the *d* orbital energies that occur through bonding.

Table 12.1 shows the number of $3d$ electrons in the di- and tri-positive ions of elements of the first transition series. None of these ions has any electrons in the $4s$ or $4p$ orbitals in its lowest energy state.

In section 12.2 we saw that the distribution of two *p* electrons amongst a set of degenerate *p* orbitals is determined by electron repulsion. The same is true for the distribution of electrons amongst the *d* orbitals of a transition metal complex. However, in surrounding the ion with ligands we have introduced more electrons into the problem and we now should include the repulsion between these and the *d* electrons when considering the distribution of the latter amongst the *d* orbitals. There is no evident reason why that arrangement of *d* electrons which is the most stable in the absence of the ligands should also be the most stable in their presence. As we shall see, a detailed analysis supports this superficial judgement. Of course, the effect of the presence of the ligands is not just additional electron repulsion, although this dominates. There is also an attraction between the *d* electrons and the ligand nuclei.

Rather than try to decompose the effect of the ligands into components we shall assume that their resultant may be described by some simple model. The simplest is to assume that each ligand may be represented by a negative point charge. The array of point charges gives rise to a potential field – the ligand field – and we shall be concerned with the effects of such fields on the electrons of the central metal ion. We see that the energies associated with these *d* electrons are subject to two main perturbations – the effect of electron repulsion and the effects of the ligand field. Transition metal complexes in which the ligand field effects dominate are said to be *strong field* complexes. Those in which electron repulsion plays the major role in determining the *d*-electron energy levels are said to be *weak field* complexes.

Although this distinction is valid for complexes of all geometries it is most usefully applied to octahedral complexes – those in which a metal ion is surrounded by six ligands at the corners of an octahedron – and so it is with octahedral complexes containing six identical electrons with which we shall first be concerned.

12.2. Octahedral complexes

Complexes in which six ligands, placed more or less at the corners of a regular octahedron, surround a metal ion are the most numerous of all complexes. The geometry of a regular ML_6^{n+} octahedral complex is shown in Figure 12.1. We have first to assess the effect of the ligand field associated with this arrangement on the d-orbital energies and it is easier to do this if we consider the molecular symmetry.

An octahedron is closely related to a cube. If lines are drawn between the mid-points of the faces of a cube an octahedron is obtained. The cube and the octahedron have the same set of symmetry elements and hence they belong to the same symmetry group which is called the octahedral group, and is given the symbol O_h. The character table of this group is given in Table 12.2. From it we see that d_{xy}, d_{yz}, and d_{zx}, which have the same symmetries as the products xy, yz, and zx respectively, belong to the species T_{2g}. The other two d orbitals, $d_{x^2-y^2}$ and d_{z^2} (which we noted in 3.23 had the symmetry of $3z^2 - r^2$) belong to the species E_g. It is customary to use lower case symbols to describe the symmetry of orbitals and we therefore refer to the two sets of d orbitals as the t_{2g} and the e_g sets.

We now have to determine the relative energies of these two sets of orbitals. Because most ligand atoms bonded to a metal ion carry a net negative charge, the lowest energy set of d orbitals will be that which enables d electrons to be as physically remote as possible from the ligands. We show in Figure 12.2 representative members of the t_{2g} and e_g sets: d_{yz} and $d_{x^2-y^2}$ respectively. From their spatial characteristics there is no doubt that the d_{yz} (and, thus, the t_{2g} set of

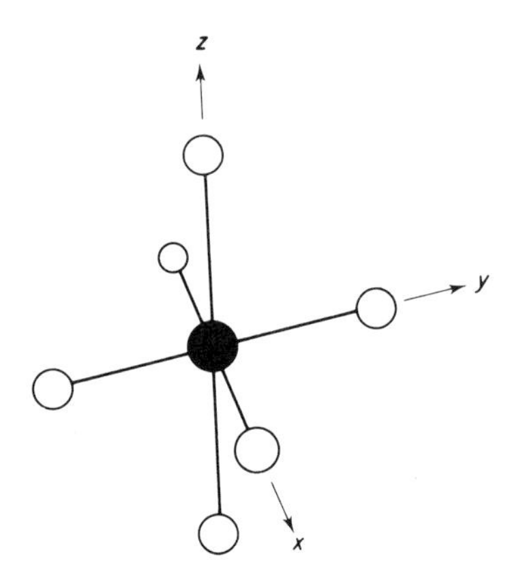

Figure 12.1 Geometry of an ML_6 octahedral complex.

TABLE 12.2 Character table of the O_h symmetry group

O_h	E	$8C_3$	$6C_2$	$6C_4$	$3C_2'$	i	$8S_6$	$6\sigma_d$	$6S_4$	$3\sigma_h$	
A_{1g}	1	1	1	1	1	1	1	1	1	1	$x^2+y^2+z^2$
A_{2g}	1	1	−1	−1	1	1	1	−1	−1	1	
E_g	2	−1	0	0	2	2	−1	0	0	2	$(x^2-y^2, 3z^2-r^2)$
T_{1g}	3	0	−1	1	−1	3	0	−1	1	−1	(R_x, R_y, R_z)
T_{2g}	3	0	1	−1	−1	3	0	1	−1	−1	(xy, yz, zx)
A_{1u}	1	1	1	1	1	−1	−1	−1	−1	−1	
A_{2u}	1	1	−1	−1	1	−1	−1	1	1	−1	
E_u	2	−1	0	0	2	−2	1	0	0	−2	
T_{1u}	3	0	−1	1	−1	−3	0	1	−1	1	(x, y, z)
T_{2u}	3	0	1	−1	−1	−3	0	−1	1	1	

orbitals) will be the more stable. This conclusion is confirmed both by more detailed calculations and by experiment. We shall therefore take the effect of an octahedral ligand field on a set of d orbitals to be that of introducing the separation in energy between the e_g and t_{2g} orbitals shown in Figure 12.3. This separation is conventionally given the symbol Δ.

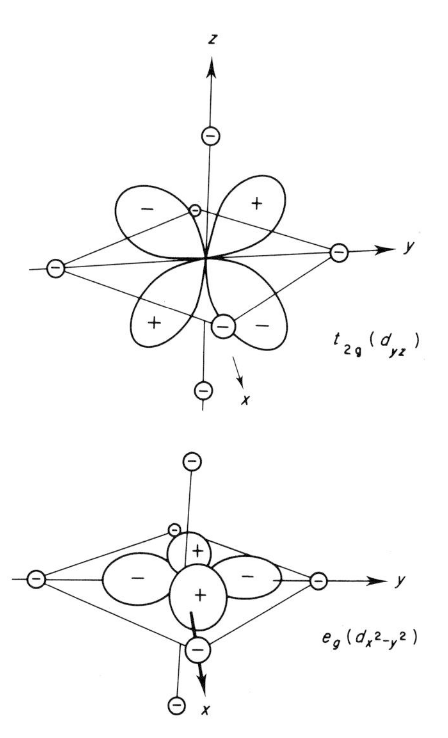

Figure 12.2 The relationship of d_{y^2} and $d_{x^2-y^2}$ orbitals to ligands on the axes.

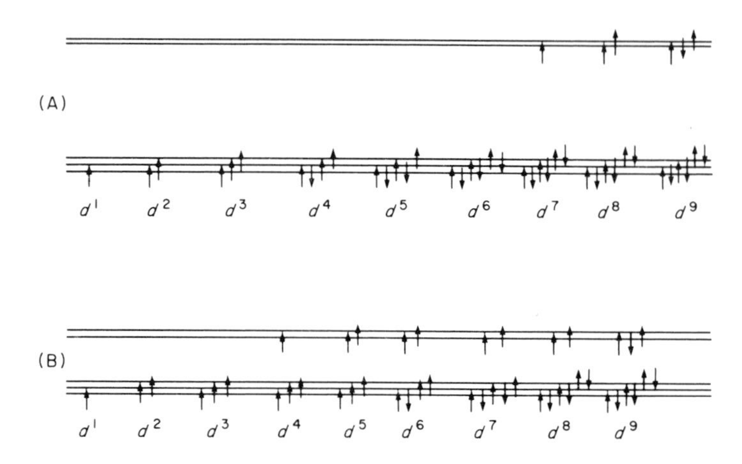

Figure 12.3 Splitting of d orbitals in an octahedral ligand field.

As the ligand field splitting of the orbitals is the dominant effect in strong field complexes, we can be very precise about the lowest energy arrangement of the d electrons. These arrangements are shown in Figure 12.4A. For weak field complexes we can, at this stage, only make qualitatively correct statements. When electron repulsion is dominant we expect that the ground state will be derived from that configuration in which the d electrons are spatially most separated. It is therefore energetically preferable to fill each of the d orbitals singly before adding a second electron to any orbital. The allocations of d electrons based on this argument are shown in Figure 12.4B. It is evident that a very clear distinction exists between weak and strong field complexes having d^4, d^5, d^6, or d^7 configurations. As the magnetic properties of complexes are dependent on the number of unpaired electrons the two cases may be distinguished by magnetic measurements.

If we ignore for the moment the contribution to the energy of any configuration from electron repulsion, then the relative energies of the configurations are

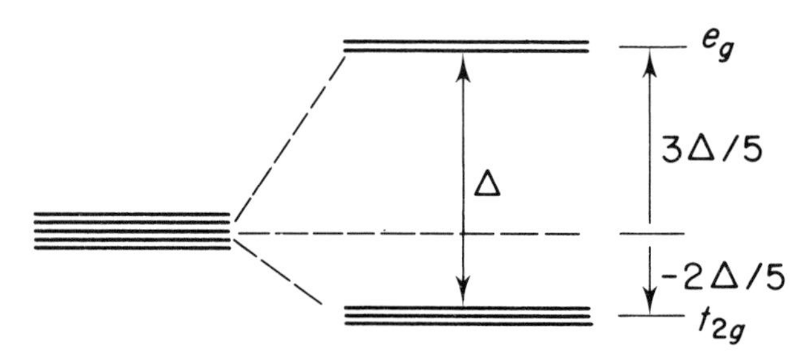

Figure 12.4 Lowest energy arrangement of electrons in the d orbitals: (A) strong ligand field, (B) weak ligand field.

obtained by adding together the ligand field contributions to the orbital energies. It is customary to define the origin of the ligand field energy so that it is zero if all the d orbitals are completely filled with electrons. This accounts for the convention shown in Figure 12.3 to have the t_{2g} orbitals stabilized by $\frac{2}{5}\Delta$ and the e_g orbitals destabilized by $\frac{3}{5}\Delta$ by the ligand field.

Table 12.3 shows the ligand field stabilization of strong and weak field complexes in terms of the parameter Δ. The appropriate value of Δ for each ligand is usually determined from electronic spectroscopy, as we shall see later. However, there is evidence to support the concept of ligand field stabilization from ground state properties.

Figure 12.5 shows the relationship between the enthalpies of hydration of dipositive transition metal ions as a function of the number of d electrons (Sc^{2+} is not included as it is unstable). There is independent evidence from spectroscopy to show that for all these ions the metal is surrounded by six water molecules and these are presumably in an approximately octahedral arrangement around the metal.

It can be seen from Figure 12.5 that the hydration energies increase with atomic number of the metal, but the increase is far from being regular. The most striking

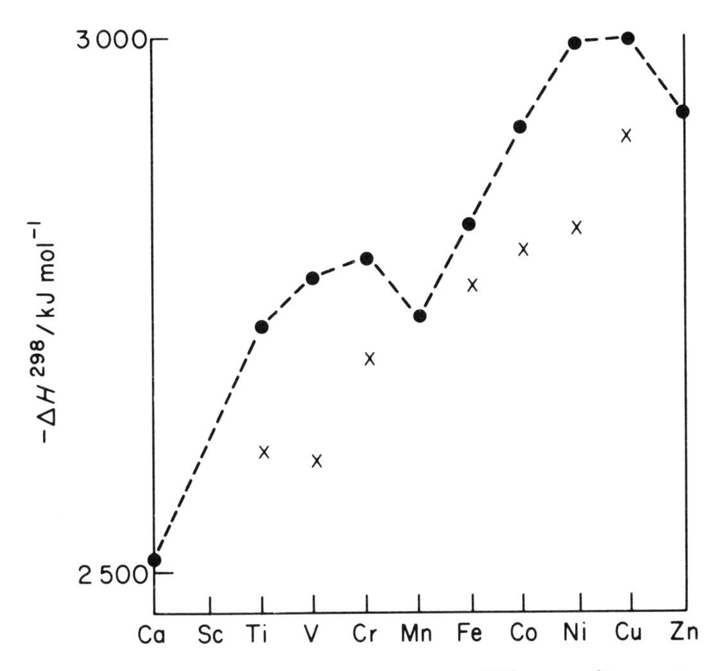

Figure 12.5 Heats of hydration ($-\Delta H^{298}$) of M^{2+} ions [P. George and D. S. McClure, *Progr. Inorg. Chem.*, **1**, 381 (1959)]. Crosses show values corrected for ligand field stabilization using $\Delta = 145$ kJ mol^{-1} (12 000 cm^{-1}). There is substantial uncertainty ($\sim$100 kJ mol^{-1}) in the values for Ti^{2+} and V^{2+}.

feature is a dip in the curve at Mn^{2+}, and this ion we note has the d^5 configuration. The pattern can in fact be qualitatively explained by supposing that there is a contribution to the hydration energy which is linearly related to the atomic number, but superimposed on this is an energy that depends on the ligand field energy of the metal d electrons. For this to be correct we must further assume that water is a *weak-field ligand*. Only with this assumption can we explain the dip for the d^5 configuration. There is again independent evidence from spectroscopy to support this. Figure 12.5 also shows the enthalpies corrected for a ligand field stabilization using the value $\Delta = 145$ kJ mol^{-1} (12 100 cm^{-1}). In fact, Δ is known to vary to a small extent from one metal ion to another so that the correction is only approximate.

Figure 12.6 shows the lattice energies of the transition metal halides of general formula MX_2. In the solid each metal ion is surrounded by an octahedron of halide ions. A comparison with Figure 12.5 shows the generality of the concept of a ligand field stabilization energy. The halide ligands, like water, are clearly behaving as weak-field ligands.

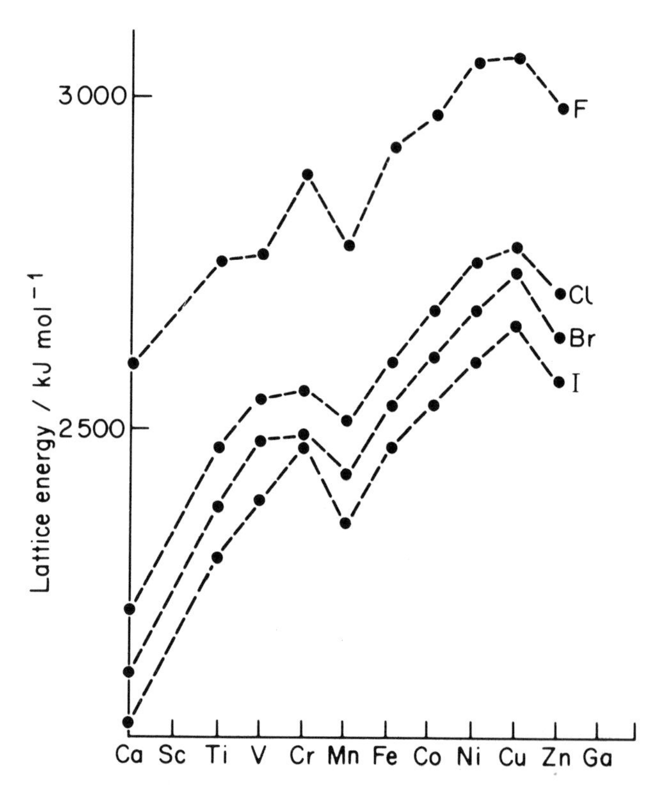

Figure 12.6 Lattice energies of M^{2+} halides [P. George and D. S. McClure, *Progr. Inorg. Chem.*, 1, 381 (1959)]. There is substantial uncertainly in the values for Ti^{2+} and V^{2+}.

Table 12.3. Ligand field stabilization energies for strong and weak-field complexes based upon the configurations shown in Figure 12.4. Units $\Delta/5$

d^n n	1	2	3	4	5	6	7	8	9	10
strong	2	4	6	8	10	12	9	6	3	0
weak	2	4	6	3	0	2	4	6	3	0

We shall now study the balance between electron repulsion and ligand field strength in more detail by considering the case of ions with two d electrons. This d^2 configuration will show features common to all configurations with more than one d electron, but is somewhat simpler than most others.

We start with the strong field case. The energies of the states are determined primarily by their electron configurations, that is, the various ways of allocating electrons to the t_{2g} and e_g orbitals. We must then consider the effect of the minor perturbation which is the repulsion between these electrons. There are three configurations for the two d electrons and in order of increasing energy, each separated by Δ from its predecessor, they are

$$t_{2g}^2, t_{2g}e_g, e_g^2. \tag{12.1}$$

We have now completed the crystal field part of the problem. What remains is the perturbation caused by electron repulsion. Just as the p^2 configuration gives rise to a variety of terms or states so also does each of the three configurations listed above. Let us consider the $t_{2g}e_g$ configuration. There is no possibility of the two electrons being in the same orbital so that no spin restrictions arise. Each orbital state that we generate may be associated with a spin singlet or with a spin triplet. The orbital states are easily determined. They correspond to all possible combinations of t_{2g} and e_g one-electron wavefunctions (to give two-electron wavefunctions in product form). The symmetries of the resulting states are obtained from the direct product of T_{2g} and E_g which can be shown, according to the rules given in section 7.3, to be

$$T_{2g} \times E_g = T_{1g} + T_{2g}. \tag{12.2}$$

Including the spin functions in an obvious way we obtain the states

$$^3T_{1g}, {}^3T_{2g}, {}^1T_{1g}, \text{and } {}^1T_{2g}. \tag{12.3}$$

The states arising from the other two configurations in (12.1) are less readily determined. This is because if we allow either spin function with any orbital arrangement we would, for some states, be allocating electrons with parallel spins to the same orbital. Fortunately, we can easily decide which spin functions are allowed for each orbital arrangement by a similar analysis to that given in section 11.2 for the p^2 configuration.

The direct product gives us the possible spatial symmetry of the wavefunction but tells us nothing about the permissible spin states. We have for the e_g^2

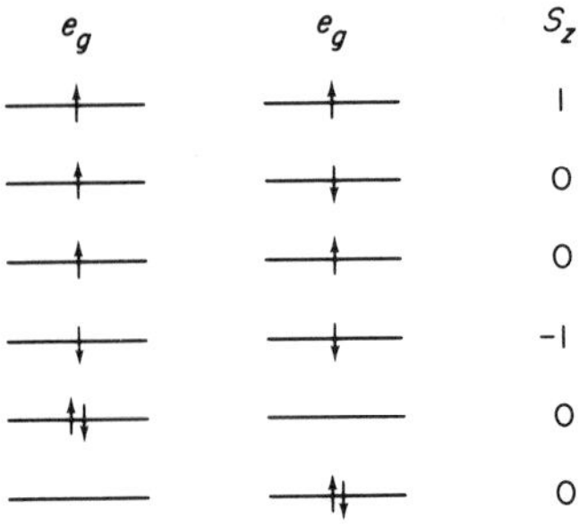

Figure 12.7 The six ways of allocating two electrons to e_g orbitals. S_z is the net component of the spin in the z direction.

configuration

$$E_g \times E_g = A_{1g} + A_{2g} + E_g. \tag{12.4}$$

Figure 12.7 shows the possible arrangement of two electrons in the e_g orbitals, six in all. There is clearly a triplet state among these, because there are arrangements with parallel spins; but only one triplet state. It must either be ${}^3A_{1g}$ or ${}^3A_{2g}$. ${}^3E_g$ would have a degeneracy of six and would not leave us any singlet states and these obviously exist by inspection of the spins of the six arrangements. The only question to decide then is whether the triplet state is unchanged by all operations of the O_h group (as is A_{1g}) or not.

It can be seen from Table 12.2 that the species A_{1g} and A_{2g} can be distinguished by their behaviour under the operations C_2, C_4, σ_d, and S_4. C_4, which is rotation about either the x, y, or z axes by $\pi/2$, is perhaps the simplest to examine.

The component of the triplet state with $S_z = 1$ has one electron in each of the e_g orbitals. Consider a C_4 rotation about the z axis. This leaves d_{z^2} unchanged and changes the sign of $d_{x^2-y^2}$. It follows that the wavefunction for the two electrons together must change sign, and hence its character under C_4 is -1.

We have therefore shown that the triplet state cannot belong to the totally symmetric species of the O_h group and hence it must, by exclusion, be ${}^3A_{2g}$. Thus the configuration e_g^2 gives rise to the states

$${}^3A_{2g},\ {}^1A_{1g},\ \text{and}\ {}^1E_g. \tag{12.5}$$

The final configuration of interest in the d^2 problem is t_{2g}^2 and this is more difficult to analyse. Firstly it can be shown that there are fifteen possible arrangements of two electrons in these three orbitals (c.f. the p^2 case) and the direct product rule gives

$$T_{2g} \times T_{2g} = A_{1g} + E_g + T_{1g} + T_{2g}. \tag{12.6}$$

We have then to decide the possible spin states that go with these orbital symmetries.

There are rigorous group theoretical techniques that can be used to decide this question but they are outside the scope of this book. There are also rather devious

Table 12.4. Terms arising from two electrons in the set of e_g and t_{2g} orbitals

e_g^2	:	${}^3A_{2g}, {}^1A_{1g}, {}^1E_g$
$t_{2g}e_g$	:	${}^1T_{1g}, {}^1T_{2g}, {}^3T_{1g}, {}^3T_{2g}$
t_{2g}^2	:	${}^1A_{1g}, {}^1E_g, {}^1T_{2g}, {}^3T_{1g}$

arguments that can be used but they have no generality. We will therefore limit ourselves to quoting the result which is that the states are

$$ {}^1A_{1g}, {}^1E_g, {}^1T_{2g}, \text{and } {}^3T_{1g}. \tag{12.7}$$

Note that the total degeneracy (the product of spin and space degeneracy) is fifteen, as it should be.

We have now obtained all of the terms arising from the configurations given in equation (12.1); they are listed in Table 12.4. These terms differ in energy because of electron repulsion; we shall discuss this immediately in the context of the weak-field limit but note that according to Hund's rules we would expect that the lowest energy state arising from each of the configurations would be a spin triplet.

We shall briefly detail the consequences of electron repulsion in the context of weak-field complexes; for these it is the dominant perturbation. The terms which are derived from the d^2 configuration are ${}^1G$, ${}^3F$, ${}^1D$, ${}^3P$, and ${}^1S$. Notice that, just as for the e_g^2 and t_{2g}^2 configurations so too for the d^2 configuration, a given orbital state appears either as a spin singlet or as a spin triplet but not as both. This is a general rule for all such two-electron configurations.

Table 12.5 shows the energy levels which arise from the d^2 configuration of V^{3+}. Spin–orbit coupling splits the ${}^3F$ and ${}^3P$ terms into several components because both have non-zero spin and orbital angular momenta (see section 12.2). However, for the weak ligand field analysis spin–orbit coupling is a small perturbation to be considered, if at all, after the ligand field, and it can be removed to give the term

Table 12.5. States arising from the d^2 configuration of V^{3+} and the resulting term energies

	J	State energy/cm^{-1}	Term energy/cm^{-1}
${}^3F$	2	0	419
	3	318	
	4	730	
${}^1D$	2	10 960	10 960
${}^3P$	0	13 121	13 344
	1	13 238	
	2	13 453	
${}^1G$	4	18 389	18 389

energies shown in the last column of Table 12.5. The term energy is the weighted average of its spin–orbit components, the weighting being the degeneracy of the state $(2J + 1)$.

Ligand field theory, being an empirical theory, contains parameters which represent the differences in term energies of the free atom. In the d^2 case the most important terms are ^3P and ^3F, and the difference in their energies we shall give the symbol B. Other parameters would be needed to specify the positions of the singlet states.

Our next task is to consider the effect of the ligand field perturbation on the ^3F and ^3P states, although our discussion will be equally applicable to the spin singlets. Indeed, since the crystal field acts directly on the orbital, but not spin wavefunctions, spin states are irrelevant in this treatment.

There is a strong similarity between the perturbation by a ligand field of orbitals having a quantum number l and states having a quantum number L. These quantum numbers determine the way in which the orbital or the state changes on rotation about an axis passing through the nucleus. Thus the characters for the rotation operations of the group will be the same for d orbitals and for D states, and the same for p orbitals and for P states. The characters for the other operations of inversion or reflection or improper rotation may, or may not, be the same.

We can see from Table 12.2 that p orbitals are not split by an octahedral field and belong in the O_h group to the T_{1u} species. T_{1u} and T_{1g} have the same characters for the rotation and hence a P state will belong to either T_{1u} or T_{1g}. For a P state formed by electrons in d orbitals the species must be T_{1g} because d orbitals are individually g.

The character table does not show how f orbitals or F states will be split in an octahedral field. There is, however, a simple general formula for the character of the operation of rotation by an angle $2\pi/n$ for a state having quantum number L and this is†

$$\chi(C_n) = \sin\,(2L + 1)\frac{\pi}{n}\Big/\sin\frac{\pi}{n}. \tag{12.8}$$

For an F state $(L = 3)$ the characters of C_2, C_3, and C_4 operations are therefore -1, $+1$, and -1 respectively. As there are seven components to the state the identity operation has character 7, and as d orbitals are symmetric to inversion each component of an F state formed from d orbitals has a character $+1$ under i so that the total set has a character $+7$. The characters for an F state under the first six

†This formula is obtained by noting that rotation of a function $e^{iM\varphi}$ by an angle α gives the new function $e^{iM(\varphi+\alpha)} = e^{iM\alpha}e^{iM\varphi}$. That is, the rotation is the same as multiplying the function by $e^{iM\alpha}$. The character of the C_n $(n = 2\pi/\alpha)$ operation for this function is therefore $e^{2iM\pi/n}$. The set of L states, $(2L + 1)$ in all, have angular functions which are $e^{iM\varphi}$ with $M = L$, $L - 1, \ldots, -L$. The sum of the characters for all components of the state are therefore

$$\sum_{M=L}^{-L} e^{2iM\pi/n}$$

and this can be summed as a geometric progression to give the answer 12.8.

operations of the O_h group are therefore

E	$8C_3$	$6C_2$	$6C_4$	$3C_2'$	i
7	1	−1	−1	−1	7 .

(12.9)

We can ignore the rest of the group, as (12,9) is sufficient to show that the symmetry species contained in this reducible representation are $A_{2g} + T_{1g} + T_{2g}$. This follows by adding the characters of these three species in the following way

	E	$8C_3$	$6C_2$	$6C_4$	$3C_2'$	i
$\chi(A_{2g})$	1	1	−1	−1	1	1
$\chi(T_{1g})$	3	0	−1	1	−1	3
$\chi(T_{2g})$	3	0	1	−1	−1	3
Sum	7	1	−1	−1	−1	7

(12.10)

Table 12.6 summarizes the application of such an analysis to the most important atomic terms that can arise from d^n configurations

Although we have shown by symmetry how the terms from the d^2 configuration are split by the octahedral field we have as yet made no decision on the relative order of the energies of the components. This is most easily done by connecting the weak and strong field limits. We shall do this only for the triplet states as these give the most important low-energy states, but a similar analysis can be made for the singlet states.

Figure 12.8 summarizes the situation reached so far in our analysis. We have taken in the weak field limit the ^3F term to lie below the ^3P (in accord with Table 12.5) and in the strong field limit the configurations are given their strong field separations. We note that there is only one $^3A_{2g}$ and one $^3T_{2g}$ state in each limit and hence these may be joined in an unambiguous way by straight lines as shown. Our justification for using straight lines is that the strong field analysis has already shown that the e_g^2 configuration is destabilized by $6\Delta/5$ and the $t_{2g}e_g$ configuration

Table 12.6. Splitting of d^n terms in an octahedral field

Term	Orbital degeneracy $(2L + 1)$	Components under O_h
S	1	A_{1g}
P	3	T_{1g}
D	5	$E_g + T_{2g}$
F	7	$A_{2g} + T_{1g} + T_{2g}$
G	9	$A_{1g} + E_g + T_{1g} + T_{2g}$
H	11	$E_g + 2T_{1g} + T_{2g}$
I	13	$A_{1g} + A_{2g} + E_g + T_{1g} + 2T_{2g}$

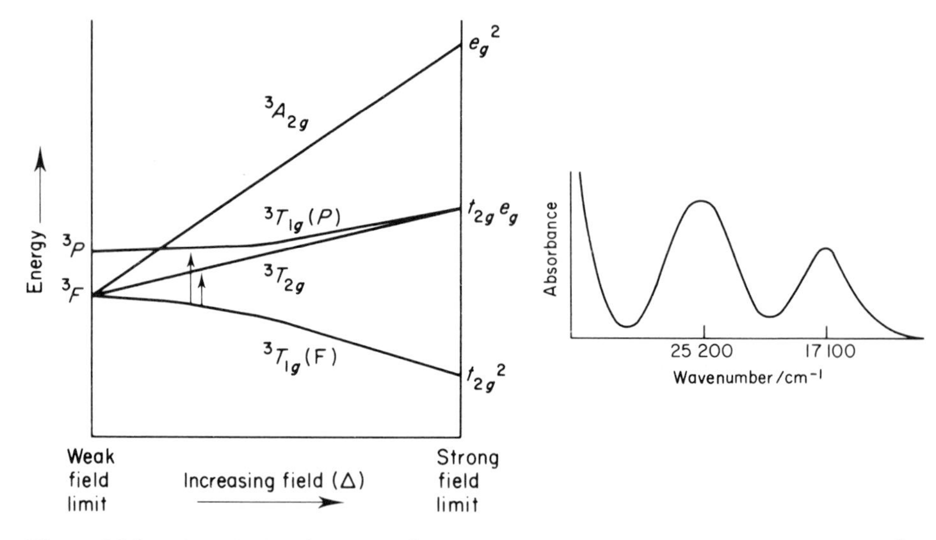

Figure 12.8 Correlation between the weak and strong field triplet states of the d^2 configuration. Arrows show the transitions that have been detected spectroscopically. The absorption spectrum of $[V(H_2O)_6]^{3+}$ is shown on the right.

destabilized by $\Delta/5$. Thus the slopes of the two lines (which are functions of Δ) should be in the ratio of 6:1.

The ambiguity that has not yet been resolved is how we connect or correlate the two $^3T_{1g}$ states. One of these originates from ^3P and one from ^3F. In the figure we have joined ^3P to $t_{2g}e_g$ and ^3F to t_{2g}^2 and we must justify this. The justification lies in an important general rule, which has wide application in quantum mechanics, called the *non crossing rule*.

Let the wavefunctions of the two $^3T_{1g}$ states in the weak field limit be Ψ_a (from ^3F) and Ψ_b (from ^3P). These two wavefunctions are orthogonal because they represent different atomic states. A general $^3T_{1g}$ wavefunction is of the form

$$\Psi = c_a\Psi_a + c_b\Psi_b \tag{12.11}$$

and we can, in principle, calculate the coefficients and energies of the two $^3T_{1g}$ states as functions of the ligand field strength Δ by the variational method, as described in Section 6.3.

The solution follows the pattern described in Section 6.3 for a two-variable problem. The energies are given as the roots of the determinant

$$\begin{vmatrix} H_{aa} - E & H_{ab} \\ H_{ba} & H_{bb} - E \end{vmatrix}, \tag{12.12}$$

which from (6.66) are

$$E = \frac{1}{2}\{(H_{aa} + H_{bb}) \pm [(H_{aa} - H_{bb})^2 + H_{ab}{}^2]^{1/2}\}. \tag{12.13}$$

H_{aa}, H_{bb}, and H_{ab} will all be functions of the ligand field parameter Δ. We note the important result that the two solutions of (12.13) will be the same only if the term involving the square root is zero; that is,

$$(H_{aa} - H_{bb})^2 + H_{ab}^2 = 0. \tag{12.14}$$

However, as this equation contains two squared terms it will be satisfied only if both of these terms are simultaneously zero;

$$H_{aa} - H_{bb} = 0, \tag{12.15}$$

and

$$H_{ab} = 0. \tag{12.16}$$

There is no reason why $H_{aa} - H_{bb}$ and H_{ab} should not be independent functions of Δ hence we conclude that (12.15) and (12.16) will not simultaneously be satisfied except by accident. In other words, the two solutions of (12.13) will never be equal, except by accident. If the two solutions start out at different energies in the weak field limit, and they are never equal for any value of Δ, the curves that join them to the strong field limits will not cross.

The non-crossing rule was first proposed for the potential energy curves of diatomic molecules and it therefore has considerable importance in spectroscopy. The proof we have given follows similar lines to one given by Teller, and although this has been criticized for lack of rigour, the non-crossing rule is well-established from experiment and from calculation.

In the case when the two basis functions have different symmetries H_{ab} will be identically zero and under these circumstances it is generally possible to find a point where (12.15) will be satisfied. Thus potential energy curves or correlation lines that are associated with states of *different* symmetry can cross.

We return to the d^2 configuration. In the strong field limit the lower $^3T_{1g}$ state, which we identify as $^3T_{1g}$ (F) has an energy which should vary as $-4\Delta/5$, and the upper $^3T_{1g}$ (P) state energy should vary as $\Delta/5$. It can in fact be shown† that the determinant (12.12) which represents the interaction between the two states has the form

$$\begin{vmatrix} -\frac{3}{5}\Delta - E & \frac{2}{5}\Delta \\ \frac{2}{5}\Delta & B - E \end{vmatrix} = 0, \tag{12.17}$$

where B is the $^3P - {}^3F$ energy separation in the free ion.

The solutions of (12.17) are

$$E = \frac{1}{2}\left[B - \frac{3\Delta}{5} \pm \left(B^2 + \frac{6}{5}B\Delta + \Delta^2\right)^{1/2}\right], \tag{12.18}$$

†J. N. Murrell, S. F. Kettle, and J. M. Tedder, *Valence Theory*, 2nd edition, p. 230, Wiley, 1970.

and by expanding the square root either in the weak-field limit ($B \gg \Delta$) or in the strong-field limit ($B \ll \Delta$), we obtain the following results:

weak-field limit	strong-field limit
$E = B$	$E = \frac{4B}{5} + \frac{\Delta}{5}$
or $E = -\frac{3\Delta}{5}$	or $E = \frac{B}{5} - \frac{4\Delta}{5}$.

(12.19)

One way of looking at this result is that in the weak-field limit the $^3T_{1g}$ (F) state has, on average a population of 9/5 electrons in the t_{2g} orbitals and 1/5 in the e_g orbitals, whereas the $^3T_{1g}$ (P) state has 4/5 electrons in e_g and 6/5 in t_{2g}. These fractional populations account for our hesitation earlier in this chapter of assigning definite orbital populations in the weak-field limit (see Figure 12.4B).

Figure 12.8 shows the absorption spectrum of $[V(H_2O)_6]^{3+}$ and this is interpreted as arising from transitions from the ground state to the first two excited triplet states. Values of $\Delta = 18\,400\ cm^{-1}$ and $B = 9\,400\ cm^{-1}$ give agreement with the frequencies of these two transitions. The value of B is considerably smaller than the $12\,925\ cm^{-1}$ determined from the free-ion spectrum so we conclude that the electron repulsion parameters also depend on the nature of the ligands. The reason for this will be examined in more detail later.

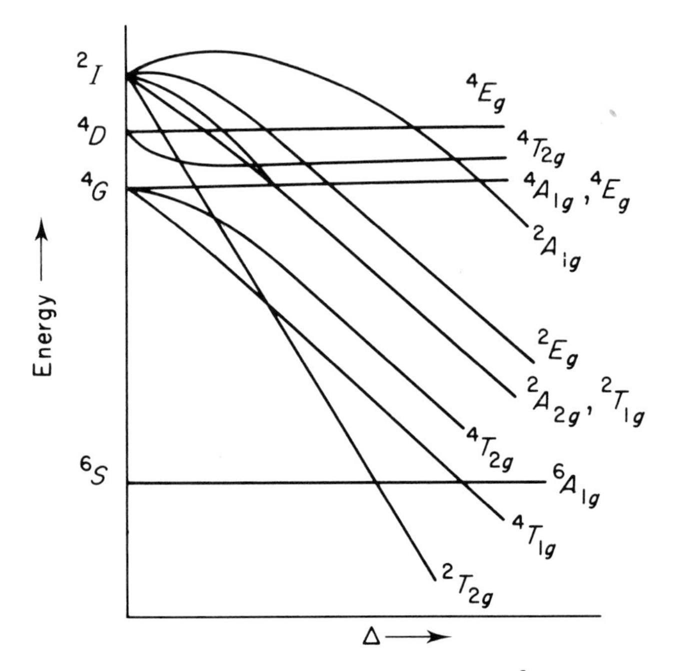

Figure 12.9 Lowest energy levels for the d^5 configuration in octahedral fields.

The type of analysis we have given for d^2 has been thoroughly investigated for all d^n configurations and figures such as 12.8 are valuable for the interpretation of spectroscopic data. Figure 12.9 shows the diagram for the lowest energy terms of the d^5 configuration (d^5 is a complicated case as it gives a total of sixteen terms).

The ground state of the d^5 free ion is a sextet state (^6S) which has five unpaired electrons and the other states are spin quartets and doublets. As this is the only sextet state in the configuration it neither splits, nor interacts with any other state in the configuration. Its energy is therefore independent of the ligand field.

Notice that in accord with our earlier analysis the ground state of a d^5 system changes from high spin in the weak-field limit to low spin in the strong-field limit. For quantitative work it is convenient to plot the energies relative to the ground state as zero, and such diagrams will show a discontinuity when the symmetry of the ground state changes. If the ground state is a curved line, then this change of origin will introduce curvature into lines which are straight when plotted as in Figure 12.8. It also makes the diagrams more useful if they are plotted as a function of Δ/B where B is an electron-repulsion parameter (and equal to $\frac{1}{15}$ of the B of 12.17 *et seq*.). If this is done one diagram can be used for several ions. However, if there is more than one important electron repulsion parameter, as there usually is, then a given diagram is only relevant to one particular set of such parameters. Diagrams of this type were first suggested by Tanabe and Sugano. Figure 12.10 shows a typical Tanabe–Sugano diagram for the d^5 configuration.

We have seen that the pattern of energy levels for the strong and weak-field limit of octahedral complexes are quite different. By measuring the magnetic susceptibility of the ground state we can determine whether the ground state is high-spin or low-spin. By measuring frequencies of the absorption bands, we can determine the value of Δ for the ligand.

It has been found that for the common oxidation states of +2 and +3, ligands can be listed in order of increasing Δ irrespective of the metal ion to which they are complexed. This list is called the *spectrochemical series*; an abbreviated list is

$$I^- < Br^- < Cl^- < F^- < OH^- < H_2O < \text{pyridine} < NH_3 < NO_2^- \text{ (N bonded)} < CN^-. \tag{12.20}$$

It might be thought that it would be possible to draw a line through the above list such that all ligands to the left are weak-field and all to the right are strong-field. This cannot, however, be done because, although Δ is roughly a constant for a given oxidation state, it differs from one oxidation state to another. Values of Δ for M^{3+} ions are larger, sometimes twice as large, as Δ values for M^{2+} ions. Moreover, the value of Δ required to switch from high-spin to low-spin ground states is different for each configuration.

It can be seen from Table 12.3 that the gain in ligand field energy on switching from a high- to low-spin state is twice as great for d^5 and d^6 than for d^4 and d^7. Thus, providing the electron repulsion energy is comparable for each case, we would expect to obtain low-spin d^5 and d^6 complexes for smaller values of Δ than for d^4 and d^7. In summary, the situation is rather complicated. The only general

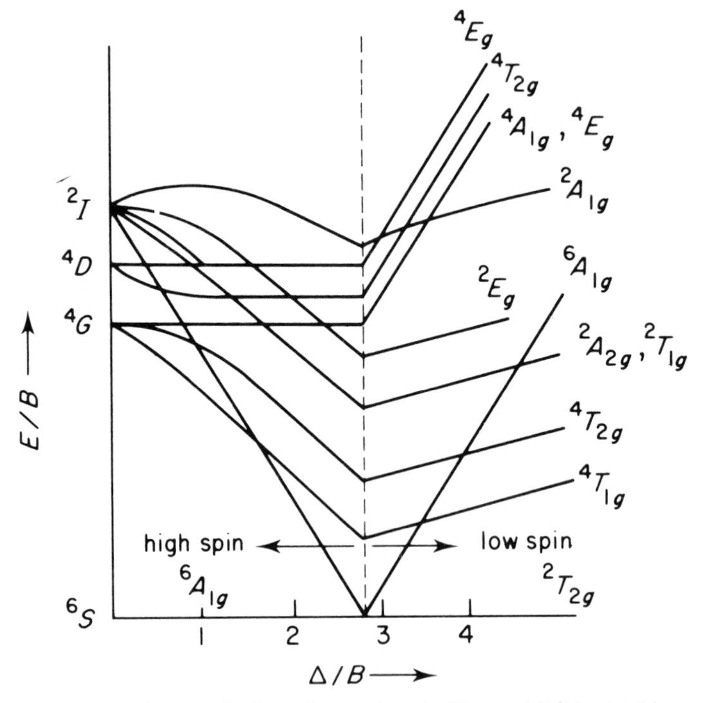

Figure 12.10 A similar plot to that in Figure 12.9 but taking the ground state to have zero energy and the ligand field parameter to be Δ/B.

rules that can be given are that CN^- is always a strong-field (low-spin) ligand for the first transition series and that the halogens are always weak-field (high-spin). H_2O is almost invariably weak-field but NH_3 can be either weak or strong depending on the metal ion.

Table 12.7. Δ/cm^{-1} values for octahedral aquo-complexes (Dunn, McClure, and Pearson, *Some Aspects of Crystal Field Theory*, Harper and Row, 1965)

M	$\Delta(M^{2+})$	$\Delta(M^{3+})$
Ti	–	20 300
V	11 800	18 000
Cr	14 000	17 600
Mn	7 500	21 000
Fe	10 000	14 000
Co	10 000	–
Ni	8 600	–
Cu	1 300	–

Table 12.8. Δ/cm^{-1} values for octahedral complexes for different ligands (after Jorgensen, Thesis, Copenhagen, 1957)

	Ligand Cl^-	H_2O	NH_3	CN^-
Cr^{3+}	13 600	17 400	21 600	26 300
Ni^{2+}	7 300	8 500	10 800	–

Table 12.7 gives some vales of Δ for octahedral aquo-complexes of the first transition series and Table 12.8 compares values for different ligands for the same metal ion. Δ is conventionally given in spectroscopic units (cm^{-1}).

The electron-repulsion integrals of second- and third-row transition metal ions are considerably smaller than those of first-row ions and hence most of their complexes are low-spin, even for halogen ligands.

12.3. Holes and electrons

It might be thought that a separate analysis is needed for each d orbital configuration. Fortunately this is not so because of the similarity between d^n and d^{10-n} configurations. We have already met the concept of positive holes in Section 10.4 when they were used to represent unfilled levels in an energy band. We use the same concept here by considering a d^{10-n} electron configuration as a d^n configuration of positive holes.

We first note that the order of term energies for a d^n electron configuration is the same as the order of term energies for a d^n positive hole configuration. This is because the order of term energies depends on electron repulsion and hole–hole interaction is, like electron–electron interaction, repulsive. Thus the lowest energy term for a d^8 ion (e.g. Ni^{2+}) is ^3F and the first-excited term is ^3P.

In Figure 12.11A we show the three strong-field d^2 configurations which we have already discussed; there are also three d^8 strong field configurations and these are shown in Figure 12.11B. It can be seen that there is a close similarity between these diagrams. This similarly is most clearly shown by the hole representation of Figure 12.11C, each hole being represented by a filled circle. The *most* stable d^2 electron configuration corresponds to the *least* stable d^8 hole configuration, and *vice versa*. Qualitatively we can say that if a ligand-field repels electrons then it will attract positive holes.

It follows that our qualitative discussion for the d^2 case is, with inversion of the ligand-field splitting, equally applicable to the d^8 case. The d^8 pattern is not in total the inverse of d^2 because the order of atomic terms is the same. Figure 12.12 shows the relationship between the two. In order of increasing energy, the d^8 energy levels are $^3A_{2g}$, $^3T_{2g}$, $^3T_{1g}$ (F), and $^3T_{1g}$ (P). Simple ligand-field arguments suggest that the two $^3T_{1g}$ curves would cross, but this is forbidden by the non-crossing rule. Hence the curvature of these two states as a function of Δ is much greater than in the d^2 case. The $^3T_{1g}$ (F) state, which has a slope of $-\frac{3}{5}\Delta$ in the weak-field limit, has a slope of $\frac{6}{5}\Delta$ in the strong-field limit: it correlates with

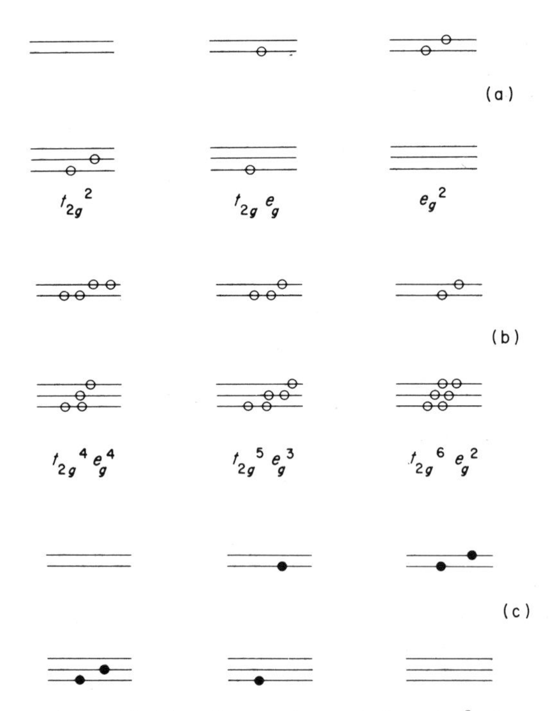

Figure 12.11 Comparison of d^2 and d^8 electron configurations, (A) and (B), and d^2 hole configuration (C).

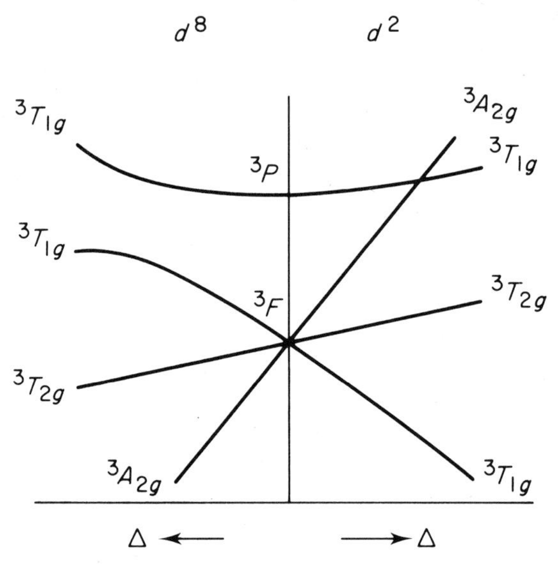

Figure 12.12 Comparison of octahedral splitting for the d^2 and d^8 configurations.

the e_g^2 hole configuration in d^8 compared with the t_{2g}^2 electron configuration in d^2.

Study of the hole–electron parallel reveals that it is of even further applicability. Because a half-filled shell is spherically symmetric, d^7 $(=d^{5+2})$ weak-field complexes follow the d^2 pattern whilst d^3 follows the d^8 pattern. Similarly, the d^1 pattern (where there is a splitting into a $^2T_{2g}$ state with energy $\frac{2}{5}$ Δ and a ^2E_g state with energy $-\frac{3}{5}$ Δ) is followed by weak-field d^6 but inverted for d^9 and d^4. There will, of course, be differences in spin multiplicity in the various cases but these follow from the spin multiplicity of the free ion terms, since the crystal field does not interact directly with electron spins. The hole–electron analogy has enabled us qualitatively to cover all of the weak-field cases in terms of the d^1 and d^2 patterns except that of d^5 which is a special case.

Although we shall not discuss them in detail here, the strong-field terms for the d^4, d^5, d^6, and d^7 configurations may be derived in a manner similar to that used above. Thus in the strong-field d^5 case, the ground state term will come from the t_{2g}^5 configuration. But this is just the t_{2g}^1 hole configuration and so, just as the d^1, t_{2g}^1, configuration gives rise to a $^2T_{2g}$ term, so too does the t_{2g}^5 configuration. The relative energies of terms arising from other configurations can either be worked out similarly or else determined by noting their correlation with weak-field states.

12.4. Complexes of other geometries

In the case of non-octahedral complexes with six-fold coordination of the central metal atom the relative d-orbital splittings (and the subsequent term splittings) may usually be readily obtained by considering small changes from the octahedral structure. Consider the case in which two *trans* ligands in an octahedral complex are moved slightly further away from the metal atom than the remaining four. This is called a tetragonal distortion. The symmetry of the complex is now D_{4h} and the character table of this is given in Table 12.9. It is evident from this character table that the t_{2g} orbitals of the octahedron (d_{xy}, d_{yz}, d_{zx}) split into an e_g degenerate pair (d_{zx} and d_{yz}) and a b_{2g} orbital (d_{xy}) in the D_{4h} complex.

Table 12.9. Character table of D_{4h}

D_{4h}	E	$2C_4$	C_2	$2C_2'$	$2C_2''$	i	$2S_4$	σ_h	$2\sigma_v$	$2\sigma_d$	
A_{1g}	1	1	1	1	1	1	1	1	1	1	x^2+y^2, z^2
A_{2g}	1	1	1	−1	−1	1	1	1	−1	−1	R
B_{1g}	1	−1	1	1	−1	1	−1	1	1	−1	x^2-y^2
B_{2g}	1	−1	1	−1	1	1	−1	1	−1	1	xy
E_g	2	0	−2	0	0	2	0	−2	0	0	(R_x, R_y), (xz, yz)
A_{1u}	1	1	1	1	1	−1	−1	−1	−1	−1	
A_{2u}	1	1	1	−1	−1	−1	−1	−1	1	1	z
B_{1u}	1	−1	1	1	−1	−1	1	−1	−1	1	
B_{2u}	1	−1	1	−1	1	−1	1	−1	1	−1	
E_u	2	0	−2	0	0	−2	0	2	0	0	(x, y)

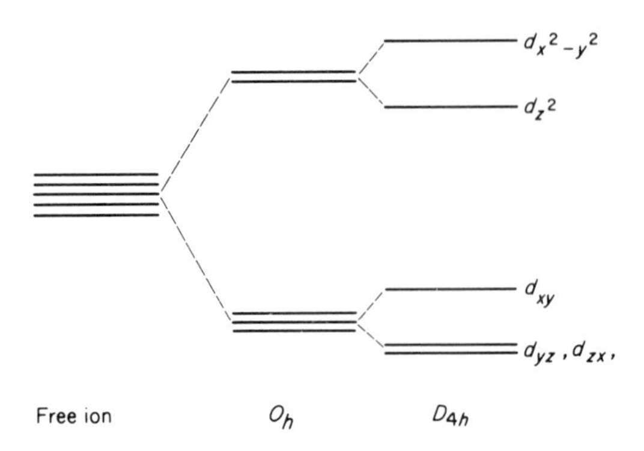

Figure 12.13 Effect of elongation, in an octahedral system, of the ligand bonds along the z axis.

Simple electrostatic arguments show that because the d_{zx} and d_{yz} pair are, in general, further from the ligands than d_{xy}, the e_g set will have a lower energy than the b_{2g}. Similarly, the orbitals which are e_g in the octahedron lose their degeneracy, the a_{1g} (d_{z^2}) being more stable than the b_{1g} ($d_{x^2-y^2}$). The resulting energy level pattern is shown in Figure 12.13.

Distortions of the octahedron of the kind we have just discussed are quite commonly observed in crystals. The reason for them lies in a general theorem, which was proved by Jahn and Teller, that if a non-linear molecule is in an orbitally degenerate state then the system will distort to relieve that degeneracy.† We conclude, for example, that weak-field octahedral complexes with ground states of E_g or T_{1g} symmetries are Jahn–Teller unstable. Thus only the weak-field d^3, d^5, d^8 configurations, whose ground states are $^4A_{2g}$, $^6A_{1g}$, and $^3A_{2g}$ respectively, will be expected to exist as regular octahedra.

If we take Cu^{2+} as an example, its configuration is d^9 and in an octahedral complex the lowest energy electron configuration will be $t_{2g}^6\ e_g^3$ which gives rise to a ^2E_g state. For such a state the hole is as likely to be in d_{z^2} as in $d_{x^2-y^2}$. However, a tetragonal extension along the z axis will, as we have seen, split this ^2E_g state into two components, the lower energy of which has the hole in $d_{x^2-y^2}$. If, on the other hand, there were to be a tetragonal compression along the z axis then d_{z^2} would have a higher energy than $d_{x^2-y^2}$ so the hole would be in d_{z^2}. The Jahn–Teller theorem does not tell us which of these will occur, or if some other distortion which removes the degeneracy will occur. X-Ray evidence shows that axial bonds in octahedral copper complexes are almost always longer than equatorial bonds. However, it is in practice difficult to obtain an unambiguous example of the Jahn–Teller theorem in operation in a ground state because in most cases there are other factors involved such as long-range bonding interactions in the

†For a proof of this see J. N. Murrell, S. F. Kettle, and J. M. Tedder, *Valence Theory*, 2nd edition, p. 248, Wiley, 1970.

crystal. The evidence of spectroscopy for John–Teller distortions in excited states is much more convincing.

If the *trans* ligands are removed completely then a square-planar complex is obtained and this also has D_{4h} symmetry. For such complexes the d_{z^2} orbital may be sufficiently stabilized that it comes below d_{xy}.

For geometries which are not closely related to the ML_6 octahedral arrangement it is necessary to carry out a subjective estimate of the d-orbital splittings from a simple electrostatic model combined with the use of the appropriate character table. It is then necessary to attempt to resolve any ambiguities in the splitting pattern experimentally; the use of a more sophisticated model of the crystal field is seldom worthwhile.

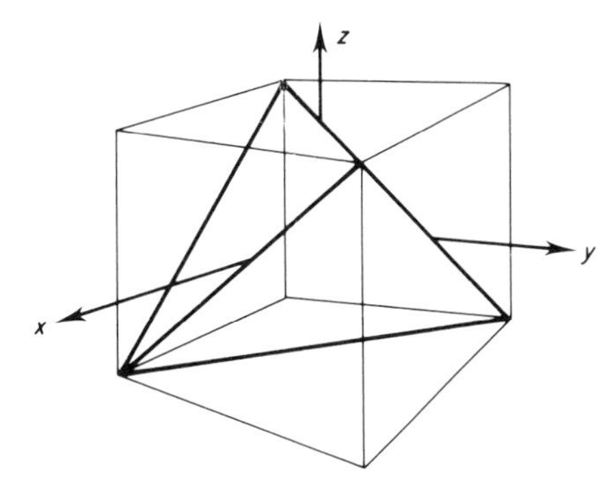

Figure 12.14 The relationship between the tetrahedron and the cube.

There is one important geometry which merits detailed attention. This is the tetrahedral, ML_4, arrangement of ligands. A tetrahedron, like the octahedron, is closely related to a cube – as shown in Figure 12.14. The d-orbital splittings in octacedral and tetrahedral complexes are therefore closely related. The tetrahedral (T_d) character table has been given in Table 8.1. Using the coordinate axes shown in Figure 12.14 it is easy to see on electrostatic grounds that the t_2 set of d orbitals (d_{xy}, d_{yz}, and d_{zx}) are *less* stable than the e set (d_{z^2} and $d_{x^2-y^2}$). The consequence of this is that in the weak-field limit the term splittings are inverted in T_d compared to O_h.

Both octahedral and tetrahedral situations may be included in the same diagram, as is done for the weak-field d^2 case in Figure 12.15. Such a figure is called an Orgel diagram. A comparison with Figure 12.12 shows that the tetrahedral d^2 diagram is essentially the same as the d^8 octahedral diagram (d^2 positive hole), so that an Orgel diagram contains information relevant to d_n and d^{10-n}, octahedral and tetrahedral complexes.

Although in principle one can distinguish between weak- and strong-field tetrahedral complexes, in practice the d-orbital splitting in tetrahedral complexes is only about half of that of the corresponding octahedral complex. This means that strong-field tetrahedral complexes are never found.

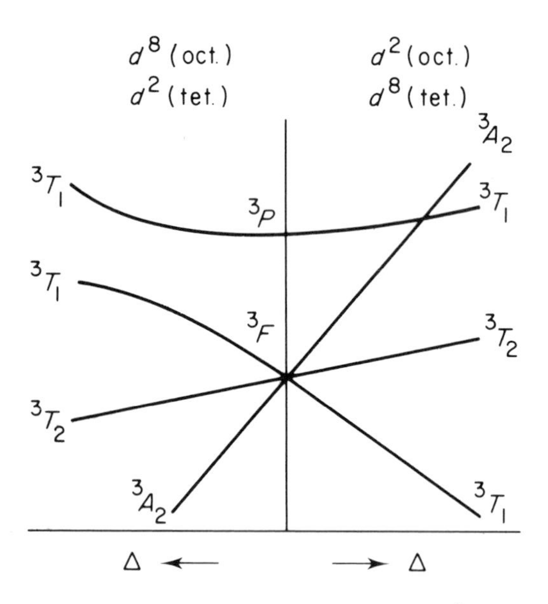

Figure 12.15 Orgel diagram for the d^2 and d^8 configurations in octahedral and tetrahedral fields.

12.5. Molecular orbital theory of transition metal complexes

Crystal field and ligand field theories are empirical theories which are successful in rationalizing experimental data in terms of a few parameters. There have been attempts to calculate Δ from the electrostatic potential of the ligand but the results have been disappointing: it has een been found difficult to obtain the correct sign in some cases.

An additional reason for being dissatisfied with the simple electrostatic model is that the electron repulsion term B, which is treated as an empirical parameter in ligand field theory, is usually appreciably smaller than its free-ion value, and the most likely reason for this is that the d electrons are delocalized into ligand orbitals. The most direct method for quantifying this delocalization is by molecular orbital theory, and this has the advantage that it may be approached by empirical or by *ab-initio* methods. *Ab-initio* calculations on transition metal complexes are, of course computationally expensive but not outrageously so with present technology, and they are likely to become much more common in the future.

The empirical molecular orbital theory of transition metal complexes uses a basis of nd, $(n + 1)s$, and $(n + 1)p$ orbitals of the metal and appropriate orbitals of the ligands. Thus for a first-row transition metal the basis orbitals are $3d$, $4s$, and $4p$. However, *ab-initio* calculations show that $3s$ and $3p$ orbitals are not far below this 'valence shell' and cannot be ignored in more accurate work.

For octahedral complexes we have already determined the symmetry species to which s, p, and d orbitals belong: they are $s(A_{1g})$, $p(T_{1u})$, and $d(E_g + T_{2g})$. We will

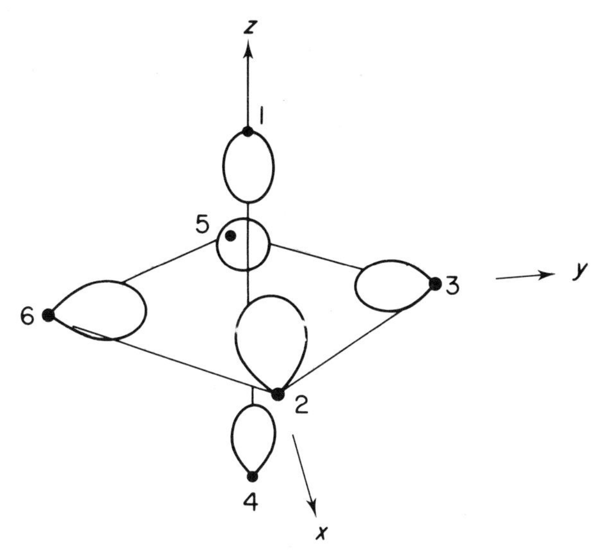

Figure 12.16 Ligand basis orbitals of σ symmetry for octahedral complexes.

start by seeing how these orbitals combine with a set of six ligand orbitals each of which is axially symmetric to the M—L bond as shown schematically in Figure 12.16. These so-called σ orbitals could be s orbitals of the ligand atoms or $p\ \sigma$ orbitals or suitable hybrids of the two. The lone-pair orbital of an NH_3 ligand would, for example, be such an orbital.

If we identify the six ligand orbitals with wavefunctions $\phi_1 \ldots \phi_6$, using the numbering of Figure 12.16 then our task is to find symmetry adapted combinations of these that belong to one of the O_h symmetry species. We follow a procedure similar to that used in Section 8.2 to find the symmetry adapted combinations of hydrogen orbitals for the methane molecule.

One obvious combination that belongs to the totally symmetric species, A_{1g}, is

$$\theta_1 = 6^{-1/2}(\phi_1 + \phi_2 + \phi_3 + \phi_4 + \phi_5 + \phi_6), \tag{12.21}$$

the $6^{-1/2}$ being a normalizing constant. Three other functions that suggest themselves are

$$\begin{aligned} \theta_2 &= \sqrt{1/2}(\phi_1 - \phi_2), \\ \theta_3 &= \sqrt{1/2}(\phi_3 - \phi_4), \\ \theta_4 &= \sqrt{1/2}(\phi 5 - \phi_6). \end{aligned} \tag{12.22}$$

Figure 12.16 shows that these functions have the same symmetry properties as the three p orbitals, p_z, p_x, and p_y respectively, so that if these three p orbitals belong to T_{1u} so do the three functions (12.22).

Having used up four of the possible six independent combinations of $\phi_1 \ldots \phi_6$ we are left with two to find and we can guess that they must belong to a doubly degenerate species, either E_g or E_u. The required combinations must be orthogonal to (12.21) and to each of the functions (12.22), and one function which fulfills these conditions is

$$\theta_5 = \tfrac{1}{2}(\phi_2 - \phi_3 + \phi_5 - \phi_6). \tag{12.23}$$

Another function can be formed with a similar pattern of coefficients

$$\theta_6 = \tfrac{1}{2}(\phi_1 - \phi_2 + \phi_4 - \phi_5). \tag{12.24}$$

However, this is not orthogonal to (12.23) and it is more convenient to choose for the last function a mixture of θ_5 and θ_6 which is orthogonal to θ_5, this is

$$\theta_6' = \frac{1}{\sqrt{3}}(2\theta_6 + \theta_5) = \frac{1}{2\sqrt{3}}(2\phi_1 - \phi_2 - \phi_3 + 2\phi_4 - \phi_5 - \phi_6) \tag{12.25}$$

Figure 12.17 shows sketches of θ_5 and θ_6'. It is evident that they resemble $d_{x^2-y^2}$ and d_{z^2} respectively, and we therefore conclude that they belong to the E_g symmetry species.

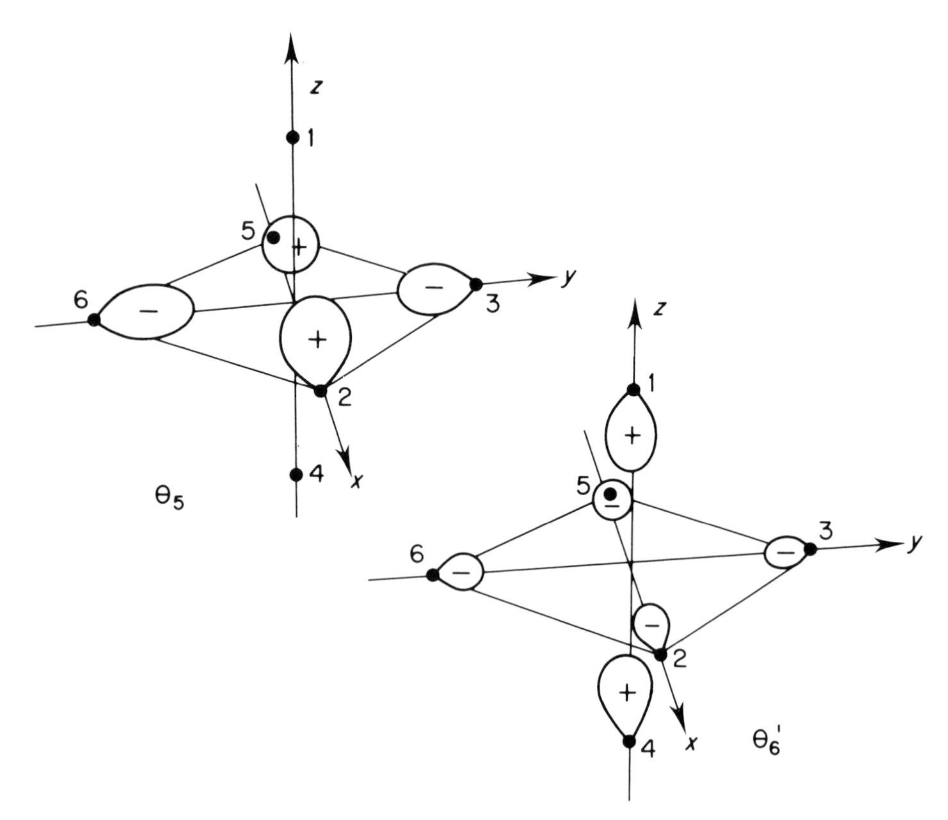

Figure 12.17 The symmetry adapted ligand orbitals θ_6' (see equations 12.23 and 12.25).

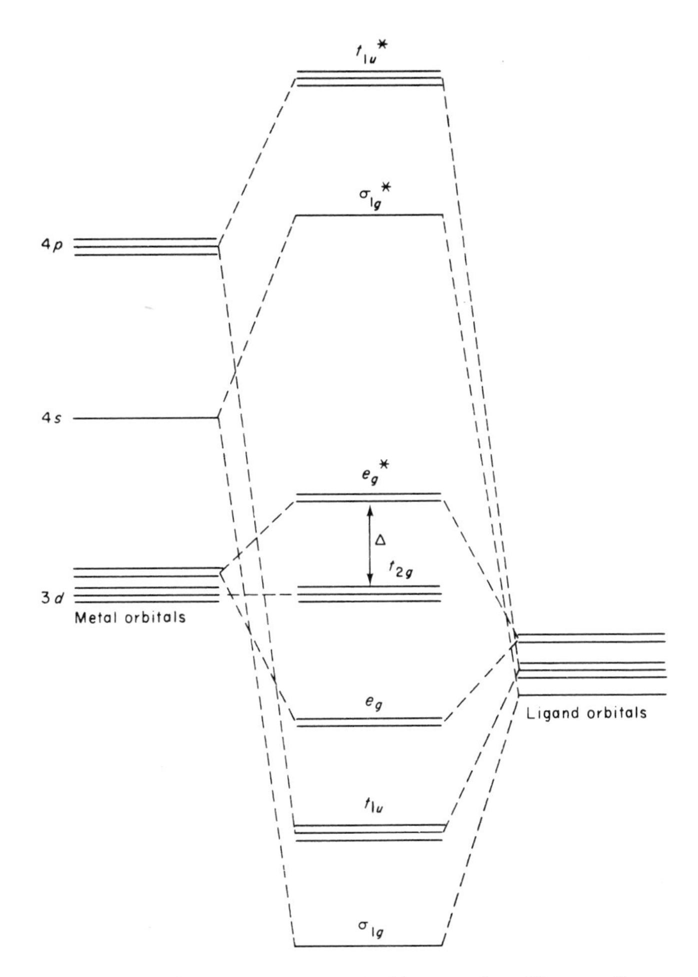

Figure 12.18 Molecular orbital interaction diagram for an octahedral complex (asterisks indicate antibonding orbitals).

The molecular orbital interaction diagram is shown in Figure 12.18. Note the important result that there are combinations of ligand σ orbitals which can interact with the 4*s*, 4*p*, and 3*d* (e_g) orbitals but there is no combination that can interact with $3d(t_{2g})$ are in this scheme non-bonding orbitals.

If we take NH_3 as a typical ligand then each of the ligand orbitals is accompanied by two electrons (the NH_3 lone-pair electrons). These twelve electrons are just sufficient to fill the six-bonding molecular orbitals, and this confirms our earlier statement that 4*s* and 4*p* orbitals of the metal are important for covalent bond formation. Any electrons in the 3*d* shell of the metal will then be placed in the non-bonding t_{2g} and the antibonding e_g orbitals (we are not implying that we can distinguish between ligand and metal electrons, but are just using a convenient accounting procedure). We therefore have a situation which is equivalent to the ligand or crystal field model if we identify the $e_g - t_{2g}$ separation with the parameter Δ.

In order that the antibonding e_g orbital be composed primarily of metal *d* orbitals it is necessary that the *d* orbitals have greater energies than the ligand orbitals. The ionization potentials of the first-row transition elements are approximately 6.5 eV, hence by Koopmans' theorem the *d*-orbital energies are approximately -6.5 eV. Thus for the zero oxidation state there is no doubt that the *d* orbitals have a higher energy than the ligand orbitals: NH_3, for example, has an ionization potential of 10.15 eV so that the lone-pair orbital energy is -10.15 eV. However, for higher oxidation states the situation is less clear because the metal ionization potential increases to ~ 16 eV for +1 oxidation states and ~ 30 eV for +2 states. The *d* orbital energy is therefore sensitive to the net charge on the metal.

Ab-initio calculations generally support the view that the e_g orbitals are mainly $3d$. For example, calculations by Moskowitz and co-workers† on $[NiF_6]^{4-}$, in which the metal has the formal oxidation state +2, showed that the net charge on the Ni was +1.8, and the e_g orbitals were made up of 98% *d* orbitals and only 2% fluorine $2p$ orbitals. In contrast the t_{2g} orbitals in this calculation were found to have an appreciable amount (23%) of fluorine *p* orbitals and these were the ligand $p\pi$ orbitals which we have not yet considered.

The symmetry adapted combinations of the twelve ligand $p\pi$ orbitals are more difficult to determine. We confine ourselves to quoting the result that they belong to the species T_{1g}, T_{1u}, T_{2g}, and T_{2u}. As there are no metal orbitals of T_{1g} and T_{2u} symmetries, these ligand orbitals are non-bonding. The T_{1u} ligand orbitals will interact with metal $4p$ and the T_{2g} with metal $3d$. These interactions are evident from Figure 12.19. It is the latter that has most relevance to the spectroscopic properties of the complexes because it partially determines the value of Δ.

If the π orbitals of the ligand are occupied by electrons and are lower in energy than the *d* orbitals then π bonding will reduce the value of Δ; halide ions are thought to be in this category. If they are higher in energy than the $3d$ orbitals and empty of electrons then they will increase the value of Δ. These two cases are illustrated in Figure 12.20.

Common ligands for which π bonding is thought to be important are CO, NO, CN^-, and ligands comprising second-row elements such as PF_3. The ligand π orbitals of CO are π molecular orbitals and representations of these have been given in Figure 6.11. Both the occupied π-bonding orbital (1π) and the vacant π-antibonding orbitals (2π) are important. The effect of the former will be to reduce Δ and of the latter to increase Δ. As such ligands always lead to low-spin complexes it can be surmised that the 2π orbitals are the more important. Note from Figure 6.11 that the 2π orbital has the larger amplitude on the carbon atom.

The mixing of $3d(t_{2g})$, 1π, and 2π orbitals leads to some redistribution of

†J. W. Moskowitz, C. Hollister, C. J. Hornbach, and H. Basch, *J. Chem. Phys.*, **53**, 2570 (1970).

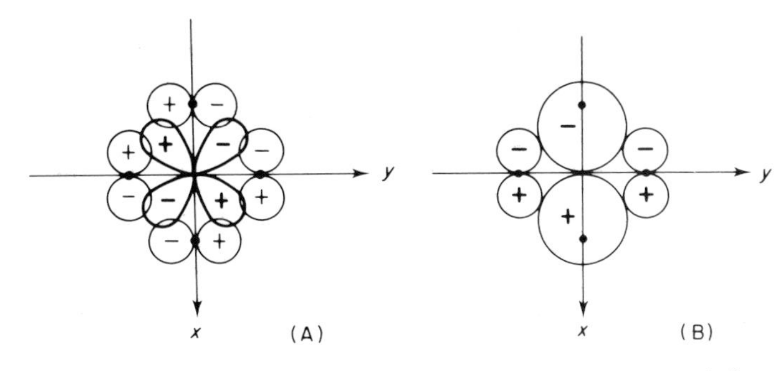

Figure 12.19 Ligand π overlap with metal orbitals of (A) t_{2g} symmetry, (B) t_{1u} symmetry.

electrons between ligand and metal, and even to a redistribution within the ligands. In the case of CO, for example, the dominant flow is thought to be from the $3d$ orbitals into the carbonyl 2π orbital and this is partly compensated by a flow the other way in the σ system. The reduction of the C=O force constant on bonding to a transition metal has been attributed to the partial occupation of the carbonyl π-antibonding orbital.

Brief mention should be made at this point to an approximate molecular orbital technique that was developed originally for solid-state band theory calculations but was taken into the molecular arena by Slater and Johnson.† This is called the SCF-Xα method. It is a molecular orbital method in the sense that it is based upon one-electron wavefunction, but it does not use the LCAO approximation. The

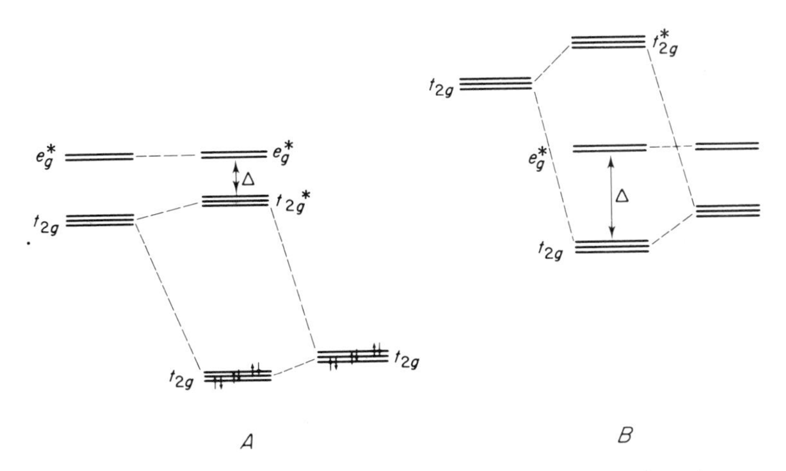

Figure 12.20 Effect of ligand to metal π bonding on Δ. (A) filled ligand orbitals, (B) empty ligand orbitals.

† See for example, K. H. Johnson, *J. Chem. Phys.*, **45**, 3085 (1966).

Hamiltonian is empirical (the factor α in the title represents an empirical parameter) but allows for the repulsion of electrons including the exchange effects associated with the antisymmetry of the wavefunction. Molecular orbitals are calculated in an SCF manner by first dividing up the space into spherical regions and obtaining wavefunctions for these, and then imposing conditions of continuity for wavefunctions that pass from one sphere across the intervening gap, into another sphere.

The SCF-Xα method is less computationally demanding than the LCAO procedure because one does not have to calculate the electron repulsion integrals for a basis of atomic orbitals. The results have been encouraging for the interpretation of spectroscopic properties, but discouraging for the calculation of molecular geometries. The method will almost certainly have continued use for solid state calculations but may be only passing fashion for molecules.

12.6. Sandwich compounds

In this section we give a brief review of one of the most interesting types of chemical bond, namely that between a transition metal and an unsaturated organic hydrocarbon. Such compounds have been known for many years although it is only since the 1950s that their special features have been appreciated.

A complex salt of ethylene and platinum was prepared by Zeise in 1827 and has the formula $[Pt(C_2H_4)Cl_3]^-$. The X-ray structure of this, which was not determined until 1954, is shown in Figure 12.21. In 1951 a complex between the cyclopentadienyl radical and iron was prepared which is commonly called ferrocene. Its X-ray structure was determined in 1955 and is also shown in Figure 12.21. The term 'sandwich' compound applied to such systems is self-evident. These two compounds are archetypes of a large family of compounds whose special feature is that the bonding appears to be due to the overlap of metal *d* orbitals and the π molecular orbitals of the hydrocarbon.

A qualitative molecular orbital picture of metal–ethylene bonding was given by Dewar and by Chatt and Duncanson in the early 1950s, and this has stood the test of time and recent *ab-initio* calculations. The essential feature is the inter-action of

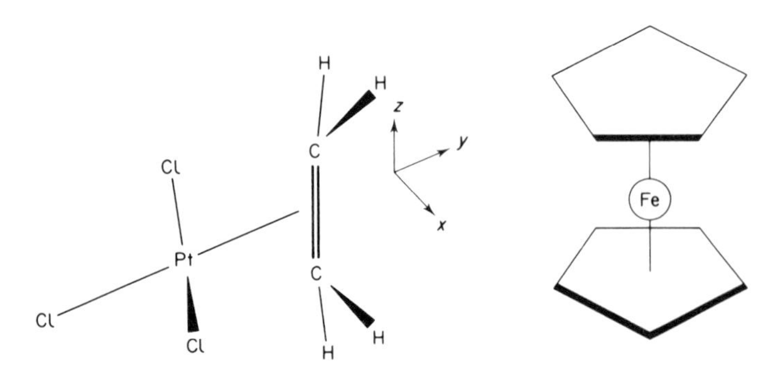

Figure 12.21 The structure of the anion of Zeise's salt and that of ferrocene.

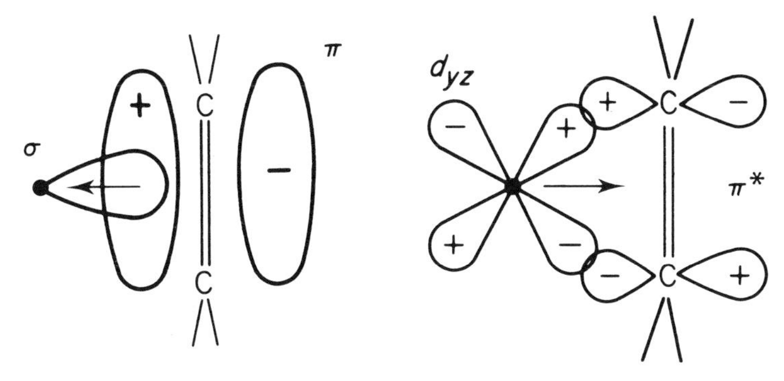

Figure 12.22 Ethylene π and π^* orbitals involved in bonding to the metal. The arrows indicate the direction of electron flow arising from each interaction.

the filled π-bonding orbital of the ethylene with vacant *d, s,* or *p* orbitals of the metal. This leads to donation of electrons from ethylene to the metal, but there is a compensated flow in the opposite direction through the overlap of filled *d* or *p* orbitals of the metal with the vacant π-antibonding orbital of ethylene: this is commonly referred to as 'back-bonding'. The orbitals involved are shown in Figure 12.22.

Ab-initio calculations have been made on $[Ag(C_2H_4)]^+$ and $[Pd(C_2H_4)]$.† Figure 12.23 shows the energies of the valence orbitals of Ag^+ and Pd, and the π bonding and antibonding (π^*) orbitals of ethylene. Also shown are the highest occupied and lowest vacant σ orbitals of ethylene. Both of the metals are d^{10} systems. From the relative energies of these orbitals it is not surprising that the net transfer of electrons is from ethylene to Ag^+ (approximately 0.15*e*) in the first complex but from Pd to ethylene (~0.20*e*) in the second. Backbonding is inhibited in the silver complex by the positive charge on the metal.

Although both of these calculations were based on structures in which the ethylene retained its planar geometry, and the X-ray evidence for Zeise's salt suggests that there will be some departure from this, it was found in both cases that there was considerable mixing of σ and π orbitals of the ethylene by virtue of the fact that they both interact with the same metal orbitals. For the Pd complex in particular the σ^* orbital gains approximately 0.1*e* at the expense of the *d* orbitals of the metal.

The molecular orbitals of ferrocene have been the subject of many calculations but the semi-empirical models failed to reach agreement on either the order of orbital energies or the charge distribution in the molecule. Two extreme views of the molecule can be taken: either it is a complex between Fe^{2+} and two cyclopentadientyl anions $C_5H_5^-$, or between neutral Fe and two cyclopentadientyl radicals. *Ab-initio* calculations by Coutiere and co-workers‡ gave a net charge of

† J. N. Murrell and C. E. Scollary, *J.C.S. Dalton*, 1034 (1977).
‡ M.M. Coutiere, J. Demuynck, and A. Veillard, *Theor. Chim. Acta*, **27**, 281 (1972).

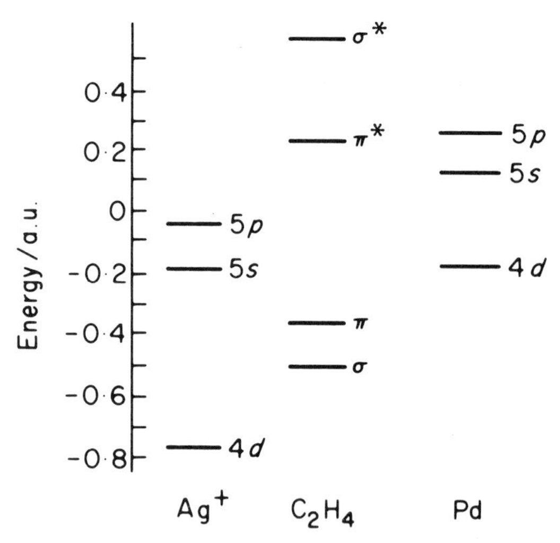

Figure 12.23 Energies of the highest occupied (π, σ, $4d$) and lowest vacant (π^*, σ^*, $5s$, $5p$) orbitals relevant in the species $[Ag(C_2H_4)]^+$ and $Pd(C_2H_4)$.

$1.23e$ on the metal, suggesting that the actual situation is almost half-way between these two extremes. The calculations support the view that the highest occupied molecular orbitals are composed mainly of Fe($3d$) and C_5H_5 π orbitals and we shall examine them on that basis.

The π orbitals of C_5H_5 are discussed in more detail in Chapter 14. They may be classified by their behaviour in the C_5 symmetry group (which consists only of the identify and rotations about the C_5 axis) as A, E_1, and E_2 in order of increasing energy. A is totally symmetric and has no sign changes on moving around the ring. The doubly degenerate orbitals E_1 have a single nodal line through the ring and E_2 have two such nodal lines. The pattern is very similar to that for the first three π orbitals of benzene (equations 9.38–9.40).

Ferrocene in the solid state has symmetry D_{5d} with the two rings in a staggered arrangement. The C_5H_5 π orbitals can be combined in plus and minus combinations in an obvious way. The two A orbitals give, in D_{5d}, the species $A_{1g} + A_{2u}$, the two E_1 orbitals give $E_{1g} + E_{1u}$, and the two E_2 orbitals give $E_{2g} + E_{2u}$.

Taking the C_5 axis as the z axis, the d orbitals belong to the following species

$$\begin{aligned} A_{1g} &: d_{z^2} ,\\ E_{1g} &: 3d_{xz}, 3d_{yz} ,\\ E_{2g} &: 3d_{xy}, 3d_{x^2-y^2} . \end{aligned} \tag{12.26}$$

Thus each of these will interact with one symmetry adapted combination of π orbitals. The interaction diagram is as shown in Figure 12.24 with the final order of bonding orbitals as obtained from the *ab-initio* calculation. The π orbitals of A_{2u} and E_{1u} symmetry would be stabilized by interaction with Fe($4p$) orbitals. Two

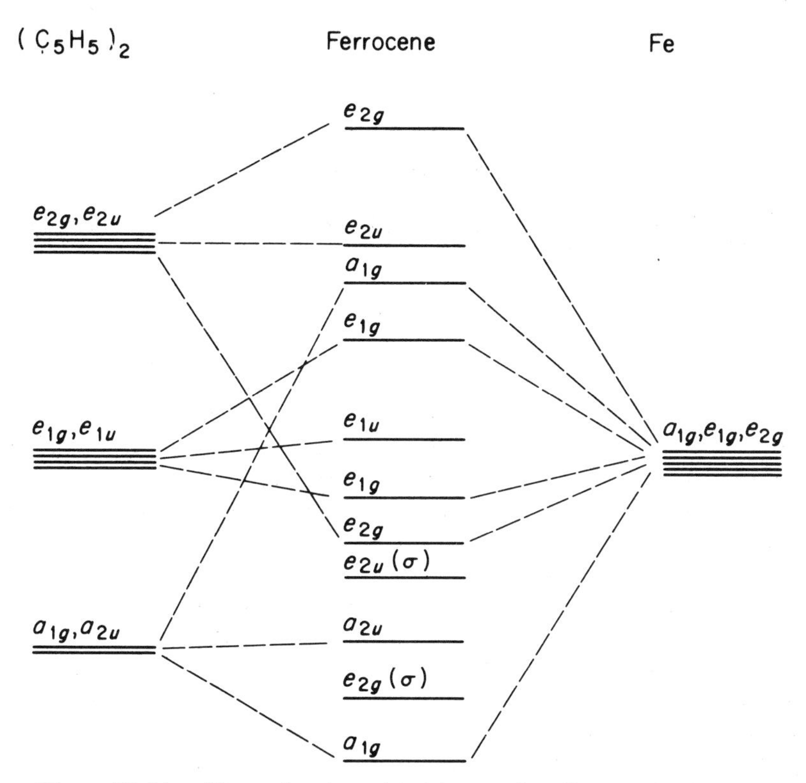

Figure 12.24 The molecular orbital interaction diagram for ferrocene. The order of the unoccupied orbitals above the wavy line is uncertain.

orbitals which originate from the σ orbitals of the C_5H_5 groups are also shown in Figure 12.24.

There are sufficient electrons to fill all orbitals up to e_{1u} (up to the wavy line in Figure 12.24) and the highest occupied orbital is largely composed of the π orbitals. The highest energy orbital which is mainly Fe(3*d*), is the e_{2g}. If we start from the ionic approximation of two $C_5H_5^-$ groups combining with Fe^{2+}, then the e_{1g} and e_{1u} orbitals of the ligands are filled with electrons. Donation of electrons to the metal then arises either through mixing with $3d$ (e_{1g}) or with $4p(e_{1u})$ orbitals. From this starting point transfer of electrons from $3d$ into the π antibonding orbitals is less likely because of the high energy of these orbitals.

The orbital pattern of Figure 12.24 could be confirmed by photoelectron spectroscopy; however, the *ab-initio* calculations show that this molecule is one of the few which Koopmans' theorem fails. The ground state of the positive ion actually arises by removing an electron from the highest e_{2g} orbital and not the e_{1u} or e_{1g} which lie above it. This can be confirmed by electron spin resonance spectroscopy because the e_{2g}, unlike the e_{1u} or e_{1g} is primarily a $3d$ orbital. The reason for this failure of Koopmans' theorem is that there is a large redistribution of electrons following ionization so that the molecular orbitals of the positive ion are very different from those of the neutral molecule.

Chapter 13
The Concepts of Valence Bond Theory

13.1. The Heitler–London wavefunction for the electron-pair bond

The first quantum mechanical theory of the electron-pair bond was that given by Heitler and London for the hydrogen molecule in 1927. This theory was developed by Pauling and others in the 1930s into a comprehensive theory of bonding called *Valence Bond Theory* that encompassed all molecules from small to large and even to solids; from aromatic hydrocarbons to transition metal complexes. In fact, for all the systems we have discussed in earlier chapters there is a valence bond analysis that parallels the molecular orbital analysis we have presented.

It may therefore appear strange, in view of its historical position and its comprehensive coverage of the field, that we have left the details of valence bond theory to such a late stage in this book, particularly when some of the most important concepts of valence such as *hybridization* and *resonance* were first formulated as part of valence bond theory. The reason for leaving the subject to this late stage is that valence bond theory *in its simplest form* gives, in general, a less satisfactory picture of bonding than does molecular orbital theory *in its simplest form*.

There are many examples to support this statement. We have seen in Section 6.2 that the properties of the homonuclear diatomic molecules, particularly their magnetic properties and their behaviour on ionization, are simply interpreted from the molecular orbital energy diagrams shown in Figures 6.6 and 6.7. There is no comparable interpretation of these properties in valence bond theory.

A second example occurs in the theory of aromaticity which will be discussed in the next chapter. Molecular orbital theory provides a simple explanation of why the cyclic ions $C_5H_5^-$ and $C_7H_7^+$ are stable but $C_5H_5^+$ and $C_7H_7^-$ are not. To find the explanation in valence bond theory one needs to know some subtle details about the Hamiltonian integrals between valence bond wavefunctions. In both examples, valence bond theory taken in its complete form will give a satisfactory explanation of the facts but these explanations cannot be fully understood without going further into quantum mechanics than would be appropriate in this book.

One reason for the decline in importance of valence bond theory in the past twenty years has been the difficulty of raising it to the *ab-initio* level for all except the simplest molecules. Computer programs for carrying out *ab-initio* molecular orbital calculations as a routine procedure have been available for many years and require only a moderately large computer. *Ab-initio* valence bond programs, however, are not widely available and those that have been written by individual groups are usually only applicable to restricted classes of molecule.

There are several reasons why *ab-initio* valence bond calculations are relatively difficult to carry out. Perhaps the most important is that there is nothing comparable to the molecular orbital Hartree–Fock limit in valence bond theory which can be represented by a single determinantal wavefunction, such as (8.9). The simplest valence bond wavefunctions for molecules having covalent bonds all consist of more than one determinantal function. A second reason is that the algebraic manipulation of the Hamiltonian and overlap integrals for valence bond determinantal functions is more difficult than it is for molecular orbital determinantal functions and this is due to the fact that the one-electron functions ($\psi_r(i)$ of 8.9) of molecular orbital theory are mutually orthogonal whereas in valence bond theory they are not orthogonal. Our estimate is that less than 5% of all *ab-initio* calculations published in the past 20 years have used valence bond theory, and none of these have been on large molecules.

The above remarks might appear sufficient reason for ignoring valence bond theory altogether in a book such as this. However, this would be to ignore the important role that the theory had in the past in forming the language of chemistry and it is necessary to look at the basic ideas of the theory in order to understand concepts such as resonance which are still widely used.

The concept of the covalent bond in the Lewis theory of valence was a sharing of electrons between two atoms such that each achieved an inert gas structure. In the simplest case, H_2, each atom shared the pair of electrons and achieved the closed shell structure of He. Heitler and London gave a quantum mechanical description of this electron sharing in the following way.

Let ϕ_a and ϕ_b be the $1s$ atomic orbitals of two hydrogen atoms. Wave functions which describe the two atoms taken together but very far apart are

$$\phi_a(1)\phi_b(2) \quad \text{or} \quad \phi_b(1)\phi_a(2) \tag{13.1}$$

where 1 and 2 label the electrons. These two functions differ only in the allocation of the two electrons to the two atoms. As electrons are indistinguishable a satisfactory wavefunction must allow for both allocations and this led Heitler and London to consider the following two functions as wavefunctions for the hydrogen molecule ($N_\pm$ is a normalizing factor that will be different in the two cases):

$$\Psi_\pm = N_\pm[\phi_a(1)\phi_b(2) \pm \phi_b(1)\phi_a(2)]. \tag{13.2}$$

Insofar as these two functions are an exact description of the system when the atoms are infinitely separated, we expect them to be approximate functions for electronic states of H_2 that dissociate to ground state hydrogen atoms, and the ground state of H_2 is one such state. We can see which function, Ψ_+ or Ψ_-, represents the ground state of H_2 by comparing the electron densities in the two states with what is already known from Chapters 5 and 6 about the electron density in the ground state of H_2.

The two-particle probability densities are obtained from (13.2) by the standard definition

$$P_{\pm}(1,2) = \Psi_{\pm}^2 = N_{\pm}^2 [\phi_a^2(1)\phi_b^2(2) + \phi_b^2(1)\phi_a^2(2) \pm 2\phi_a(1)\phi_b(1)\phi_a(2)\phi_b(2)] . \tag{13.3}$$

To obtain the one-electron density we must average over the position of the second electron thus

$$P_{\pm}(1) = \int P_{\pm}(1,2)\, dv_2 = N_{\pm}^2 [\phi_a^2(1) + \phi_b^2(1) \pm 2S_{ab}\phi_a(1)\phi_b(1)] , \tag{13.4}$$

where S_{ab} is the overlap integral. The overlap density $\phi_a(1)\phi_b(1)$ has its largest value where both $\phi_a(1)$ and $\phi_b(1)$ are large and that is between the two nuclei. For the state Ψ_+ the overlap density *increases* the density between the nuclei and for Ψ_- it *decreases* it. As we have already shown (c.f. Figure 5.9) that in a bonding state electrons are drawn into the region between the nuclei and in an antibonding state they are withdrawn, we can conclude that Ψ_+ is the wavefunction for the ground state of H_2 that we are seeking and Ψ_- represents an antibonding state.

The functions (13.2) represent only the spatial parts of the electronic wavefunctions and they must be multiplied by spin functions for a complete description of the electronic states. Moreover, we know (chapter 8) that the *total* wavefunction must be antisymmetric to exchange of the two electrons. Ψ_+ is a function that is symmetric to exchange and hence it must be multiplied by a spin function that is antisymmetric. For two electrons which individually have the spin functions α or β (see Chapter 8) there is only one antisymmetric spin combination, namely

$$\alpha(1)\beta(2) - \beta(1)\alpha(2). \tag{13.5}$$

Three spin functions can be constructed which are symmetric to exchange

$$\begin{aligned} &\alpha(1)\alpha(2), \\ &\alpha(1)\beta(2) + \beta(1)\alpha(2), \\ &\beta(1)\beta(2). \end{aligned} \tag{13.6}$$

The function (13.5) clearly has no net z-component of spin angular momentum because the α and β components cancel. As there is no other antisymmetric spin state, (13.5) must be a singlet spin state ($S=0, M_s=0$) as defined in Section 11.2. The set of functions (13.6) make up a triplet spin state with spin quantum numbers $S=0, M_S=1,0,-1$.

Combining (13.2) and (13.5) we obtain the electronic wavefunction for the ground state of H_2 as derived by Heitler and London

$$\Psi = N_+[\phi_a(1)\phi_b(2) + \phi_b(1)\phi_a(2)]\,[\alpha(1)\beta(2) - \beta(1)\alpha(2)] . \tag{13.7}$$

That this represents a bonding state can be confirmed by evaluating its energy, according to (5.14), and comparing it with that of two hydrogen atoms. As the Hamiltonian does not depend on electron spin we can ignore the spin part of the wavefunction in evaluating the Hamiltonian integral. Substituting (13.2) into (5.14) gives

$$E_{\pm} = N_{\pm}^2 \int [\phi_a(1)\phi_b(2) \pm \phi_b(1)\phi_a(2)]\, \mathscr{H} [\phi_a(1)\phi_b(2) \pm \phi_b(1)\phi_a(2)]\; dv_1 dv_2. \tag{13.8}$$

On multiplying out the integrand in (13.8) we obtain four separate integrals, one of which is

$$\int \phi_a(1)\phi_b(2)\, \mathscr{H} \phi_a(1)\phi_b(2)\; dv_1 dv_2. \tag{13.9}$$

It can be seen that the functions on both sides of the Hamiltonian are the same in this integral and hence it represents the energy of the system having electron (1) on one atom with density $\phi_a^2(1)$ and electron (2) on the other atom with density $\phi_b^2(1)$. The energy is just the energy of two separate hydrogen atoms except for the electrostatic (Coulomb) interaction between the two.

To evaluate the integral (13.9) it is convenient to divide up the total Hamiltonian into three parts as follows

$$\mathscr{H} = \mathscr{H}_A(1) + \mathscr{H}_B(2) + \mathscr{H}', \tag{13.10}$$

where

$$\mathscr{H}_A(1) = -\frac{1}{2}\nabla^2{}_1 - \frac{1}{r_{a1}}, \tag{13.11}$$

$$\mathscr{H}_B(2) = -\frac{1}{2}\nabla^2{}_2 - \frac{1}{r_{b2}}, \tag{13.12}$$

$$\mathscr{H}' = -\frac{1}{r_{a2}} - \frac{1}{r_{b1}} + \frac{1}{r_{12}} + \frac{1}{R}. \tag{13.13}$$

$\mathscr{H}_A$ (1) represents the Hamiltonian for an isolated hydrogen atom with electron (1) associated with nucleus a : r_{a1} is the distance of electron (1) from this nucleus. $\mathscr{H}_B$ (2) is similar in that it represents the Hamiltonian for a second hydrogen atom; electron (2) being associated with nucleus b. $\mathscr{H}'$ collects all the extra potential energy terms in the Hamiltonian of H_2 : the potential of electron (1) with nucleus b, electron (2) with nucleus a, electron (1) with electron (2) and nucleus a with nucleus b (R is the internuclear distance).

If we now make use of the fact that $\phi_a(1)$ is a solution of the Schrödinger equation for atom A, that is

$$\mathscr{H}_A(1)\phi_a(1) = \epsilon_H \phi_a(1), \tag{13.14}$$

and

$$\mathscr{H}_B(2)\phi_b(2) = \epsilon_H \phi_b(2), \tag{13.15}$$

where ϵ_H is the energy of the $1s$ orbital of a hydrogen atom, then the integral

(13.9) can be written

$$\int \phi_a(1)\phi_b(2)[\mathcal{H}_A(1) + \mathcal{H}_B(2) + \mathcal{H}']\phi_a(1)\phi_b(2)\, dv_1\, dv_2$$

$$= \int \phi_a(1)\phi_b(2)\,[2\epsilon_H + \mathcal{H}']\,\phi_a(1)\phi_b(2)\, dv_1\, dv_2. \qquad (13.16)$$

For normalized wavefunctions this is equal to

$$2\epsilon_H + Q, \qquad (13.17)$$

where

$$Q = \int \phi_a(1)\phi_b(2)\mathcal{H}'\phi_a(1)\phi_b(2)\, dv_1\, dv_2, \qquad (13.18)$$

is called the Coulomb integral.

If we write Q out in full using (13.13) and integrate over any coordinates not involved in the operators, we obtain

$$Q = -\int \phi_a(1)\frac{1}{r_{b1}}\phi_a(1)\, dv_1 - \int \phi_b(2)\frac{1}{r_{a2}}\phi_b(2)\, dv_2$$

$$+ \int \phi_a(1)\phi_b(2)\frac{1}{r_{12}}\phi_a(1)\phi_b(2)\, dv_1\, dv_2 + \frac{1}{R}. \qquad (13.19)$$

The first term in (13.19) is equal to the attraction of an electron in ϕ_a to the nucleus of atom B. The second term is the attraction of an electron in ϕ_b to the nucleus of A. The third term is the repulsion between electron (1) with density $\phi_a^2(1)$ and electron (2) with density $\phi_b^2(2)$, and the last term is the repulsion between the two nuclei. It is clear that Q represents the total Coulomb interaction energy between the two atoms.

Because a hydrogen atom is a neutral spherical system, there is no electrostatic interaction between the two atoms until the two electron densities $\phi_a^2(1)$ and $\phi_b^2(2)$ overlap significantly. When this first happens the electrostatic energy is attractive (negative), but if the atoms are pushed closer together the repulsion of the two nuclei becomes dominant and the electrostatic term is repulsive. Figure 13.1 shows how Q varies with the internuclear distance.

Another component integral in (13.8) is identical to (13.9) except for the exchange of the labels (1) and (2) on both sides of the Hamiltonian. As $\mathcal{H}$ is symmetric with respect to the coordinates of the two electrons this integral must have the same value as (13.9). However, there are two other integrals in the expansion which are quite different from (13.9), and these, after making a similar reduction to that leading to (13.17), involve the integral

$$A \equiv \int \phi_a(1)\phi_b(2)\mathcal{H}'\phi_b(1)\phi_a(2)\, dv_1\, dv_2, \qquad (13.20)$$

and an equivalent integral obtained by exchanging (1) and (2).

A is called the *exchange integral* because the function on the right-hand side

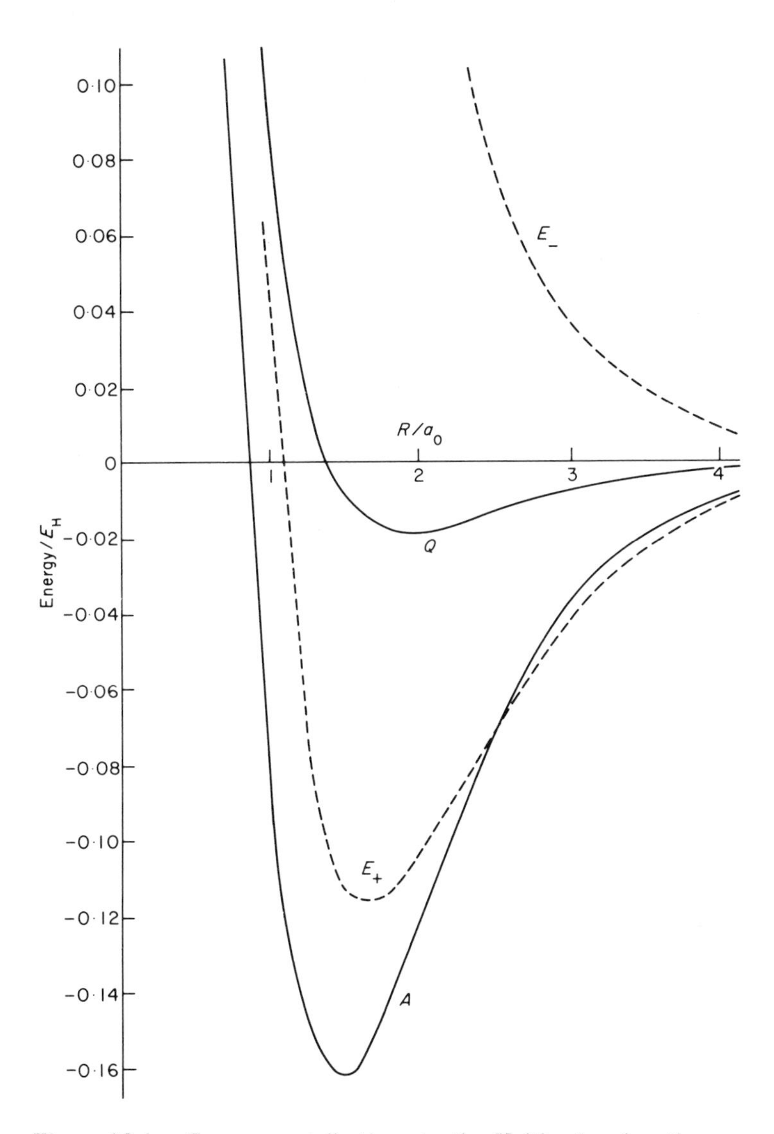

Figure 13.1 Energy contributions to the Heitler-London theory. A, Q, E_+, and E_- are explained in the text.

of the Hamiltonian is obtained by exchanging the electrons between the orbitals on the left-hand side of the Hamiltonian. A is the energy of a system of two nuclei and two electrons both of which have the probability density $\phi_a\phi_b$. This density, which is the overlap density occurring in (13.4), must become zero at infinite separations of the two atoms, hence A must become zero in this limit. Figure 13.1 shows how A varies with internuclear distance. Its R dependence is roughly the same as that of S_{ab}^2 (c.f. Figure 6.2) because (13.20) has two overlap densities in the integrand, and it is a more rapidly varying function than Q. The most important fact is that A is a *negative* energy.

On substituting (13.18) and (13.20) into (13.8) we obtain the energies of the two states as

$$E_{\pm} = 2\epsilon_{\mathrm{H}} + 2N_{\pm}^2(Q \pm A), \tag{13.21}$$

where the factor 2 arises from the two equivalent Coulomb and exchange integrals that occur. Because A is negative, E_+ will have the lower energy of the two states, as shown in Figure 13.1. By imposing the normalization condition, that the integral of (13.4) is unity, it can be shown that the normalization constants have the values

$$N_{\pm} = (2 \pm 2S_{\mathrm{ab}}^2)^{-\frac{1}{2}} \tag{13.22}$$

Although E_+ is far from being an exact representation of the ground state potential curve of H_2, it is *better* than the corresponding curve obtained from molecular orbital theory and, more important, it has the correct dissociation limit. We have not mentioned in earlier chapters that molecular orbital theory fails in this respect. To prove this we must reexamine the molecular orbital wavefunction for H_2.

The molecular orbital wavefunction for H_2 has two electrons in the bonding molecular orbital (equation 6.18)

$$\psi_g = (\phi_{\mathrm{a}} + \phi_{\mathrm{b}}), \tag{13.23}$$

with opposite spins. The antisymmetric wavefunction which describes this situation is (c.f. equation 8.6)

$$\begin{aligned}\Psi &= N[\psi_g(1)\alpha(1)\psi_g(2)\beta(2) - \psi_g(1)\beta(1)\psi_g(2)\alpha(2)] \\ &= N\psi_g(1)\psi_g(2)[\alpha(1)\beta(2) - \beta(1)\alpha(2)],\end{aligned} \tag{13.24}$$

N being a normalizing factor. On substituting (13.23) into (13.24) and expanding the orbital product we find

$$\begin{aligned}\Psi = N[&\phi_{\mathrm{a}}(1)\phi_{\mathrm{b}}(2) + \phi_{\mathrm{b}}(1)\phi_{\mathrm{a}}(2) + \phi_{\mathrm{a}}(1)\phi_{\mathrm{a}}(2) + \phi_{\mathrm{b}}(1)\phi_{\mathrm{b}}(2)] \\ &\times [\alpha(1)\beta(2) - \beta(1)\alpha(2)].\end{aligned} \tag{13.25}$$

A comparison of this and the Heitler–London function (13.7) shows that (13.25) contains (13.7) but it has an equal component of another function whose spatial part is

$$[\phi_{\mathrm{a}}(1)\phi_{\mathrm{a}}(2) + \phi_{\mathrm{b}}(1)\phi_{\mathrm{b}}(2)] \tag{13.26}$$

This is a symmetric combination of functions in which *both* electrons are in the same atomic orbital, and would be represented by an ionic valence structure for H_2 which is $H^- H^+$.

The molecular orbital wavefunction (13.24) therefore contains an equal mixture of covalent (Heitler–London) and ionic structures. At the dissociation limit this is of course incorrect. The ground state of H_2 dissociates to atoms and not to a mixture of atoms and ions. At the equilibrium bond length of H_2 we do not *a priori* know which of the molecular orbital and Heitler–London functions is the better. If we specify that one electron is on one atom (in ϕ_a say) then we can ask where the second electron will be. From the Heitler–London function we deduce that the second electron is in ϕ_b but from the molecular orbital function we deduce that there is equal probability of it being in ϕ_a and ϕ_b because the molecular orbital function gives equal weighting to the ionic and covalent representations.

We can use the variational theorem (Section 6.3) to determine a function that must be better than either the Heitler–London or the molecular orbital functions of H_2. We do this by allowing a free mixing of covalent and ionic contributions to the wavefunction, as follows:

$$\Psi = \{[\phi_a(1)\phi_b(2) + \phi_b(1)\phi_a(2)] + \lambda[\phi_a(1)\phi_a(2) + \phi_b(1)\phi_b(2)]\} \times [\alpha(1)\beta(2) - \beta(1)\alpha(2)]. \qquad (13.27)$$

Figure 13.2 shows the energy of the resulting state, and compares it with that of the Heitler–London and simple molecular orbital functions. At $R=\infty$ the Heitler–London function is exact ($\lambda = 0$) and at the equilibrium distance, $R=2a_0$, the optimum value of λ is about half way between the Heitler–London ($\lambda = 0$) and molecular orbital ($\lambda = 1$) limits.

The success of the Heitler–London treatment of the hydrogen molecule might seem to refute the discouraging comments we made about valence bond theory at the beginning of this chapter. Single determinantal molecular orbital wavefunctions such as (8.9) often give incorrect dissociation limits for the potential energy curves and *always* do so for homonuclear bonds when the separate atoms have incomplete shells of electrons (e.g. $F_2 \rightarrow 2F$). However, this can be corrected by taking a wavefunction with more than one determinant; this is called *configuration interaction* in molecular orbital theory. In valence bond theory, however, all representations of the covalent bond require more than one determinant. For example, the Heitler–London function can be written as two determinants

$$\Psi = N\left\{\begin{vmatrix} \phi_a(1)\alpha(1) & \phi_b(1)\beta(1) \\ \phi_a(2)\alpha(2) & \phi_b(2)\beta(2) \end{vmatrix} - \begin{vmatrix} \phi_a(1)\beta(1) & \phi_b(1)\alpha(1) \\ \phi_a(2)\beta(2) & \phi_b(2)\alpha(2) \end{vmatrix}\right\} \qquad (13.28)$$

as can be confirmed by expansion of these determinants and comparison with (13.7). Moreover, for n covalent bonds the valence bond wavefunction consists of 2^n determinants as it is a product of n functions such as (13.7), one for each covalent bond, the whole function being antisymmetrized. There is always, however, a single determinantal molecular orbital function which at least gives a satisfactory picture of the equilibrium molecule. It is for this reason that valence bond theory gets rapidly more difficult to handle as the number of covalent bonds in a molecule increases.

If a number of approximations are made in evaluating the Hamiltonian and overlap integrals between valence bond wavefunctions then it is possible to formulate rules for evaluating these integrals which do not require a full expansion of the wavefunction. This is the basis of the empirical valence bond theory

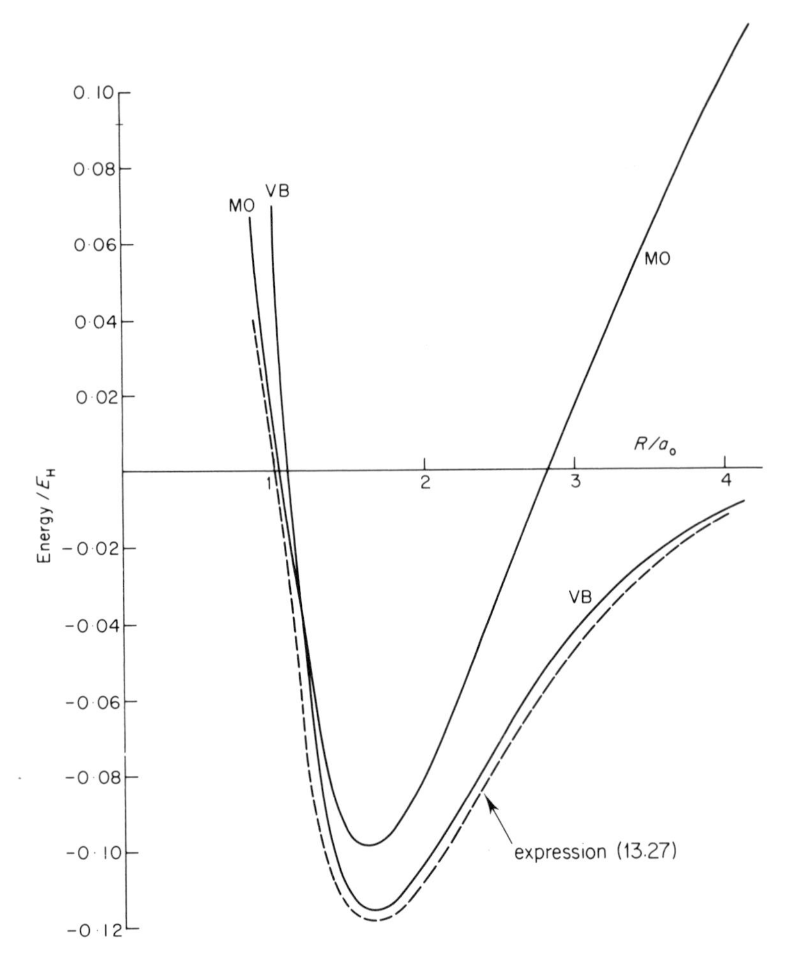

Figure 13.2 Comparison of Heitler-London (VB) and simple molecular orbital (MO) energies with that of wavefunction (13.27).

developed by Pauling and others in the 1930s. In such a theory it is sufficient to represent a wavefunction (the so-called bond eigenfunction) by the chemical bond structure from which it is derived. Thus the wavefunction (13.7) can be represented by Ψ(H—H), and the two Kekulé structures would be represented symbolically by wavefunctions

$$\Psi\left(\text{Kekulé structure I}\right) \quad \text{and} \quad \Psi\left(\text{Kekulé structure II}\right). \qquad (13.29)$$

The energies of such structures can be represented by the so-called *perfect-*

pairing approximation which for closed-shell molecules takes the form

$$E = E_{\text{atoms}} + \sum_{\substack{\text{all} \\ \text{a b}}} Q_{\text{ab}} + \sum_{\substack{\text{paired} \\ \text{a b}}} A_{\text{ab}} - \frac{1}{2} \sum_{\substack{\text{non-paired} \\ \text{a b}}} A_{\text{ab}}. \tag{13.30}$$

Q and A are Coulomb and exchange integrals such as (13.18) and (13.20). Paired orbitals are those indicated by lines in the valence structure which are coupled to give a Heitler–London type of function and unpaired orbitals are those that are not so coupled. For a molecule such as benzene, the Q and A integrals can be determined empirically in the same way as are the Hückel α and β parameters of molecular orbital theory (Chapter 9).

The rules for calculating the Hamiltonian integrals between different structures are more difficult and in view of the lack of current activity in this field would seem to have no place in this book. The reader is referred to a paper by Pauling for the details.† We will, however, return to the question of the interaction between structures in the last section of this chapter.

It is clear that valence bond theory, at least in its simplest empirical form, relies heavily on chemical intuition for finding suitable valence structures. In some respects this may seem an advantage in that chemical experience can be directly introduced into a quantum mechanical calculation. In another respect, however, it is a weakness because it assumes some knowledge of the answer to the problem at the start. In other words, the results of such a valence bond calculation will be wrong if chemical intuition is wrong. The most valuable calculations are often those whose results refute chemical intuition.

In contrast, molecular orbital theory does not make any assumption about the way orbitals are paired in a molecule: for a specified molecular geometry this will be found from the results by examining bond orders and overlap populations (Chapter 9).

Most stable molecules can be represented by a single valence structure and it should be possible to represent this by a single valence bond wavefunction. However, to do this it is necessary to introduce two new concepts into chemical bond theory. One of these is the *valence state* of an atom and the other is *hybridization* of orbitals. The second of these concepts has already been discussed within a molecular orbital framework but we shall look at it also in a valence-bond context as that is where it was first introduced.

13.2. Valence states and hybridization

These concepts can be very well illustrated by the bonding of carbon compounds but any other atom other than hydrogen or an alkali metal would serve as well. The lowest energy configuration of the carbon atom is $1s^2 2s^2 2p^2$ and in this only the two $2p$ electrons can have their spins uncoupled. From this configuration we would expect, from valence bond theory, that carbon would be divalent with a bond angle

† L. Pauling, *J. Chem. Phys.*, **1**, 280 (1933).

close to 90°. This would follow by coupling the $2p$ electrons, one in $2p_x$ and one in $2p_y$ say, with unpaired electrons on the combining atoms. However, carbon only rarely shows bonding of this type. CF_2 (bond angle 105°) is one example of a divalent carbon species which is sufficiently long-lived for its microwave spectrum to be studied.

Carbon is almost universally tetravalent in its compounds and in valence bond theory four covalent bonds can only be formed by first uncoupling the spins of the two electrons in the $2s$ orbital. This can be done with least expenditure of energy by promoting an electron from the $2s$ to the $2p$ sub-shell to give the configuration $1s^2 2s2p^3$. It is now possible for all the electrons in the $2p$ orbitals and that in the $2s$ to have unpaired spins.

It requires a considerable amount of energy to promote an electron from the $2s$ to a $2p$ orbital. From atomic spectroscopy we know that it is at least 400 kJ mol^{-1}. However, this energy can be retrieved by the ability of the atom to form two more bonds. The CH bond energy for example is itself ~400 kJ mol^{-1} so that there is plenty of energy to spare for the electron promotion.

We have seen in Chapter 11 that the lowest energy configuration of the carbon atom will give rise to the spectroscopic terms ^3P, ^1D, and ^1S. The excited configuration $1s^2 2s2p^3$ will also give rise to a number of terms (^5S, ^3S, ^3P, ^1P, ^3D, ^1D) the lowest energy of which is ^5S. However, the bonding in carbon compounds cannot be represented by a single valence bond structure based upon the ^5S term because this would imply that three bonds were of one kind (formed from the $2p$ orbitals) and the fourth was different (formed from the $2s$ orbital). We know from chemical and spectroscopic evidence that the four bonds in CH_4 are identical. We therefore conclude that the state of the carbon atom in CH_4 does not resemble a single spectroscopic state (term) but must correspond to a mixture of spectroscopic states. Moreover this mixture is not necessarily confined to the states arising from one configuration. Such a state is called a *valence state*.

A valence state is not a spectroscopic state and it is therefore not directly observable. It is a concept introduced to describe the nature of the atom as it is bound in the molecule, this bonding being represented by a valence bond wavefunction. For the carbon atom in CH_4 it has been shown that the valence state has a wavefunction†

$$\frac{1}{8}\,[2\sqrt{5}(sp^3\!:\!{}^5S) + 3\sqrt{2}(sp^3\!:\!{}^3D) - \sqrt{2}(sp^3\!:\!{}^1D) + 3(p^4\!:\!{}^3P) - \sqrt{3}(p^4\!:\!{}^1D) + \sqrt{3}(s^2p^2\!:\!{}^1D) - 3(s^2p^2\!:\!{}^3P)] \tag{13.31}$$

and its energy, which may be inferred from its spectroscopic components, is approximately 670 kJ mol^{-1} above the ground state energy of the carbon atom.

It is beyond the scope of this book to show how the wavefunction for a valence state such as (13.31) is determined, but one of the points that has to be specified is

†H. H. Voge, *J. Chem. Phys.*, **4**, 581 (1936); **16**, 984 (1948).

the type of orbital of the atom that is being used to form the electron pair bonds, and that leads us to the concept of hybridization.

In order that a valence bond structure for CH_4 implies four equivalent CH bonds, the four carbon orbitals that are used to form the electron pair bonds must be equivalent. It is possible to take combinations of the $2s$ and the three $2p$ orbitals which satisfy this property; such combinations are called hybrid orbitals and in this particular case sp^3 hybrids.

Wavefunctions have been given in (8.23) for sp^3 hybrids, this particular set of functions being appropriate for the choice of coordinate axes shown in Figure 8.2. A general wavefunction for a sp^3 hybrid is, from equation (8.28),

$$\psi = (2s + \sqrt{3}.2p)/2 \tag{13.32}$$

where $2p$ is any normalized combination of the three $2p$ orbitals. Wavefunctions for other important hybrid combinations of s and p orbitals are given by (8.26) and (8.27).

It is seen that the concept of hybridization is essential in valence bond theory if one wants to represent the wavefunction of a molecule by a single valence bond structure. Thus the wavefunction for CH_4 can be represented by

$$\overset{\frown}{\psi_1 h_1}\overset{\frown}{\psi_2 h_2}\overset{\frown}{\psi_3 h_3}\overset{\frown}{\psi_4 h_4} \tag{13.33}$$

where $\psi_1 \ldots \psi_4$ are the four sp^3 hybrids defined by (8.23), $h_1 \ldots h_4$ are the four hydrogen $1s$ orbitals and the connections indicate the pairing of electrons in these orbitals. By such a function we imply that there is a pairing of orbitals ψ_1 with h_1, ψ_2 with h_2 etc., such that each pairing is represented by a Heitler–London function such as (13.28), the whole wavefunction being anti-symmetrized. Without hybridization one could still construct functions like

$$\overset{\frown}{2sh_1}\overset{\frown}{2p_x h_2}\overset{\frown}{2p_y h_3}\overset{\frown}{2p_z h_4} \tag{13.34}$$

but many such pairing schemes would have to be combined in the total wavefunction $\overset{\frown}{2sh_2}$ etc.) for the four bonds to be equivalent.

It is perhaps worth reminding the reader at this point that hybridization does not have an important role in molecular orbital theory. A hybridization scheme may be inferred by analysis of a molecular orbital wavefunction, as we showed in Section 8.3, but there is little to be gained by using hybrid orbitals as an LCAO basis for molecular orbitals rather than atomic orbitals themselves unless one intends to make severe approximations in the secular equations by neglecting many of the off-diagonal Hamiltonian integrals.

13.3. Resonance theory and canonical forms

There are many molecules and ions which cannot be represented by a single valence structure of single or multiple bonds. For example, in acetic acid the two carbon–oxygen bonds have different lengths and this can be understood in terms of

the single valence structure

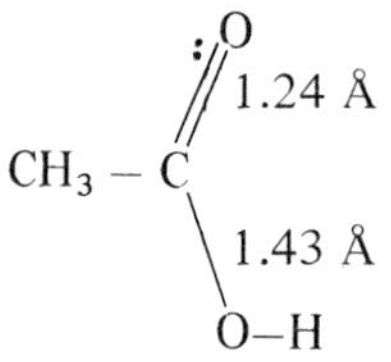

$$\mathrm{CH_3-C}\begin{matrix} //\ddot{\mathrm{O}} \quad 1.24\ \text{Å} \\ \backslash \mathrm{O{-}H} \quad 1.43\ \text{Å} \end{matrix}$$

In the acetate ion however the two carbon–oxygen bonds are equal in length and the negative charge is divided equally between the two oxygen atoms. Clearly the 'classical' structure (A) does not give a correct picture and we must either use the 'non-classical' structure (B) or develop an alternative symbolism.

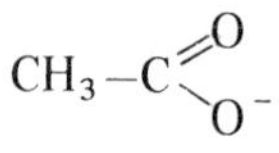

$$\underset{\text{A}}{\mathrm{CH_3-C}\begin{matrix} {=}\mathrm{O} \\ \backslash \mathrm{O^-} \end{matrix}} \qquad \underset{\text{B}}{\mathrm{CH_3-C}\begin{matrix} {\stackrel{...}{=}}\mathrm{O^{-1/2}} \\ {\stackrel{...}{=}}\mathrm{O^{-1/2}} \end{matrix}}$$

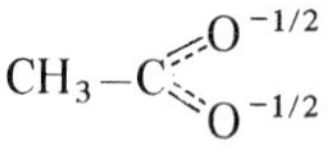

We have already seen that improved valence bond wavefunctions can be obtained by combining wavefunctions for different valence structures, as in (13.27). This mixing of structures is called *resonance*, the term originating from the mathematical formalism which is similar to that used to describe classical resonating systems (e.g. coupled pendulums). Pauling and Wheland developed a pictorial theory called resonance theory which found extensive application in describing the properties and reactions of molecules, especially in organic chemistry. At the present time resonance theory is giving way to a new symbolism, that of a pictorial molecular orbital theory. However it will be many years, if ever, before resonance theory disappears completely from the literature so a brief description will be given here.

In resonance theory the acetate ion is said to be a resonance hybrid of two classical structures or *canonical forms*. A canonical form is a valence structure that obeys the laws of classical valence theory. This is symbolised as follows

$$\mathrm{CH_3-C}\begin{matrix} {=}\mathrm{O} \\ \backslash \mathrm{O^-} \end{matrix} \longleftrightarrow \mathrm{CH_3-C}\begin{matrix} /\mathrm{O^-} \\ {=}\mathrm{O} \end{matrix}$$

The double headed arrow is a specific symbol meaning that the ion cannot be correctly described by either structure, but is in between the two. Unfortunately the term resonance suggests that there is oscillation between the two structures and we must emphasize that this is not the case. A mule is a hybrid; the result of crossing a horse with a donkey. It is not a horse one moment and a donkey the next, but something whose properties are always between the two.

As well as providing a symbolism for representing molecules which cannot be adequately represented by a single classical structure, resonance theory also introduced the concept of *resonance energy*. If we take a wavefunction which is a linear combination of more than one structure then, by applying the variation theorem, we obtain a state which has an energy equal to or lower than that of the

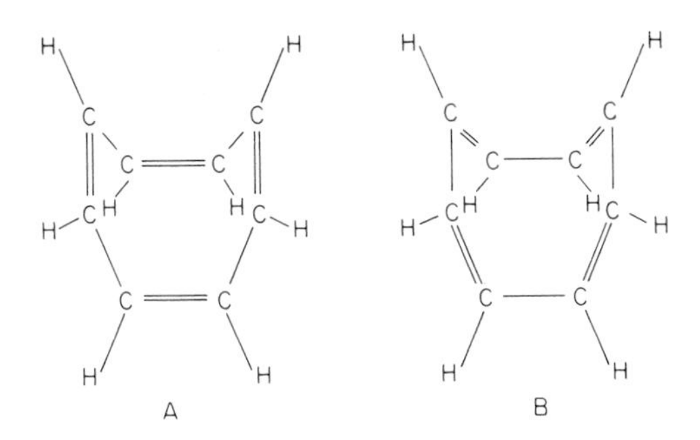

Figure 13.3 Covalent structures for cyclo-octatetraene.

lowest energy structures in the linear combination. This lowering of energy by mixing structures is called the resonance energy.

We can define a set of rules for resonance to occur between two classical valence structures: (i) the nuclei are in the same positions in all structures, that is the structures differ only in electron distribution; (ii) the energies of the contributing structures must be similar; and (iii) the number of electrons whose spins are paired must be the same in all structures.

Some care must be taken over the second point because the energy of a structure depends on several factors. For example, the energy of a covalent bond depends on the overlap of the two orbitals making up the bond. Thus, although we can formally write two covalent structures for cyclooctatetraene, as shown in Figure 13.3, structure A has a much lower energy than B because the π orbital overlap is more favourable.

We can compare the acetate ion with possible resonance structures for ethyl acetate and acetamide:

$$CH_3{-}C\begin{matrix} {=}O \\ {-}O{-}C_2H_5 \end{matrix} \longleftrightarrow CH_3{-}C\begin{matrix} {-}O^- \\ {=}\overset{+}{O}{-}C_2H_5 \end{matrix}$$

$$CH_3{-}C\begin{matrix} {=}O \\ {-}NH_2 \end{matrix} \longleftrightarrow CH_3{-}C\begin{matrix} {-}O^- \\ {=}\overset{+}{N}H_2 \end{matrix} .$$

In both these hybrids the second canonical form is dipolar. In ethyl acetate the ionic form is of very high energy and makes little contribution to the resonance hybrid. In acetamide the dipolar structure, although still of high energy, represents transfer of charge from nitrogen to the more electronegative oxygen and therefore it makes a small but significant contribution to the resonance hybrid. In the acetate ion, shown earlier, the two forms are of equal energy and contribute equally to the hybrid. In conformity with this picture acetamide shows polar properties not exhibited by ethyl acetate.

Probably the most misunderstood concept of the theory is resonance energy. From resonance theory we deduce that a resonance hybrid will have a lower energy than any single canonical form. The theory tells us nothing about the absolute stability of one species relative to another. The assumption that it does is a very common mistake. The acidity of phenol is often attributed to the large number of possible canonical forms one can draw for the phenoxide ion

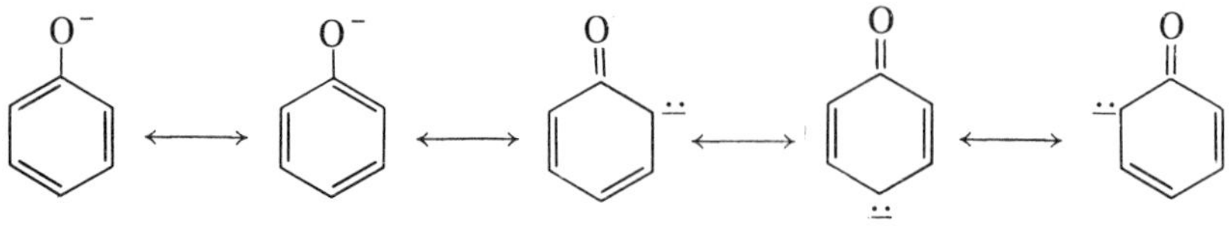

Although the delocalization of the π electrons over the benzene ring will lower the electronic energy it is probable that the spreading of the negative charge (correctly represented in the hybrid) affects the entropy of ionization, and that this rather than the electronic energy is the governing factor.

It is important to realize that the number of canonical forms tells us nothing about the absolute stability of a molecule. In other words, just counting structures is not always a sufficient recipe for solving a problem. In particular, the signs of the Hamiltonian integrals can be important. This can be shown by considering the case of three equivalent structures.

Let the wave functions of the three structures be Ψ_1, Ψ_2, Ψ_3, and the integrals be

$$\begin{aligned} &\int \Psi_i \mathscr{H} \Psi_i \, dv = B \ (i = 1, 2, 3) \\ &\int \Psi_i \mathscr{H} \Psi_j \, dv = b \ (i \neq j) \end{aligned} \tag{13.35}$$

We will assume that the wavefunctions are normalized and orthogonal ($S_{ij}=\delta_{ij}$). The secular determinant obtained by applying the variation theorem to this problem is (equation 6.55)

$$\begin{vmatrix} B-E & b & b \\ b & B-E & b \\ b & b & B-E \end{vmatrix} = 0 \tag{13.36}$$

and on expansion this gives

$$(B-E)^3 - 3b^2(B-E) + 2b^3 = 0, \tag{13.37}$$

which has roots

$$E = B - b, B - b, B + 2b. \tag{13.38}$$

If b is negative the lowest energy state is $B + 2b$ and the resonance stabilization is $2b$. If b is positive the lowest energy state is doubly degenerate with energy $B - b$ and the resonance stabilization is b.

The importance of this result can be shown in the case of the cyclic polyene ions. For both $C_3H_3^+$ and $C_3H_3^-$ three equivalent resonance structures can be drawn

which are of the type

(both C^- and C^+ are trivalent species) and simply by counting structures one would infer equal resonance stabilization. However, for the cation b is negative and for the anion b is positive so that the cation has the larger resonance stabilization. Moreover the degeneracy of the ground state of the anion would imply instability to distortion from the D_{3h} geometry by virtue of the Jahn–Teller theorem (Section 12.4).

Applying such an analysis to the stability of the C_nH_n cyclic ions leads to the same predictions as molecular orbital theory (Section 14.7) that those with $(4r + 2)\,\pi$ electrons ($r = 0,1,2 \ldots$) are stable: just counting structures is not sufficient as this would also imply stability for the ions with $4r\ \pi$ electrons and, at least for the small rings, this is not true.

Chapter 14
The Chemical Bond and Reactivity

14.1. Potential energy surfaces

Valency theories are concerned with explaining why atoms are held together in a particular way to form molecules. They were developed initially to explain the arrangement of atoms in molecules and the strength of the chemical bonds. The reactions which molecules undergo raise much wider issues, but quantum mechanics has proved extremely useful in providing a theoretical background into which empirical studies of reactions and reactivity can be fitted. There are two approaches to the theory of the rate of chemical reactions which can be distinguished by the terms microscopic and macroscopic. In the microscopic approach we attempt to calculate a rate for an individual molecular collision with the reactants and products in defined quantum states. Such rates can in principle be determined experimentally in molecular beam experiments. It is possible to average these rates over quantum states to get the rate for a macroscopic sample.

In the macroscopic approach we attempt to go directly to the rate for a macroscopic sample by using the methods of thermodynamics and statistical mechanics. Only the macroscopic method is feasible for the reactions of complicated molecules because there are so many individual quantum states of the reactants and products, and they are so close in energy, that there is little hope of studying them individually by theory or by experiment. Certainly all reactions that take place in solution are of this type. There is one unifying principle linking the microscopic and macroscopic approaches and that is the concept of the potential energy surface, which we have discussed in Section 5.1 in the context of the Born–Oppenheimer approximation. London in 1929 was the first to see the implications of this to chemical reactions and postulated that most chemical reactions are adiabatic in the sense that they occur over a single electronic potential energy surface. In other words if we were to follow the electronic energy and electronic wavefunction of a molecule or groups of molecules as they undergo a reaction it would be found that these quantities change smoothly on going from reactants to products. There are some reactions where this is not so (non-adiabatic) but not many.

The role of valence theory in the theory of chemical reactivity is to calculate those regions of the potential energy surface which are important for an understanding of the reaction. In the microscopic approach one actually calculates the detailed dynamics of the atoms as they move over this surface and hence we need to know all regions of the surface that are likely to be reached in the reaction.

In the macroscopic theories, the most important of which is called *transition state theory*, we need to know the form of the surface only in the region of the equilibrium configuration of the reactants (this in any case can be deduced by spectroscopic study of the reactants) and at certain other crucial regions of the surface associated with the transition state.

Figure 5.3 shows contours of the potential energy surface for one of the simplest chemical reactions which is the exchange of atoms in the collision of H and H_2 (a reaction that can be followed by examining the *ortho-para* hydrogen interconversion or isotopic exchange). This reaction is known to proceed most easily (require least input of energy) if the three atoms are collinear, and Figure 5.3 shows the surface for such configurations.

The reactants and products are associated with the valleys at either large R_1 or large R_2 and the transition state is the col, or saddle point on the surface which has been marked with a cross. This saddle point is the maximum on the minimum energy route that passes from reactants to products.

Transition state theory is based on the idea that in the transition state we have a quasi-molecular species called the activated complex in equilibrium with reactants. From this assumption the chance of having the system pass through the transition state to reaction can be calculated by statistical mechanics. Of course the transition state does not represent an equilibrium position for a molecule, not even a metastable one, because it is not a true minimum on the surface. At the saddle point the potential is at a minimum with respect to all dimensions except one, and that is the direction of the reaction coordinate which is the curved line passing along the reactant valley, through the transition state, and ending at the product valley. In this direction the surface has positive curvature at the saddle point.

Let us consider the reaction of an atom or radical X with the diatomic molecule YZ to yield the new molecule XY and the atom or radical Z. The experimental fact is that all bimolecular reactions of this type go faster as the temperature is raised. This effect of temperature finds expression in the empirical Arrhenius equation. The rate of formation of XY is proportional to the product of the concentrations of X and YZ

$$d[XY]/dt = k[\mathrm{X}]\,[\mathrm{YZ}], \tag{14.1}$$

and k, the rate constant, is given by the Arrhenius equation

$$k = A\mathrm{e}^{-E/RT} \tag{14.2}$$

A and E are experimentally determined constants, T is the absolute temperature and R is the gas constant. The important feature of this equation is that E (which must be in energy units) represents the energy barrier which the reacting species must equal or exceed if the reaction is to take place. Reactant molecules colliding with an energy less than this will not react, at least according to classical mechanics, and as atoms are relatively heavy particles (in contrast to electrons) their quantum behaviour, which finds expression in the so-called tunnel effect, is small.†

†The tunnel effect in quantum mechanics is associated with the fact that the wavefunction is not necessarily zero at a point where the total energy is less than the potential energy. In other words there is a non-zero (but usually small) probability of finding a particle 'inside' a potential barrier.

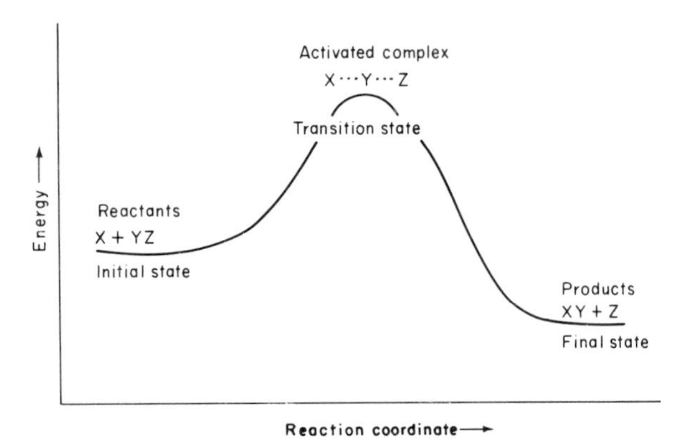

Figure 14.1 Energy against reaction coordinate diagram for a reaction X + YZ → XY + Z.

London was the first person to account for this barrier in terms of quantum mechanics. In the above reaction the atoms Y and Z are initially joined by a single bond, that is by two electrons with opposed spins, while atom X has an uncoupled electron. As X is brought up to Y—Z the interaction between the three electrons causes a decrease in the bond strength of Y—Z so that Y and Z tend to separate. Hence the approach of X to Y—Z is accompanied by an increase in potential energy of the system which is only partly offset by the formation of an X—Y bond. Eventually a point is reached where the X—Y attraction becomes dominant and Z departs with a resulting drop in the overall potential energy of the system. We can demonstrate this diagramatically by plotting the overall potential energy of the system against the extent of reaction as in Figure 14.1.

The experimental energy barrier, called the activation energy, is the difference in potential energy of the initial state and the transition state. However, we have not particularized the approach of the atom X to the molecule YZ nor the path of atom Z when leaving the new molecule XY. X could approach at right angles to the bond Y—Z towards its centre or directly towards Y; alternatively it could approach Y along the axis of the bond Y—Z. We can in theory calculate the energy of the three-electron system for every possible relative position of X, Y, and Z within the terms of the Born—Oppenheimer approximation, that is assuming the nuclei are at rest. If we do this for every possible position of X, Y, and Z we will obtain a potential energy surface with low potential energy areas where X is far away and Y—Z are a normal bond lengths apart, and again where Z is far away and X—Y are a normal bond length apart. Between these areas of low potential energies there will be a barrier when X, Y, and Z are all fairly close together. The lowest point on the barrier between the region representing X and YZ, and the region XY and Z is the transition state and the arrangement of the three nuclei at this point the activated complex.

For one of the simplest reactions, which we have already mentioned,

$$H + H_2 \longrightarrow H_2 + H$$

there have been many attempts to calculate the complete potential energy surface and the most accurate calculation gives an activation energy 41 kJ mol^{-1} compared with the experimental value of 38 kJ mol^{-1}. Most three-atom reactions of this type have a collinear transition state i.e. the lowest energy approach for atom X in the system X + YZ is along the Y—Z molecular axis with Z departing along the same line. Results for these collinear processes are the easiest to compute because we can treat them as a two dimensional problem. However if we consider a slightly more complicated system, the addition of a hydrogen atom to ethylene for example, the determination of the complete potential energy surface is a matter of much greater difficulty. Not only can the hydrogen atom approach from above or in the plane of the ethylene molecule, but as it approaches so the planar nature of the ethylene molecule will be disturbed and if the attack centres on one carbon atom changes in bond lengths and bond angles around the other carbon atom will occur. In other words for every position of the incoming hydrogen atom relative to the carbon atom it is attacking we have to optimize the positions of all the other atoms in the system. That is, we have to move all the other atoms until we obtain a minimum energy for that particular carbon–hydrogen bond distance. Calculations of this kind can be made either using semi-empirical or *ab-initio* methods. Either way they require a lot of computer time and as yet have only proved practicable in the simplest of cases.

Even if we can construct a complete energy surface we are still a long way from being able to calculate the absolute rate of a chemical reaction. As we have seen reactions in general involve reacting species with a wide range of thermally distributed energies and we have to employ the methods of statistical mechanics to calculate the relevant partition functions for the reactants and for the activated complex, the latter being treated exactly as a normal molecule except for the vibration frequency along the coordinate of reaction. Thus in transition state theory we first calculate the complete potential energy surface and so determine the shape of the activated complex. We then use the bond lengths, bond angles, and force constants we have obtained to calculate the appropriate partition functions. The reactants are assumed to be in equilibrium with the activated complex, which decomposes to products at a fixed rate.

If the calculation of absolute rates for reactions occurring in the gas phase is difficult, the calculation of absolute rates for reactions in solution is almost impossible at the present time. Not only is there a problem of diffusion of reactive species through the solution, but each species including the activated complex will interact to a greater or lesser extent with solvent molecules. The models of solvation in mixed solvents are too crude at the present time to enable meaningful calculations to be attempted. Furthermore we must remember that for reactions at room temperature ($RT \approx 2.5$ kJ mol^{-1}) an uncertainty of the activation energy of only 5 kJ mol^{-1} represents an uncertainty in rate of a factor of seven. Since chemists are often interested in rates which vary by a factor of two or less some idea of the magnitude of the task of trying to replace kinetic experiments by the computer can be appreciated.

Although quantum mechanics cannot, as yet, be used to determine absolute rates for any but the simplest reactions, it can prove of great value in providing a semi-quantitative framework in which to correlate results. Even simple procedures like those of Hückel theory have proved useful in this way. There are three possible approaches. The first is to look at the properties of the ground state of one of the reactants, such as charge density and try and correlate these with reactivity. The second approach is to try and calculate the loss in electronic energy associated with forming the activated complex. The third is to combine these approaches and to investigate the interaction between the highest occupied molecular orbital of one species with the lowest unoccupied molecular orbital of the other (see Section 9.5). These approaches called respectively isolated molecule theory, localisation theory, and frontier orbital theory will be discussed briefly in the next three sections.

14.2. Isolated Molecule Theory of Reaction Rates

In the isolated molecule treatment we examine the orbitals of the molecule whose reactivity we wish to study, and attempt to relate some quantity calculable from the orbitals with reactivity. In the 1950s a large number of parameters were suggested such as free valence, self polarizability, etc., but experience has shown that most of these are of limited value and probably the only parameter used widely today is the charge density. The following example will illustrate the approach.

The coupling of aromatic diazonium salts with phenols often involves the attack by the diazonium cation (RN_2^+) with the phenoxide anion. As a model of the phenoxide anion we can take the benzyl anion (i.e. exchanging CH_2 for the isolelectronic O). Benzyl is an odd alternant radical and using the procedure in Section 9.4 we can readily determine the Hückel coefficients. According to equation (9.61) the sum of the coefficients around any one atom is zero and the smaller (unstarred) set have zero coefficients. If we put the coefficient for the non-bonding orbital at the *para*-carbon atom equal to a, the values for the non-bonding orbitals at each carbon atom are as in Figure 14.2.

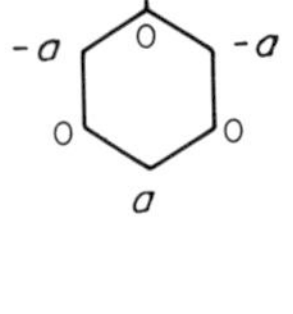

2a
-a 0 -a
0 0
a

Figure 14.2 Coefficients of the non-bonding orbital of the benzyl radical.

The normalization condition requires $7a^2 = 1$, hence $a = 1/\sqrt{7}$. Because the π-electron density is unity for all atoms for a neutral alternant, the distribution of negative charge in the benzyl anion is equal to the density of the extra electron that occupies the non-bonding orbital. The charge distribution is shown in Figure 14.3.

Similar calculations on (CH_2) substituted naphthalenes as models for the charge distribution in the 1-naphthol and 2-naphthol anions give the results shown in Figure 14.4.

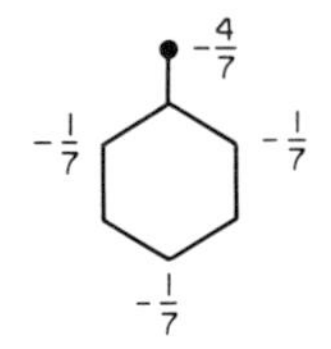

Figure 14.3 Charge densities in the benzyl anion.

The common red dye formed when benzenediazonium chloride couples with 2-naphthol is the result of attack by the diazonium cation on the 1-position. When 1-naphthol is used a mixture of dyes resulting from attack at both the 2- and 4-positions is formed. Furthermore no attack occurs at all on 2-naphthol derivatives if the 1-position is blocked, (e.g. as in 1-methyl-2-naphthol).

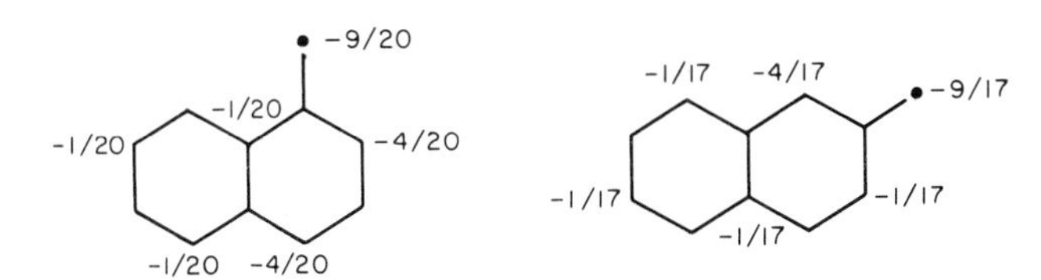

Figure 14.4 Charge densities in the 1- and 2-naphthol anions.

Alternant aromatic hydrocarbons are non-polar and do not couple with diazonium salts. However azulene, a non-alternant aromatic compound, is polar and does couple with the benzenediazonium cation. We can calculate the charge densities for all the positions in azulene using Hückel theory (see Figure 14.5) and the site carrying the highest negative charge, the 1-position, is indeed the site at which coupling occurs. However, we have already seen that the dipole moment calculated for azulene, from the Hückel charge distribution, is far too large (see Table 9.2). Because no allowance is made for electron repulsion in such calculations.

Several methods have been developed which account to some degree for specific electron repulsion. The first of these was due to Wheland and Mann who suggested that the Coulomb integral of Hückel theory (α) should be related to the π-electron density on the atom (q) through the expression

$$\alpha_r = \alpha_0 + (1 - q_r)\omega\beta. \tag{14.3}$$

If $q_r = 1$ the Coulomb integral has the value α_0 of a neutral atom. ω is a dimensionless parameter and the value 1.4 appears to give, overall, the best results. The C—C resonance integral β provides the necessary dimensions of energy. For $q_r > 1$, α_r will be greater than α_0 because β is a negative quantity, and this is in accord with the principle that an excess of electrons on an atom will make it less electron attracting.

The use of expression (14.3) to calculate molecular orbitals requires an iterative procedure. Firstly we must guess the value of q_r and calculate an improved value of

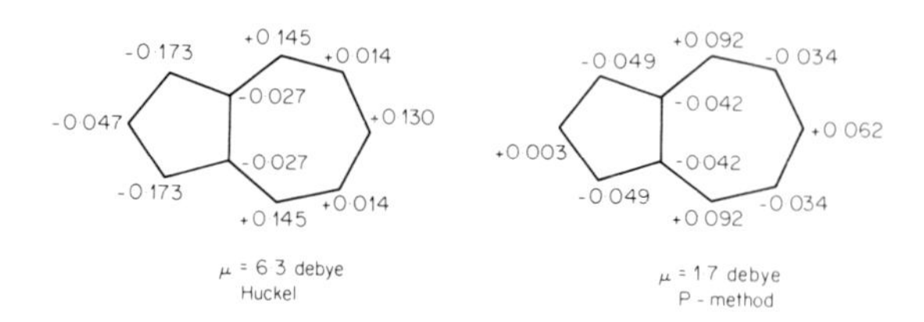

Figure 14.5 The net atom charges and dipole moment of azulene as calculated by the Hückel and P-methods.

α_r. The cycle is continued until a constant result is obtained. The Wheland-Mann method therefore belongs to the self-consistent-field class of calculation (Section 3.4).

In 1953 a much more elaborate SCF treatment of π electrons was introduced separately by Pariser and Parr and by Pople, and this is often referred to as the P-method. It extends the Wheland–Mann method firstly in modifying the Coulomb integrals α_r according to the electron density on all the atoms, and not (as in equation 14.2) just by the density on atom r. Secondly it modifies the resonance integral for electron repulsion; β_{rs} is corrected by a factor proportional to the product of the bond order (equation 9.49) p_{rs} and the repulsion energy between an electron on r and one on s. The details of the P-method do not appear appropriate for this book,† but it is perhaps interesting to examine a typical result. Figure 14.5 shows the net atom charges $(1 - q_r)$ produced by Hückel theory and by the P-method for a typical non-alternant hydrocarbon, azulene. The SCF procedure greatly reduces the polarity of the molecule due to the fact that atoms carrying net negative charge are less electron attracting than those carrying net positive charges. There is therefore an inbuilt tendency to equalize charges.

14.3. Localization energy

If we know the geometries of the reactants and of the transition state then it is a straightforward procedure to calculate the activation energy by molecular orbital theory. One simply calculates the total energy of the activated complex and the total energy of the reactants and the difference between them is the quantity of interest. To have relevance to the experimental situation we must assume that molecular orbital theory, which, it must be emphasized, does not lead to an exact solution of the Schrödinger equation, is in error by the same amount for reactants and products, and also any complications from the surrounding medium, such as solvation, must be allowed for.

The most difficult aspect of such calculations is to determine the geometry of the transition state. This cannot be determined by spectroscopic or diffraction

†See for example, *Semi-Empirical SCF Theory*, J. N. Murrell and A. J. Harget, Wiley-Interscience, 1972.

measurements because the activated complex does not exist for finite times. We must therefore determine the geometry by calculating the energy for various configurations over the region of the potential energy surface where the transition state is suspected, and finding the point which is a minimum in all directions except one, and a maximum in that one, which is the direction of the reaction coordinate.

If we are interested in the relative reactivities of a family of similar molecules under similar experimental conditions then it may be justified to assume a common structure for the activated complex. This assumption is usually connected with the use of simplified methods of calculating the activation energy, as we shall see. Providing that pre-exponential A-factors are constant for the series of reactions we can write, from equation (14.2)

$$E - E_0 = RT \log \frac{k}{k_0} = \Delta E^{\neq}, \tag{14.4}$$

where E_0 and k_0 are the appropriate values for one member of the series chosen as a standard and all other members are measured against this. $\Delta E^{\neq}$ is therefore the difference in activation energy of one reaction from that of the standard reaction, which we may hope to calculate and correlate with the logarithm of the ratio of the rate constants.

Many reactions in organic chemistry involve transient intermediates of high energy which undergo further reaction to yield the products. A particularly important example is the addition-with-elimination reactions of aromatic compounds which result in the substitution of a hydrogen atom by a group like NO_2. The intermediate in this particular case is called the Wheland intermediate and it is represented by the structure shown in Figure 14.6. The geometry of this intermediate has been calculated for the simplest case of hydrogen atom exchange and is as shown in the figure.

If we draw a reaction coordinate diagram for such reactions we see that there are two transition states (Figure 14.7). Whichever is the higher of the two will determine the activation energy for the overall reaction.

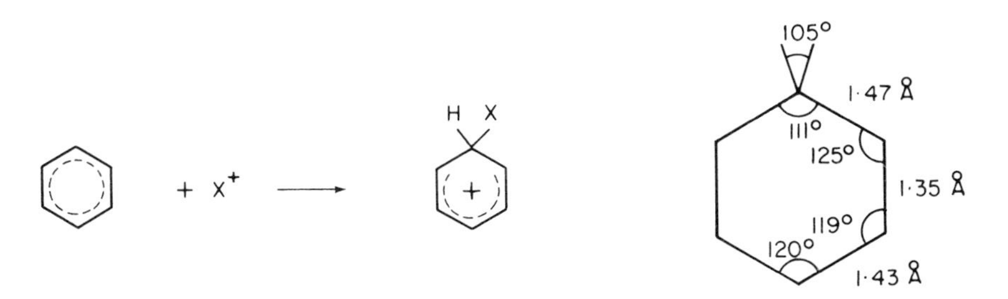

Figure 14.6 The formation of the Wheland intermediate and its structure as determined by *ab-initio* calculations for the situation where X = H.

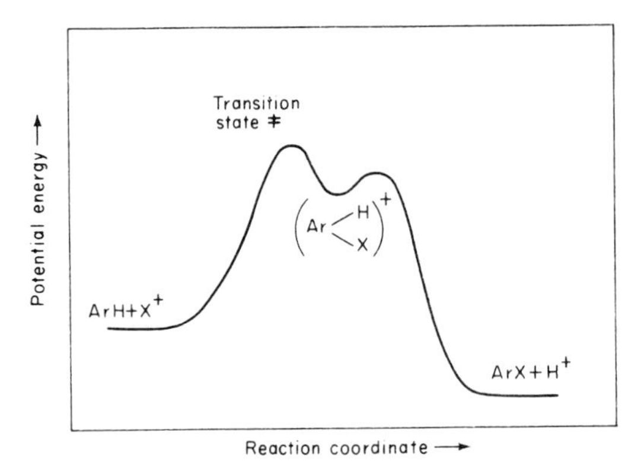

Figure 14.7 The potential energy diagram for electrophilic substitution.

It can be assumed that the Wheland intermediate is near in energy to the transition state and it can therefore be used as a model on which to calculate the activation energy. This is done in Hückel theory by calculating the energy lost if we localize two of the π electrons in the aromatic nucleus to form a bond with the incoming electrophile (NO_2^+ say). If the energy of the parent hydrocarbon is $a\alpha + b\beta$ and that the Wheland intermediate is $(a-2)\alpha + b_m^+\beta$ then the localization energy for electrophilic attack is defined as

$$L_m^+ = (b - b_m^+)\beta. \tag{14.5}$$

This represents only that part of the total energy, due to the π electrons, which changes from one site to another in the molecule or from one aromatic molecule to another. The assumption we make is that the remainder of the energy is the same for all sites. For example, the π-electron energy for benzene computed by Hückel theory is $6\alpha + 8.0\beta$ (see equation 9.44) while the π-electron energy for the pentadienyl cation (i.e. the Wheland intermediate shown in Figure 14.6) is $4\alpha + 5.46\beta$. The localization energy for attack on benzene is thus 2.54β.

Although such calculations are now straightforward, by standard computer programs at several levels, Hückel or *ab-initio* for example, this has only recently been the case. For this reason there was considerable activity in the field of organic reactivity during 1940–1970 to obtain approximate methods of estimating the molecular orbital activation energy. The names of Dewar, Fukui, Brown, Hudson, and Klopman are a few of the many connected with this area of work. Because this work led to some important concepts, in particular to the so-called *frontier orbital* concept, it retains its value even in this computer age. The basis for all this work is perturbation theory which was developed in Chapter 11 and which we will now make use of.

14.4. The perturbation MO theory of chemical reactions

Let us consider the change in molecular orbitals and their energies when two molecular species R and S come together. A molecular orbital calculation on the super molecule RS differs from that of its components in two ways. Firstly there will be non-zero Hamiltonian integrals involving the atomic orbitals of R and those of S which have to be considered. In a Hückel-type theory, for example, we would identify these as

$$\int \phi_m^{\mathrm{R}} \mathscr{H} \phi_n^{\mathrm{S}} \, \mathrm{d}v = \beta_{mn}^{\mathrm{RS}} \tag{14.6}$$

and this would be non-zero if there was a bond between the atoms associated with atomic orbitals ϕ_m^{R} of R and ϕ_n^{S} of S. Secondly, the integrals such as

$$\int \phi_m^{\mathrm{R}} \mathscr{H} \phi_m^{\mathrm{R}} \, \mathrm{d}v = \alpha_m^{\mathrm{R}} \tag{14.7}$$

or

$$\int \phi_m^{\mathrm{R}} \mathscr{H} \phi_n^{\mathrm{R}} \, \mathrm{d}v = \beta_{mn}^{\mathrm{R}} \tag{14.8}$$

would have a different value for RS than they have for R on its own because $\mathscr{H}$ will contain terms not only for the nuclei and electrons of R but also for the nuclei and electrons of S.

In some theories integrals like (14.7) and (14.8) are assumed not to change as the two components are brought together. This is quite a good approximation if R and S are non-polar species, such as alternant hydrocarbons, because the changes in $\mathscr{H}$ due to the presence of S can be attributed to the electrostatic field of S and this is negligible outside the electron density of a neutral, non-polar molecule. In such theories the change in the orbital energies is entirely due to delocalization effects brought about by the integrals such as (14.6). It can be shown moreover that even if the other integrals do change on bringing the two components together then the effects of this on the total energy can be approximated by just calculating the electrostatic (Coulomb) energy between the two components and adding this energy to whatever is calculated from the delocalization effect. This is by no means an exact procedure but it has been widely adopted and gives, as we shall see, useful answers.

The problem we pose is therefore what is the delocalization energy arising from the interaction between R and S? We shall calculate this within a Hückel-type theory assuming that the interaction is sufficiently small for perturbation theory to be valid. In effect a small perturbation is one in which R and S retain their separate characteristics to a large extent even when they are joined together.

To apply perturbation theory to the mixing of molecular orbitals of the two components we must distinguish two cases. The easiest situation to analyse is one in which there is no degeneracy between the two sets of molecular orbitals (no orbital of R has the same energy as an orbital of S). We can then apply the equation of second-order perturbation theory (11.27) to calculate the energy change. The more difficult case is when there is degeneracy because we then have to solve the secular equation (6.74) for each set of degenerate orbitals, with the perturbation term as Hamiltonian. We will deal with this situation first.

Suppose that there is degeneracy between an orbital of R whose wavefunction is

ψ^{R} and one of S of wavefunction ψ^{S}. These wavefunctions will be linear combinations of the atomic orbitals ϕ_m^{R} and ϕ_n^{S} of R and S respectively and in the Hückel approximations they will be orthogonal because they are made up of different sets of atomic orbitals

$$\psi^{\mathrm{R}} = \sum_m c_m \phi_m^{\mathrm{R}},$$

$$\text{and} \quad \psi^{\mathrm{S}} = \sum_n c_n \phi_n^{\mathrm{S}}, \tag{14.9}$$

The molecular orbitals of the combined system will have the form

$$\psi = a\psi^{\mathrm{R}} + b\psi^{\mathrm{S}}, \tag{14.10}$$

and the coefficients (a,b) are found by solving the secular equations for which the secular determinant (equation 6.55) is as follows:

$$\begin{vmatrix} H'_{\mathrm{RR}} - E & H'_{\mathrm{RS}} \\ H'_{\mathrm{RS}} & H'_{\mathrm{SS}} - E \end{vmatrix} = 0, \tag{14.11}$$

where

$$H'_{\mathrm{RR}} \equiv \int \psi^{\mathrm{R}} \mathscr{H}' \psi^{\mathrm{R}}\, \mathrm{d}v; \quad H'_{\mathrm{SS}} \equiv \int \psi^{\mathrm{S}} \mathscr{H}' \psi^{\mathrm{S}}\, \mathrm{d}v, \tag{14.12}$$

$$H'_{\mathrm{RS}} \equiv \int \psi^{\mathrm{R}} \mathscr{H}' \psi^{\mathrm{S}}\, \mathrm{d}v. \tag{14.13}$$

The integrals H'_{RR} and H'_{SS} are made up of terms like (14.7) and (14.8) and can be taken as zero if we are just calculating the delocalization energy. In this case the two solutions of (14.11) are

$$E = \pm H'_{\mathrm{RS}}, \tag{14.14}$$

which on substituting (14.9) into (14.13) can be expanded to give

$$\begin{aligned} E &= \pm \sum_m \sum_n c_m c_n \int \phi_m^{\mathrm{R}} \mathscr{H}' \phi_n^{\mathrm{S}}\, \mathrm{d}v \\ &= \pm \sum_m \sum_n c_m c_n \beta_{mn}^{\mathrm{RS}}. \end{aligned} \tag{14.15}$$

Although it is not relevant to our discussion at this point we note that on substituting these solutions back into the secular equations we arrive at the two wavefunctions

$$\psi^{+} = \psi^{\mathrm{R}} + \psi^{\mathrm{S}},$$

$$\text{and} \tag{14.16}$$

$$\psi^{-} = \psi^{\mathrm{R}} - \psi^{\mathrm{S}}.$$

One orbital has a lower energy than that of the two component orbitals, by $|H'_{RS}|$ and the other has an energy higher by $|H'_{RS}|$.

We must now consider the allocation of electrons to these orbitals. In the separate components the orbitals ψ^R and ψ^S may be empty or may contain one or two electrons. In the combined system we therefore have 0. 1, 2, 3, or 4 electrons to go into these orbitals. If there are no electrons there is no gain in energy. If there are four electrons there is no gain in energy because we occupy both the stabilizing and the destabilizing orbital. However if we have 1, 2, or 3 electrons there will be net stabilization because we have more electrons in the stabilizing than in the destabilizing orbital.

Dewar has made extensive use of a situation in which there is stabilization by the interaction of degenerate orbitals. Let us consider the union of two alternant radicals R and S. We have seen that there is no net stabilization from the pairwise interaction of completely filled bonding orbitals or completely empty antibonding orbitals. Stabilization *will* arise from the pairwise interaction of the half-filled non-bonding orbitals whose wavefunctions we represent by ψ_0^R and ψ_0^S. This interaction gives rise to two molecular orbitals of the combined molecule RS, whose wavefunctions are $\sqrt{½}(\psi_0^R \pm \psi_0^S)$. The lower energy of these orbitals in Hückel theory is, from (14.15),

$$\left| \sum_{m,n \to m} c_{0m}^R c_{0n}^S \right| \beta_{mn}^{RS}, \tag{14.17}$$

where $n{\to}m$ indicates that atoms m (in R) and n (in S) are joined. C_{0m}^R and C_{0n}^S are the LCAO coefficients of ψ_0^R and ψ_0^S respectively. The total energy of the resultant molecule RS is given in the above approximation by

$$E_{RS} = E_R + E_S + 2\beta \left| \sum_{m,n \to n} c_{0m}^R c_{0n}^S \right|, \tag{14.18}$$

the factor of 2 arising because two electrons will occupy this orbital. We shall see in Section 14.7 that this approach can be used to assess the aromaticity of a molecule but at present we are concerned with reactivity.

We can use Dewar's first-order perturbation treatment to determine the localization energy in a very simple way. We can regard the Wheland intermediate in an electrophilic substitution as a single carbon atom R($c_{0m}^R = 1$) and a pentadienyl cation S. The energy lost in isolating two π electrons at one atom is therefore given by

$$E_{RS} - E_R - E_S = 2\beta \left| \left(\sum_n c_{0n}^S \right) \right|, \tag{14.19}$$

where $\Sigma_n c_{0n}^S$ is the sum of the coefficients of the non-bonded orbital of S on atoms which are joined to the single atom of R. The sum $N_i = |\Sigma_n c_{0n}^S|$ is called the *Dewar number*. The smaller the Dewar number, the smaller the interaction energy between R and S, and the easier it is to isolate one carbon atom (R) from the π-electron system of the molecule RS.

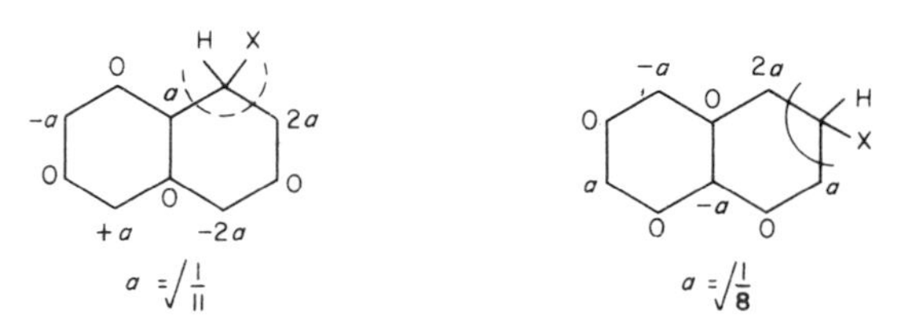

Figure 14.8 Coefficients of the non-bonding orbital of the Wheland complexes.

We can illustrate this method by computing the Dewar numbers for electrophilic attack on naphthalene making use of the procedure for determining the coefficients of the non-bonding orbital as defined in Hückel theory (equation 9.61). The two non-bonding orbitals are shown in Figure 14.8. From the normalization conditions $a = 1/\sqrt{11}$ for attack at the 1 position and $1/\sqrt{8}$ for attack at the 2 position. The Dewar numbers are therefore $N_1 = 6/\sqrt{11} \doteq 1.81$ and $N_2 = 6/\sqrt{8} = 2.12$. N_i is smaller for attack at the 1-position so that electrophilic attack would be expected to occur preferentially at the 1-position, as indeed it does.

14.5. Frontier orbital theory

We have seen from the above analysis of the degenerate case that there is no delocalization energy from the interaction of degenerate orbitals which are both fully occupied or both empty. This result is true also for the second-order mixing of non-degenerate filled orbitals or empty orbitals as can be seen from the following analysis.

Two orbitals ψ_i^{R} and ψ_j^{S} of energies E_i and E_j respectively, which are not degenerate will mix in perturbation theory to give new orbitals whose energies are, to second order, from equation (11.27)

$$E_i'' = \frac{(H_{ij}')^2}{E_i - E_j}, \tag{14.20}$$

and

$$E_j'' = \frac{(H_{ij}')^2}{E_j - E_i} = -\frac{(H_{ij}')^2}{E_i - E_j}. \tag{14.21}$$

If $E_i > E_j$ then the energy E_i'' is greater than E_i and E_j'' is less than E_j, the difference being equal in magnitude in the two cases. It follows that if both ψ_i^{R} and ψ_j^{S} are empty orbitals and the perturbed orbitals are also both empty, then there is no gain in energy. If they are both doubly occupied then there is again no gain in energy. Only if there are 1, 2, or 3 electrons shared between the two is there some gain.

The most important situation for our purpose is when one of ψ_i^{R} and ψ_j^{S} is doubly occupied in the separate compounds and the other is empty as this gives the maximum gain in second-order delocalization energy by placing the two electrons in the lower of the two orbitals. We can sum up all second-order interactions of the form (14.21) to obtain the expression

$$E'' = -2\left\{\sum_{i(\mathrm{R})}\sum_{j(\mathrm{S})} + \sum_{i(\mathrm{S})}\sum_{j(\mathrm{R})}\right\}\frac{(H'_{ij})^2}{E_i - E_j}. \tag{14.22}$$

The factor 2 arising because there are two electrons in each occupied orbital (j); the unoccupied orbitals are labelled (i).

The individual integrals H'_{ij} can also, in the Hückel model, be expressed in β units by inserting an expression such as

$$H'_{ij} = \sum_m \sum_n c^{\mathrm{R}}_{im}\, c^{\mathrm{S}}_{in}\, \beta^{\mathrm{RS}}_{mn}. \tag{14.23}$$

The total interaction energy between the two components R and S can now be obtained by summing three contributions: the first order energy arising from the mixing of incompletely filled degenerate orbitals (if any); the second-order energy arising from the mixing of filled and empty orbitals (or incompletely filled orbitals if there are any); and finally a Coulomb energy which may be expressed as the sum of all pair interactions between the atom charges of R and the atom charges of S as follows

$$E_{\mathrm{Coul}} = \sum_m \sum_n \frac{q^{\mathrm{R}}_m\ q^{\mathrm{S}}_n}{R_{mn}}, \tag{14.24}$$

R_{mn} being the distance between atoms m and n. We write (E' being the first-order energy)

$$E_{\mathrm{Total}} = E_{\mathrm{Coul}} + E' + E'' \tag{14.25}$$

and can now investigate the implications of this expression.

We first examine the case where there is a large energy gap between all occupied and unoccupied orbitals so that the energy terms $E_i - E_j$ that appear in the denominator of (14.22) are all large. In this case it may be a valid approximation to replace the individual energy differences by an average for the whole set

$$\Delta E = \overline{E_i - E_j}. \tag{14.26}$$

Expression (14.22) with this approximation and with the insertion of (14.23) becomes

$$E'' = \frac{-2}{\Delta E}\left\{\sum_{i(\mathrm{R})}\sum_{j(\mathrm{S})}\left[\sum_m \sum_n c^{\mathrm{R}}_{im}\, c^{\mathrm{S}}_{jn}\, \beta^{\mathrm{RS}}_{mn}\right]^2 + \sum_{i(\mathrm{S})}\sum_{j(\mathrm{R})}\left[\sum_m \sum_n c^{\mathrm{R}}_{jm}\, c^{\mathrm{S}}_{in}\, \beta^{\mathrm{RS}}_{mn}\right]^2\right\}. \tag{14.27}$$

Suppose now that R and S are joined through only one bond which links specific atoms m in R and n in S. There will only be one non-zero β and expression (14.27) becomes

$$E'' = \frac{-2(\beta_{mn}^{RS})^2}{\Delta E}\left\{\sum_{i(R)}(c_{im}^{R})^2\sum_{j(S)}(c_{jn}^{S})^2 + \sum_{i(S)}(c_{in}^{S})^2\sum_{j(R)}(c_{im}^{R})^2\right\}. \quad (14.28)$$

The summations over the occupied orbitals (j) can now be replaced by the atom charges using expression (9.45); for example

$$\sum_{j(S)}(c_{jn}^{S})^2 = q_n^S/2. \quad (14.29)$$

From the normalization condition of the coefficients (9.58) we have

$$\sum_{j(S)}(c_{jn}^{S})^2 + \sum_{i(S)}(c_{in}^{S})^2 = 1, \quad (14.30a)$$

hence

$$\sum_{i(S)}(c_{in}^{S})^2 = 1 - q_n^S/2. \quad (14.30b)$$

Thus all the summations over orbitals in (14.28) can be replaced by net atom charges to give the expression

$$E'' = \frac{-(\beta_{mn}^{RS})^2}{\Delta E}\left\{q_n^S + q_m^R - q_n^S q_m^R\right\}. \quad (14.31)$$

As q is a positive quantity $q_n^S + q_n^m - q_n^S q_m^R$ will be maximized when one of q_n^S and q_m^R is maximized and the other minimized. Hence with the approximations we have made the maximum stabilization through the term E'' will occur when we join two atoms, one having a large electron density and the other a small electron density. This condition is also the one which maximizes E_{Coul}, hence we conclude that the limiting case of a large energy difference between occupied and vacant orbitals leads to reactions that are *charge controlled*.

We now turn to the case where the energies of the highest filled orbital of one fragment (ψ_r of R say) is very close to the energy of the lowest vacant orbital of the other (ψ_s of S). In this case one of the denominators in (14.22) will be very small ($E_s - E_r$) and the corresponding term $(H'_{rs})^2/(E_s - E_r)$ will, in general, dominate expression (14.22). Two orbitals of this type are called *frontier orbitals*. The interaction between them will be strong and lead to a partial transfer of charge from R (the donar) to S (the acceptor). Such a situation will be considered in more general terms in Section 15.3.

The interaction of the two frontier orbitals will be maximized when $(H'_{rs})^2$ is maximized, so that we are concerned with the factor

$$\left[\sum_m\sum_n c_{rm}^{R}\, c_{sn}^{S}\, \beta_{mn}^{RS}\right]^2. \quad (14.32)$$

If there is only one bond between the two fragments, then the most favourable configuration will be when this is between the two atoms having the largest coefficients in the two frontier orbitals.

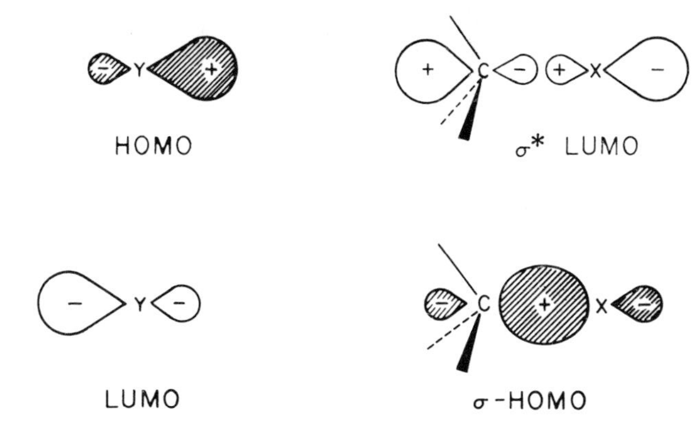

Figure 14.9 Highest occupied and lowest unoccupied molecular orbitals in the initial state for nucleophilic substitution.

Frontier orbital theory provides a very useful pictorial theory into which much of organic chemistry can be fitted. One of the most important reactions is the displacement of a halogen from an aliphatic compound by an anion, so called nucleophilic substitution. The nucleophile is the donor and it will therefore seek to interact with the lowest unoccupied molecular orbital (LUMO) of the carbon

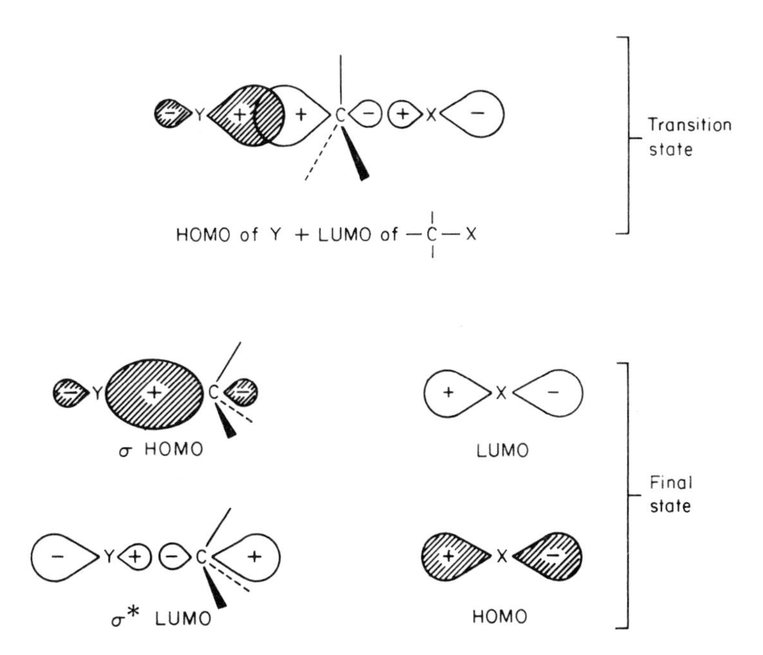

Figure 14.10 Highest occupied and lowest unoccupied molecular orbitals in the transition and final state for nucleophilic substitution.

halogen bond (Figure 14.9). The donation of two electrons from the highest occupied molecular orbital (HOMO) of the nucleophile to the LUMO of the carbon–halogen bond must be followed by the complete transfer of the original bonding pair to the halogen (Figure 14.10).

The attack at a carbonyl group provides a second example. In contrast to ethylene, for which the π orbitals have equal contributions from the two carbon atomic orbitals, the bonding π orbital of C=O (the HOMO) has a larger coefficient from the oxygen atomic orbital while the π^* orbital (the LUMO) has a larger coefficient of the carbon atomic orbital (see Figure 9.12). Thus nucleophiles, which are donors, attack the carbon atom but electrophiles, which are acceptors, attack the oxygen.

14.6. Cycloaddition reactions and the Woodward–Hoffmann rules

If the reaction between R and S involves two or more orbitals then for the reaction to proceed readily all the interacting orbitals must match, that is the relative signs of the wavefunctions must be such that they are all bonding throughout the addition processes. For the interaction between a diene and an alkene, the so-called Diels--Alder reaction, the frontier orbitals are dipicted in Figure 14.11.

It can be seen from the figure that the HOMO orbital of one molecule and the LUMO orbital of the other have the same symmetry in both cases. As the energy difference $E_{LUMO} - E_{HOMO}$ is the same, within the Hückel approximations, for both interactions, both pairs of orbitals can be considered as frontier orbitals and both interactions will facilitate the addition reaction. The addition should occur even more readily if either of these energy differences is reduced and this can be done by putting an acceptor group on either molecule (which lowers E_{LUMO}) or a donor group on either (which raises E_{HOMO}). In particular, acceptor groups on the alkene, as in maleic anhydride, lead to a much more facile addition. Care must be taken to distinguish between the term concerted (which the Diels–Alder reaction certainly is) and synchronous (which it may not be).

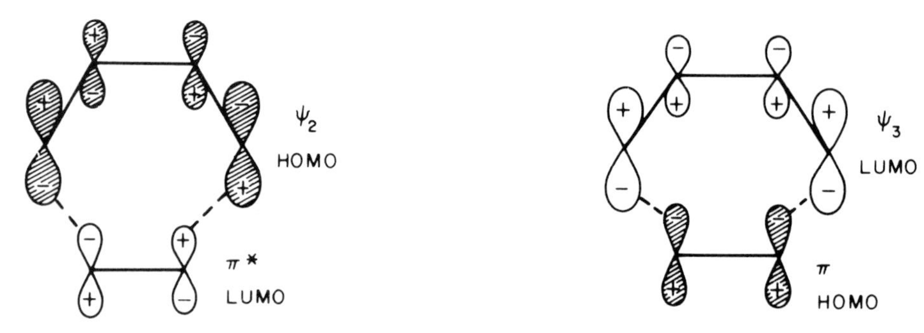

Figure 14.11 Frontier orbitals for the Diels-Alder addition of an alkene to a diene.

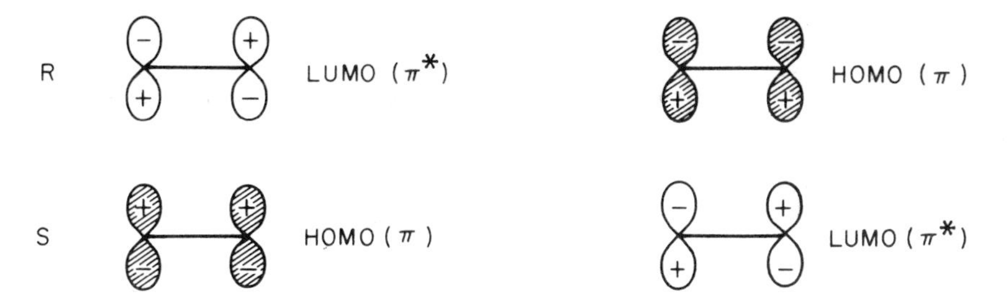

Figure 14.12 Frontier orbitals for the dimerization of ethylene.

The possible interaction between two ethylene molecules to form cyclobutane does not have matching frontier orbitals as can be seen from Figure 14.12. However, if one of the ethylene molecules (S) is raised to its excited state so that it has one electron in the π orbital and one in the π^* then the HOMO frontier orbital of S can be considered to be π^* and this has the same symmetry as the LUMO (π^*) orbital of R. Thus ethylene molecules dimerize when irradiated. Moreover, a study of substituted ethylenes (e.g. *cis*- or *trans*-butene) shows from the products that the reaction is concerted, that is, both new bonds are formed simultaneously.

Ethylenes containing powerful electron attracting or accepting groups will form cyclobutanes without irradiation. Study of the stereochemistry of these additions shows that they proceed in a two step, non-concerted process with a rotation occurring after the addition at one end of each ethylene so that the interacting orbitals can match as in Figure 14.13.

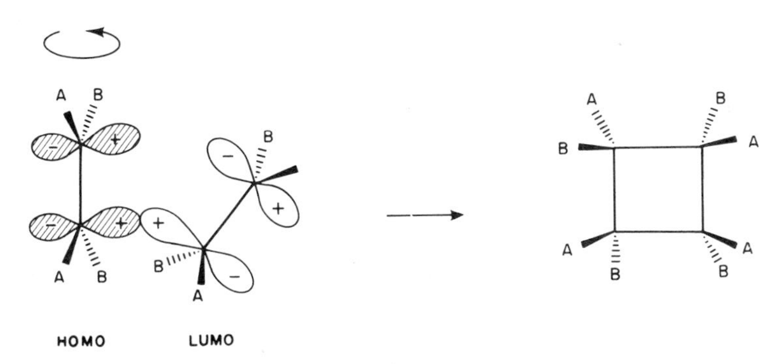

Figure 14.13 Dimerization of alkenes by a non-concerted pathway.

A very important group of cycloadditions are the so-called 1, 3-dipolar reactions in which dipolar substances such as diazomethane $H_2C^-{-}N^+{=}N$, isocyanates $R{-}C{\equiv}N^+{-}O^-$, or azides $R{-}N{=}N^+{=}N^-$ add to ethylenic double bonds. The 1, 3-dipolar compounds have three π molecular orbitals, like those of the allyl radical, which contain four electrons. The frontier orbitals are shown in Figure 14.14. The stereochemistry of products from these additions confirms that the reaction is concerted.

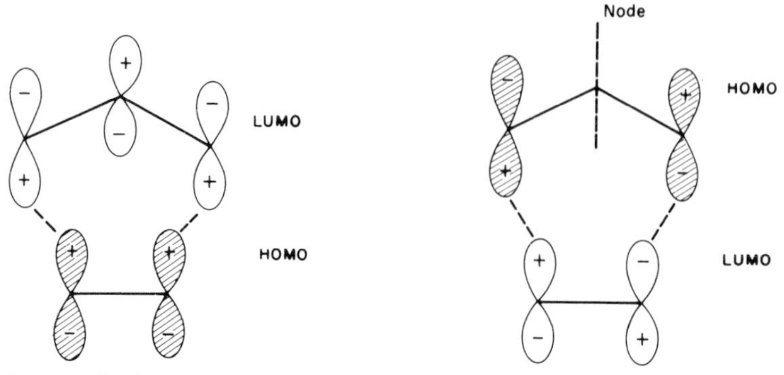

Figure 14.14 Frontier orbitals for 1,3-dipolar addition reactions to alkenes.

The Cope rearrangement (Figure 14.15) can also be interpreted in terms of frontier orbital theory.

Figure 14.15 The Cope rearrangement.

There is a large range of cyclic-addition, rearrangement, and decomposition reactions which can be interpreted by frontier orbital theory. As we have seen, an important feature of the theory is that the orbitals should match during a reaction process. The simplest way of confirming this is to compare the symmetry of the orbitals of the reactants with those of the products.

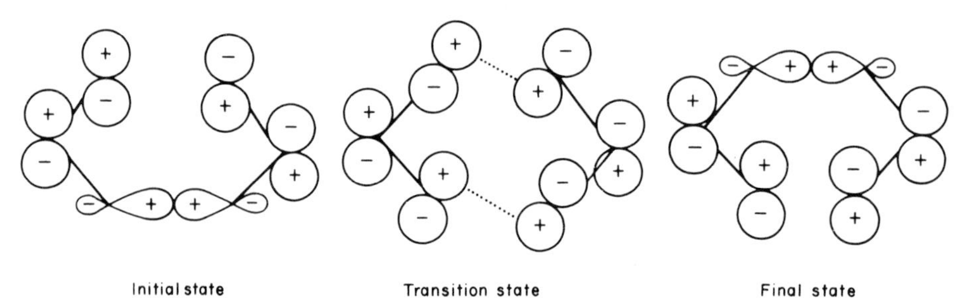

Figure 14.16 Frontier orbitals for the Cope rearrangement.

Although the precise structure of the transition state is uncertain, there is good evidence that it has a chair conformation. We have depicted the bonding orbitals of the two hexa-1,5-dienes and shown how the orbitals can match through the transition state if it involves two allyl fragments. Notice that the non-bonding orbitals of the allyl fragments are shown, in other words we are looking at the frontier orbitals and they match throughout the reaction (Figure 14.16).

We will take as an example the ring opening of cyclobutene to give butadiene. If we concentrate our attention on the bonds that change in the reaction then these are one σ and one π bond of the reactants which become two π bonds of the product. The product σ and π bonds will be associated with bonding and antibonding orbitals as shown in Figure 14.17.

The π orbitals of butadiene are four in number as has been shown in Figure 9.3 and they are shown schematically again in Figure 14.17. The cyclobutene ring can open in two ways which are referred to as conrotatory and disrotatory. These are illustrated in Figure 14.18.

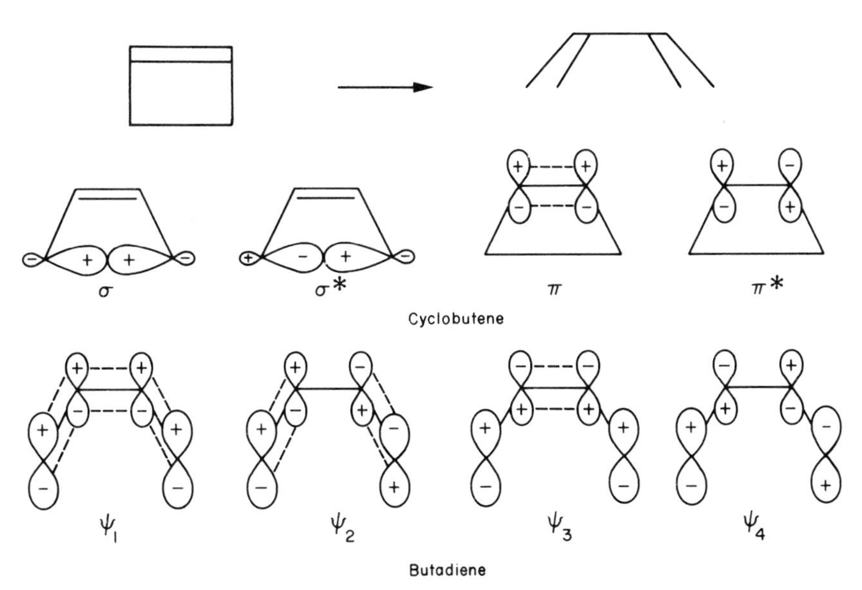

Figure 14.17 Orbitals of cyclobutene and of butadiene which change during the ring opening reaction.

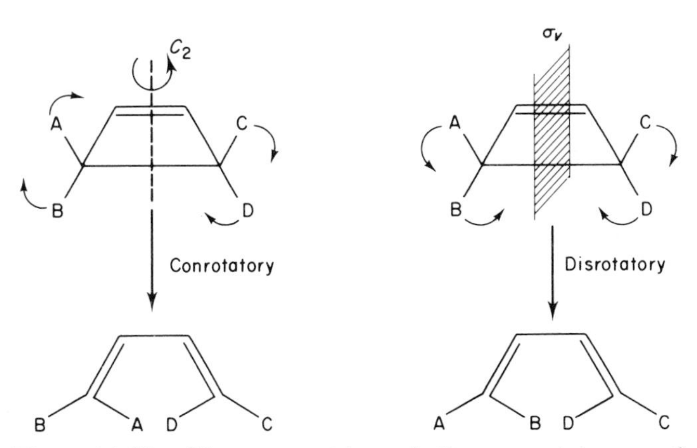

Figure 14.18 The symmetries of the conrotatory and disrotatary ring opening process of cyclobutene.

If we assume that both of these modes occur in such a way that the two end CH_2 groups maintain their equivalence throughout the reaction then in the disrotatory case there is, for all configurations, a plane of symmetry, as shown. The disrotatory mode also retains an element of symmetry, in this case a C_2 rotation axis. In Figure 14.18 the terminal hydrogen atoms have been labelled to emphasize the fact that with isotopes or substituents the products of these two modes can be distinguished.

The orbitals shown in Figure 14.17 can be classified as symmetric (S) or antisymmetric (A) under the rotation and reflection operations. These classifications are shown in Table 14.1.

Table 14.1. Classification of the orbitals shown in Figure 14.17

C_2		σ_v	
S	A	S	A
σ	σ^*	σ	σ^*
π^*	π	π	π^*
ψ_2	ψ_1	ψ_1	ψ_2
ψ_4	ψ_3	ψ_3	ψ_4

Figure 14.19 shows a correlation diagram for the orbitals under the two modes of reaction. For the conrotatory mode bonding orbitals of the reactant correlate with bonding orbitals of the product but for the disrotatory mode this is not the case: one bonding orbital of the reactant correlates with an antibonding orbital of the product.

The ground state of cyclobutene has the configuration σ^2 π^2 and if electrons remain in these orbitals throughout the reaction it is only for the conrotatory mode that they produce the ground state configuration of the product. Experimentally the thermal ring opening of cyclobutene is found to give exclusively conrotatory products.

In contrast, if we raise cyclobutene to the lowest excited state, having the configuration $\sigma_2\pi$ π^*, by absorption of u.v. radiation, then for the disrotatory mode this configuration has symmetry S^2 SA under σ_v and this matches the lowest excited configuration of the product $\psi_1^2\psi_2\psi_3$ (S^2 AS). For a conrotatory mode these first excited states do not match in their individual orbital symmetries (S^2 AS $\rightarrow$ A^2 SA), although the total symmetries of the excited configurations are indeed the same $S \times S \times A \times S = A \times A \times S \times A = A$. Photochemical ring opening of cyclobutenes has been shown to give disrotatory products.

A cyclo-addition like the Diels–Alder reaction can be analysed in the same way if we maintain a reflection plane at all stages of the reaction, as in Figure 14.20. We can treat this in exactly the same way as we developed for the ring opening of cyclobutene and if we do, we find that the ground state orbitals of the two

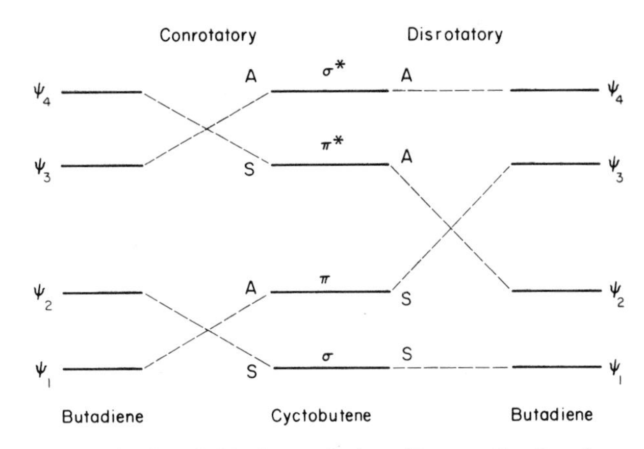

Figure 14.19 Orbital correlation diagram for the ring opening of cyclobutene.

reactants and those of the products match. We can also apply exactly the same treatment to the Cope rearrangement. Notice that as we have depicted in Figure 14.16 it is a *disrotatory* process and a mirror plane is maintained throughout.

This treatment is completely general and because of the pairing properties of Hückel orbitals we find that if n is the total number of carbon atoms in the product (n is necessarily even in a reaction of this kind), then when $n/2$ is odd the reactant and product orbitals will match. If $n/2$ is even a concerted addition will only occur through excited states. The discussion presented earlier for the ethylene dimerization represents another specific case of the general rule. A similar rule can also be derived for ring opening and ring closing reactions including rearrangements of the Cope and Claisen type. If $n/2$ is odd the ground state process is matching for the disrotatory mode, but if $n/2$ is even the matching ground state process is conrotatory. The exact reverse applies to reactions involving the first excited states. The experimental verification of these rules is well illustrated by the ring opening

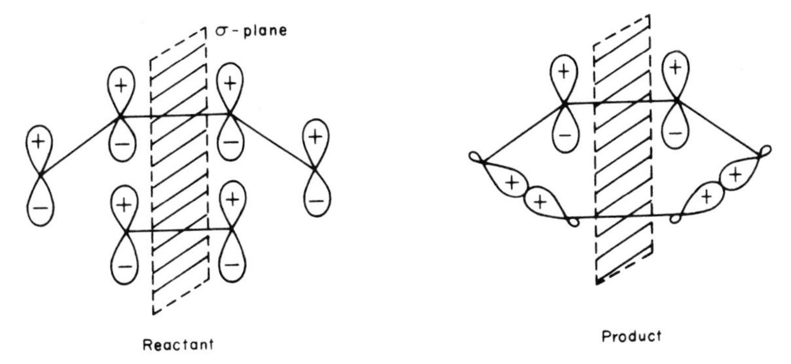

Figure 14.20 Symmetry plane for configurations on the reaction pathway of a Diels–Alder addition.

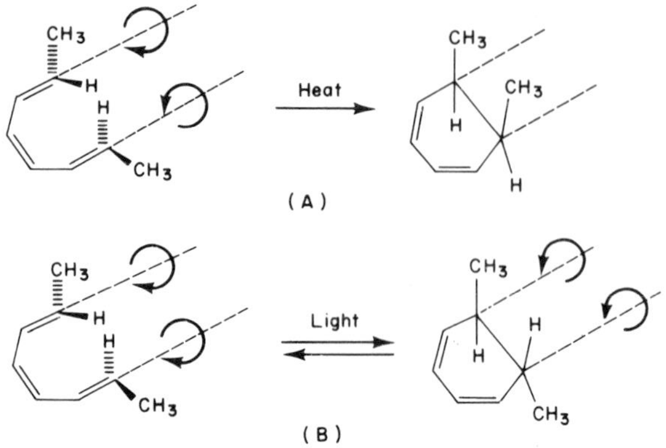

Figure 14.21 Cyclization of dimethylhexatriene: (A) disrotatory-thermal cyclization; (B) conrotatory-photochemical cyclization.

reactions of 1,2-dimethylcyclohexa-3,5-diene to give 1,6-dimethylhexa-1,3,5-triene (Figure 14.21).

The above symmetry treatment is only applicable when the reaction is concerted, but it has proved extremely successful in correlating a wide range of experimental data.

14.7. Aromatic reactivity and Hückel's rule

Although benzene is often represented by a single Kekulé (cyclohexatriene) structure its chemical properties are very different from those expected for such a formula. We have seen that in molecular orbital theory the ground state of benzene has six π electrons occupying three bonding orbitals, the electrons being delocalized over the whole ring, and it is this delocalization (or strictly non-localizability, c.f. Chapter 8) which gives the molecule its characteristic properties. There are clearly an infinite number of possible cyclic molecules with the empirical formula $C_{2k}H_{2k}$ ($2k$=number of C atoms in the ring) with Kekulé structures containing alternating double and single bonds (e.g. cyclobutadiene, cyclooctatetraene, etc.). The Hückel secular determinant for a cyclic molecule of this kind is (assuming all bonds to have equal length)

$$\begin{vmatrix} x & 1 & 0 & 0 & \dots & 1 \\ 1 & x & 1 & 0 & \dots & 0 \\ 0 & 1 & x & 1 & \dots & 0 \\ 0 & 0 & \dots & \dots & \dots & \\ \dots & \dots & \dots & \dots & \dots & \\ 1 & 0 & \dots & \dots & \dots & x \end{vmatrix} = 0. \qquad (14.33)$$

It can be shown that the roots of the above equation ($2k$ in all) are given by the expression

$$x = -2 \cos (l\pi/k) \text{ [where } l = 0, \pm 1, \pm 2 \ldots . k] . \quad (14.34)$$

Substituting $(\alpha - E)/\beta$ for x(equation 9.8) we get

$$E_l = \alpha + 2\beta \cos (l\pi/k). \quad (14.35)$$

Now since $\cos(-\vartheta) = \cos \vartheta$ the orbitals with quantum numbers l and $-l$ are degenerate.Thus according to Hückel theory the π orbitals of a cyclic polyene $C_{2k}H_{2k}$ will occur in degenerate pairs except for the lowest (l=0) and highest orbital (l=k) which are unique. In the ground state of these molecules the π orbitals will be filled by $2k$ electrons, two electrons into each orbital filling the lowest energy orbitals first. If there are two degenerate orbitals and only two electrons to go into them, then by Hund's rule (which was discussed in detail in Chapter 11) the lowest energy arrangement will be that with one electron in each orbital with their spins parallel, to form a triplet state. Thus planar cyclic alternant hydrocarbons ($C_{2k}H_{2k}$) for which k is odd, such as C_6H_6, $C_{10}H_{10}$, $C_{14}H_{14}$, and $C_{18}H_{18}$ will, according to Hückel theory, have a closed shell of $2k$ electrons delocalized over the ring in k bonding orbitals. In contrast, planar cyclic alternant hydrocarbons for which k is even, such as C_4H_4 and C_8H_8 should, according to Hückel theory, exist in a triplet state, and be reactive species.

Although cyclic odd alternant hydrocarbons can only exist as radicals or ions, they can be treated in exactly the same way. The secular determinant will be the same and the general roots the same, the only difference is that the values l can take, for an odd alternant ($C_{2k-1}H_{2k-1}$) are l=0, ±1, ±2 . . . ±(k – 1). Thus a cyclic odd alternant also has its orbitals in degenerate pairs, except for the lowest energy orbital (l=0) which is unique (notice the highest orbitals in a cyclic odd alternant are a degenerate pair).

Figure 14.22 shows the energy levels of the π orbitals cyclic alternant hydrocarbons for C_3 up to C_8. We see that the only neutral molecule in this series with all its π electrons in pairs in bonding orbitals is benzene. The C_3, C_5, and C_7 radicals can be converted into closed shell systems by removing the unpaired electron from the C_3 and C_7 species and by adding an electron to the C_5 species. The cyclopropylium ion $C_3H_3^+$ and the tropylium ion $C_7H_7^+$ have been prepared and are very stable and unreactive compared with other carbocations. Similarly the cyclopentadienyl anion $C_5H_5^-$ is readily prepared and is less reactive than other hydrocarbon anions.

According to Hückel theory planar cyclooctatetraene would have a triplet ground state. However the total π-electron energy for the planar molecule is almost the same as that of four isolated ethylenic bonds. Since considerable energy would be lost in making the molecule planar (i.e. by distorting the σ framework from the preferred bond angles of 120°), it remains puckered and has chemical properties similar to those of linear unsaturated polyenes (Figure 14.23).

Cyclobutadiene is also predicted to have a triplet ground state by Hückel theory. In this case the molecule would be expected to be flat, but again there would be no gain in energy by delocalizing the electrons compared with a

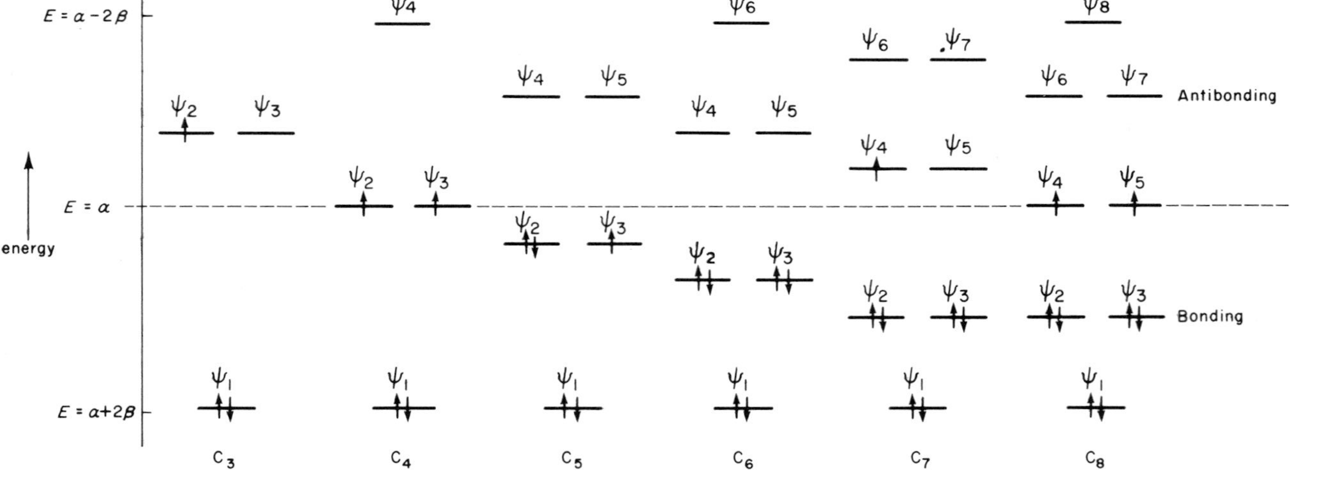

Figure 14.22 Energy levels and occupancy of the π orbitals of planar neutral cyclic molecules C_nH_n ($n = 3-8$).

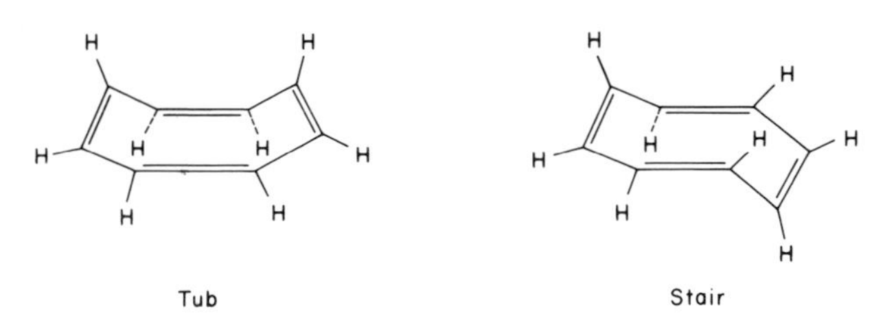

Figure 14.23 Configurations of cyclooctatetraene.

rectangular molecule with two isolated ethylenic bonds. In addition, from the Jahn–Teller theorem (Section 12.4) we know that molecules with degenerate states undergo a distortion to remove the degeneracy, so that the ground state of cyclobutadiene is almost certainly rectangular and has singlet spin.

Cyclic hydrocarbons $C_{2k}H_{2k}$ where $k>4$ are known as annulenes. As we have seen Hückel theory predicts that cyclic compounds $C_{2k}H_{2k}$ where k is odd should have closed shells of π electrons, with total π-electron energies lower than that expected for linear polyenes ($C_{2k}H_{2k+2}$), whilst cyclic compounds $C_{2k}H_{2k}$ where k is even should have triplet ground states. In practice the smallest annulenes with k odd, [10] annulene (k=5) and [14] annulene (k=7), are unable to achieve planarity because of the repulsion between endo-cyclic hydrogen atoms. Even [18] annulene is not completely planar, but bisdehydro[14] annulene, which is planar and which has seven filled delocalised π orbitals (obeying Hückel's Rule) undergoes the substitution reactions, nitration, sulphonation, and Friedel–Crafts acylation characteristic of benzenoid aromatic compounds. Notice that in the bisdehydro-[14] annulene the π_y orbitals of the acetylenic bond and of the allene bond are orthonormal to the π orbitals which form the completely delocalized system over the 14 atoms of the ring.

In the presence of a powerful magnetic field, cyclic delocalized π orbitals behave like a super conducting ring and the electrons flowing round the ring set up a magnetic field inside the ring in opposition to the applied field. This means that protons on the outside of the ring experience an increased magnetic field (the induced field plus the applied field). This ring current has an important effect on the magnetic resonance of the protons; those in the augmented field absorb at

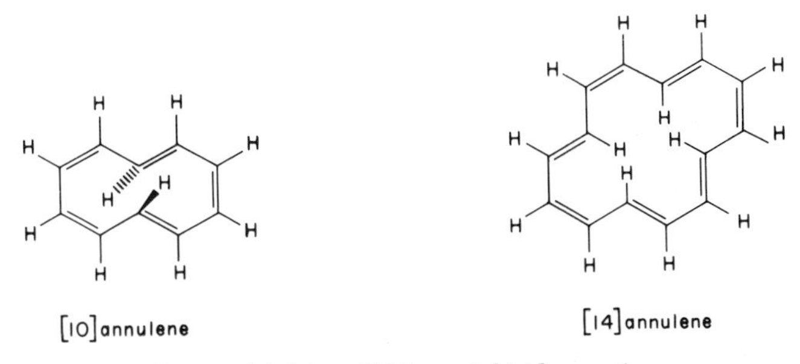

Figure 14.24 [10]- and [14]-annulene

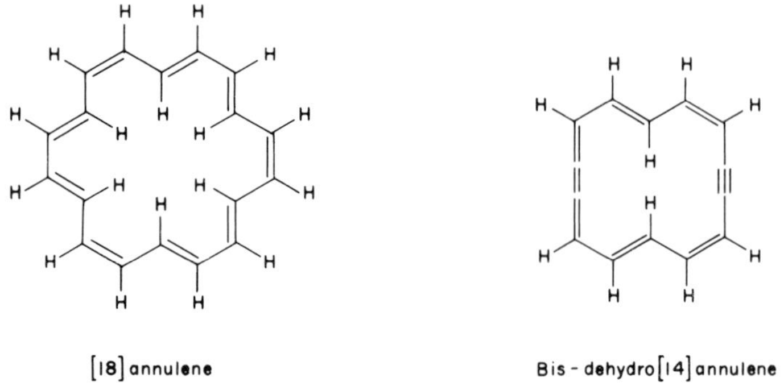

[18] annulene   Bis-dehydro[14]annulene

Figure 14.25 [18] annulene and bisdehydro[14] annulene.

lower field than protons in an otherwise similar environment (e.g. in a linear polyene), while those in the centre of the ring absorb at higher field. The n.m.r. spectrum of 1,18-bisdehydro[14] annulene is shown in Figure 14.26.

[18] Annulene is nearly planar in the crystalline state and in solution its n.m.r. spectrum shows absorption at low fields (ca. 1.1τ) due to the outside protons and at high fields (ca. 11.8τ) due to the internal protons. In sharp contrast [24] annulene is a very unstable compound, decompositing in the air and showing only one broad band (ca. 3.16τ) in its n.m.r. spectrum due to all the protons.

Dewar has shown how the perturbation MO theory developed in Section 14.2 can be used to estimate the π-electron stability of cyclic hydrocarbons ($C_{2k}H_{2k}$) relative to their linear polyene equivalent ($CH_2{=}[CH]_{2k-2}{=}CH_2$). We can use equation (14.19) to evaluate the energy lost by isolating one π electron on a carbon atom of the annulene or on a terminal atom of the polyene. In both cases the fragment remaining is an odd alternant chain of $2k{-}1$ atoms and the coefficients of the non-bonding orbital of this are (from equation 9.63) $k^{-1/2}$, 0, $-k^{-1/2}$, 0, $k^{-1/2}$,

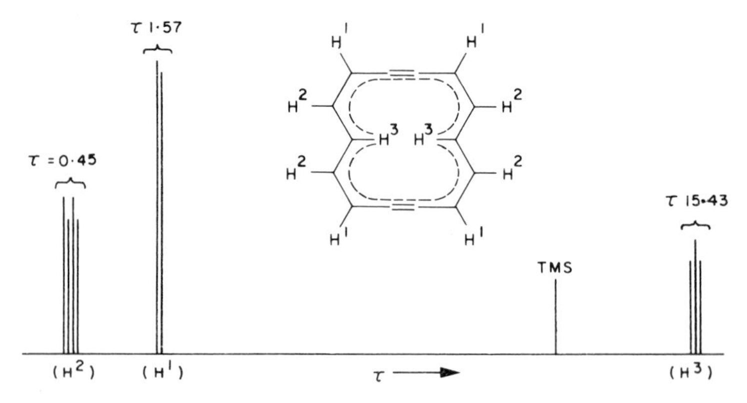

Figure 14.26 Nuclear magnetic resonance spectrum (60 MHz) of 1,8-bisdehydro[14] annulene.

etc. down the chain. The Dewar number for the terminal atom of the polyene is $k^{-1/2}$ and for an atom of the annulene is $k^{-1/2}[1+(-1)^{k-1}]$. When k is odd the annulene has the greater Dewar number, that is, more energy is lost in isolating one π electron, and hence it has a greater π-electron energy than the linear polyene and is said to be aromatic. When k is even the polyene has the greater Dewar number and hence has the greater π-electron energy. The annulene in this case is said to be anti-aromatic because delocalization of the π electrons leads to destabilization, that is the molecule in its ground state will have localized π orbitals.

The above prediction is the same as that based on Hückel's rule, that is, neutral cyclic hydrocarbons $C_{2k}H_{2k}$ where k is odd are exceptionally stable.

14.8. Substituent effects

The concepts of a substituent and of a functional group are much used in chemistry, especially in organic chemistry. Thus in benzoic acid the carboxyl group is regarded as the functional group since a majority of the chemical reactions undergone by benzoic acid involve the carboxyl group. In *p*-chlorobenzoic acid, the chlorine atom effects the relative rate of many of the reactions undergone by the carboxyl group, but usually does not effect the actual course of the reaction. The chlorine atom is therefore regarded as a substituent. It would not be surprising if the chlorine atom had a similar effect on the relative rates of a variety of reactions involving the carboxyl group. The comparison of the effects of the same series of substituents on the relative rates of a number of fairly similar reactions leads to the so called *free energy relationships*, of which the best known is the Hammett equation. If we compare the logarithms of the rates of hydrolysis of esters with the logarithm of the ratio of the dissociation constants of the corresponding acids we find that the *meta*- and *para*-substituted acids and esters lie on a straight line while the *ortho*-substituted compounds and the aliphatic acids and esters lie off the line. The straight line is given by

$$\log k_{\mathrm{h}} = \log K_{\mathrm{D}} + \text{const.}, \tag{14.36}$$

where k_{h} is the rate constant for hydrolysis and K_{D} is the dissociation constant. Hammett suggested that the dissociation constants of the *meta*- and *para*-benzoic acids should be used as standards, because so many relevant data are available. He therefore defined a substituent constant σ_i, characteristic of substituent i by the relation

$$\sigma_i = \log K_i - \log K_0, \tag{14.37}$$

where K_0 is the dissociation constant of benzoic acid and K_i is the dissociation constant of the appropriately substituted benzoic acid. The Hammett equation has the form

$$\log k_{ij} - \log k_{0j} = \rho_j \sigma_i, \tag{14.38}$$

where k_{ij} is the rate or equilibrium constant of reaction j when the substituent is i, k_{0j} is the corresponding constant for the unsubstituted compound and ρ_j is called the reaction constant of reaction j. Correlations have been obtained for a very large number of reactions. Perfect correlation is not to be expected, and is very rarely found. The best correlations are those which relate to reactions involving functional groups on a side chain attached to the benzene ring. Poor correlation is obtained when there is complete delocalization between the substituent and the functional group. Attempts to improve the correlation has led to the adoption of a whole range of σ values to suit particular reactions. The most important feature of a Hammett type of treatment is that it relates two sets of free energy data, and for reactions in solution the governing factors may well be interaction between the solvent and reactants rather than any property of the activated complex which we could calculate from valence theories alone. However, the very existence of free-energy relationships of the Hammett type shows that the concept of a substituent has a sound empirical basis.

In a substituted alkane the bonds are formally localized and the influence of substituent is called the *inductive effect*. The inductive effect is the term given to any electron displacement due to the difference in electronegativity between the substituent and the rest of the molecule. For example, the bond attaching an electronegative substituent like a fluorine atom will be polar with a higher electron density at the fluorine end of the bond. The polarity will necessarily have some effect on the adjacent bonds, both through bonds and through space (i.e. purely electrostatic). It is convenient to divide substituents into those which are more electronegative than carbon which are called *electron attracting* groups and those which are less electronegative which are called *electron repelling* groups. These terms are relative and a methyl group, for example, is an electron repeller when attached to a benzene ring. In a saturated alkane derivative the inductive effect falls off very rapidly, and becomes almost negligible beyond two carbon atoms.

When we come to delocalized bonds, that is, those associated with π orbitals, we find substituent effects can be transmitted along the length of the delocalized system. The major effects are of two kinds, those due to substituents with filled orbitals of symmetry such that they can interact with the delocalized system (such substituents are called *donors*) and those due to substituents with low lying vacant orbitals with a symmetry which permits interaction with the delocalized system; such substituents are called *acceptors*.

We can illustrate the effect of a donor by considering ethylene with one hydrogen atom replaced by CH_2^-: this would be an extreme example. As Figure 14.27 shows this gives the allyl anion which has three molecular orbitals, one bonding (ψ_1), one non-bonding (ψ_2) and one anti-bonding (ψ_3). In the allyl anion we have four electrons which occupy ψ_1 and ψ_2 and we know from the properties of the non-bonding orbitals of odd alternant hydrocarbons (Section 9.4) that the electron density in the non-bonding orbital is restricted to the starred atoms, i.e. the terminal atoms, thus the charge distribution in the allyl anion is

$$\overset{-\frac{1}{2}}{\overset{*}{C}H_2}-CH-\overset{-\frac{1}{2}}{\overset{*}{C}H_2}.$$

The extreme example of an acceptor would be CH_2^+ and we will again consider

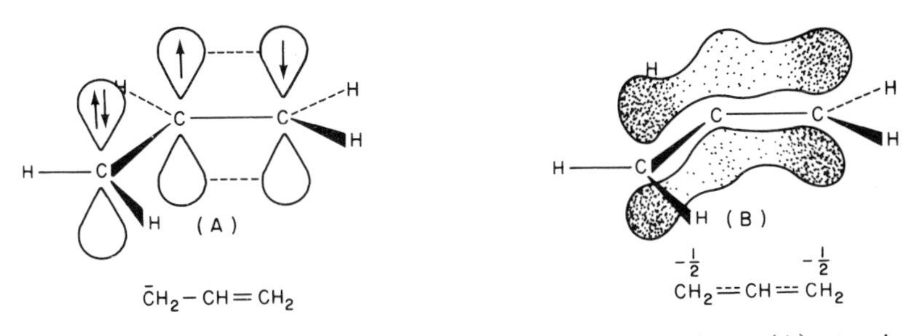

Figure 14.27 Interaction of a CH_2^- group with ethylene: (A) atomic orbital interactions (arrows representing electrons); (B) the final electron distribution.

ethylene so substituted (Figure 14.28). We again have the allyl system but there are now only two electrons to go into the π orbitals. Since we know that in a neutral alternant the electron density is unity at every atom the removal of an electron from the non-bonding orbital (i.e. from the allyl radical) will result in a charge distribution in the allyl cations of $CH_2^{+\frac{1}{2}}-CH-CH_2^{+\frac{1}{2}}$.

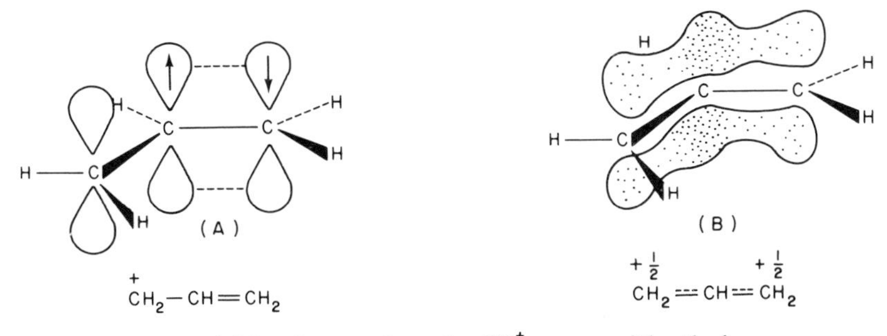

Figure 14.28 Interaction of a CH_2^+ group with ethylene.

The ionic substituents CH_2^- and CH_2^+ represent extreme examples of donors and acceptors. A more typical example of a donor is a dimethylamino group. If such a group is adjacant to a double bond, its lone pair orbital will behave as the π orbital of CH_2^- and contribute to the π molecular orbitals. This contribution will be maximized, as shown in Figure 14.29, if the three nitrogen–carbon bonds are coplanar, because the lone pair orbital is then a pure $p\pi$ atomic orbital rather than some hybrid of s and p. An interaction of this type will lead to the migration of electrons from the nitrogen lone pair to the carbon atoms, and by analogy with the allyl system, we expect to find these mainly on the end carbon atom. Organic chemists represent this transfer by the use of curly arrows as in the following:

$$(CH_3)_2\ddot{N}-CH=CH_2.$$

Electron pairs are represented as being transferred from atoms to bonds in such a way that a stable octet of electrons (cf. Lewis theory) is retained around all the

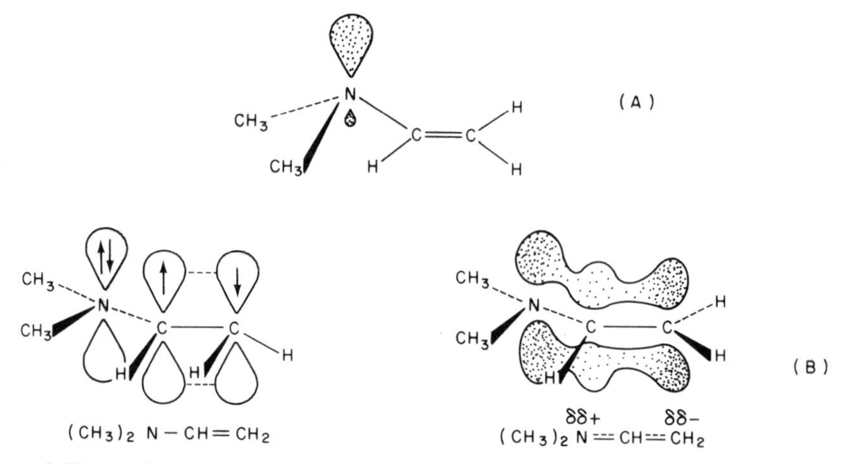

Figure 14.29 (A) Pyramidal nitrogen-weak interaction. (B) The atomic orbitals in the planar situation giving strong interaction, the resulting π-electron distribution being shown in (C).

participating atoms. Donor groups include any substituent with non-bonding atomic or filled molecular orbitals, which have the correct symmetry to interact with a suitable π-orbital system (e.g. CH_2^-, O^-, H_2N, HO, Cl, Br, etc).

Although the extreme example of an acceptor group is CH_2^+ the most common are groups with low lying anti-bonding orbitals (e.g. O=C—, N≡C—, O_2N—, O_2S— etc. Because oxygen is more electronegative than carbon both the bonding and anti bonding π orbitals of O=C— have a lower energy than those of CH_2=CH—. Thus whereas CH_2=CH— is neutral, neither a donor nor an acceptor when attached to an unsaturated hydrocarbon, O=C— will behave as a net acceptor of π electrons.

If we consider an amide group we have R_2N— as a donor group attached to the carbonyl which is an acceptor group (Figure 14.30). The difference in electronegativity between nitrogen and oxygen encourages the migration of charge and the experimental fact is that the amide group is planar and has a large dipole.

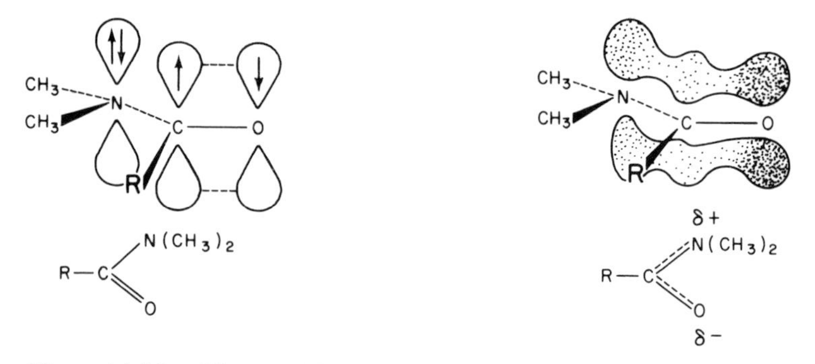

Figure 14.30 The amide group which consists of a donor $[(CH_3)_2N-]$ joined to an acceptor (>C=O). There is very substantial transfer of charge from nitrogen to oxygen.

If the amino and carbonyl groups are not directly linked but are connected through a double bond, a similar transfer of charge will be observed (Figure 14.31).

If there were no hetero-atoms this would be the odd alternant pentadienyl system and in the anion the negative charge would be on the starred atoms (1,3, and 5). In a similar way we can represent the transfer of charge with the curved arrows of the organic chemists

$$R_2\ddot{N}-CH=CH-CH=O.$$

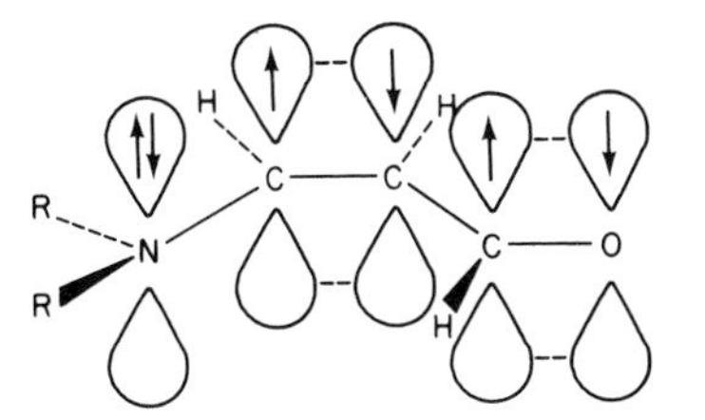

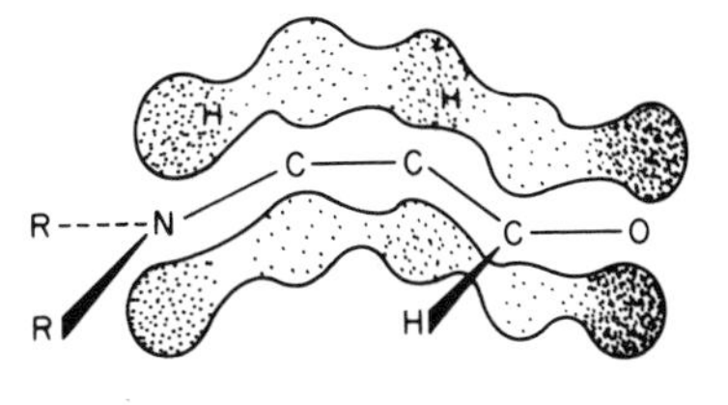

Figure 14.31 The vinyl anologue of an amide $R_2N-CH_2=CH-CHO$.

Notice that if there was complete transfer of charge (which there is not), all the atoms in the dipolar molecule would have complete valence shells.

$$R_2\overset{+}{N}=CH-CH=CH-O^-$$

In terms of resonance theory the dipolar structure is said to make a small contribution to the ground state of the molecule.

The electron attracting or repelling properties of a substituent, which depend on its relative electronegativity, are independent of its acceptor or donor characteristics. The halogen atoms have filled *p* atomic orbitals which can interact with adjacent π molecular orbitals (e.g. $Cl-CH=CH_2$ so that the chlorine atom would be expected to behave as a donor group. At the same time chlorine is more electronegative than carbon and therefore also behaves as an attracting group. These opposing effects are frequently encountered in organic chemistry, and as we shall see the donor/acceptor properties govern orientation in addition or substitution reactions, whereas repelling/attracting properties may govern the relative rates of the reactions.

We have earlier considered aromatic substitution reactions, in which a cation such as NO_2^+ (or an incipient cation such as SO_3) adds to the benzenoid ring to form a Wheland intermediate. Although the intermediate is not the transition state, it is probably closer in energy and structure to the activated complex than either the reactants or the products. The π-electron system of the Wheland intermediate is the same as that of the odd alternant pentadienyl cation, thus the charge distribution in the Wheland intermediate will be as shown in Figure 14.33.

We can see at once that a donor substituent *ortho* or *para* to the site of the

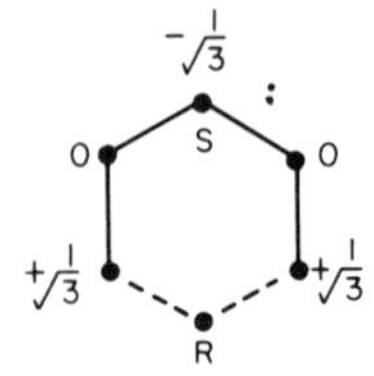

Figure 14.32 Coefficients of the non-bonding orbital of pentadienyl.

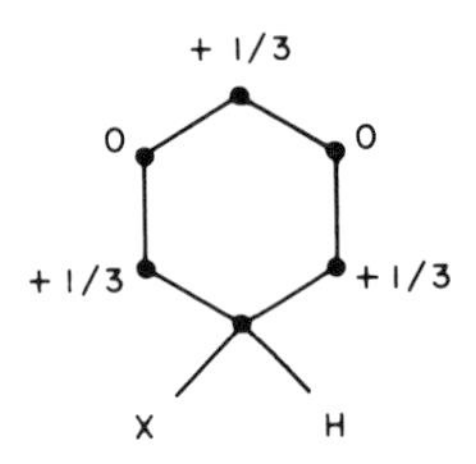

Figure 14.33 Charge distribution in the Wheland complex.

original attack will reduce the positive charges on the ring by spreading the charge to the substituent. Probably in solution the spreading of charge is as important a factor in determining the course of the reaction as any gain in electronic energy due to further delocalization. An acceptor group in the *ortho* or or *para* positions, on the other hand, would increase the charge on the carbon atom. We can therefore understand the *ortho-*, *para*-directing and activating characteristics of a donor substituent, and the deactivating, *meta*-directing characteristics of an acceptor group. Similarly a repelling group is weakly activating, *ortho-*, *para*-directing, and an attracting group weakly deactivating, *meta*-directing. When there are opposing effects, orientation is always decided by the donor/acceptor properties, but if there is a strong inductive effect the rate of reaction may be determined by the repelling/attracting properties (e.g. chlorobenzene is nitrated more slowly than benzene, but substitution occurs in the *ortho-* and *para*-positions).

The repelling/attracting and donor/acceptor characteristics can be simulated in Hückel π-electron theory in various ways. Firstly, direct calculations using the hetero-atomic parameters described in Section 9.6 will lead to changes in electron density and orbital energy. Two new parameters enter the calculation, the Coulomb integral of the substituent (α_X) and the resonance integral of the C—X bond (β_{CX}).

To introduce repelling/attracting properties in Hückel theory it is necessary to change the Coulomb integral of the substituted carbon atom and possibly also of more distant atoms in the chain.

In Hückel theory donor/acceptor properties depend on at least *two* factors, one is the energy of the orbitals of π symmetry of the substituent and the other is the magnitude of the appropriate resonance integral. In some molecules the importance of these two can be evaluated separately by twisting the substituent groups out of the plane of the conjugated hydrocarbon so that α_X remains constant but β_{CX} varies, being roughly proportional to φ the angle of twist. The introduction of bulky groups R and R′ *ortho* to a dimethylamino substituent, as in Figure 14.34 will produce the effect, and both the u.v. spectra and the chemistry of such molecules are sensitive functions of the angle φ.

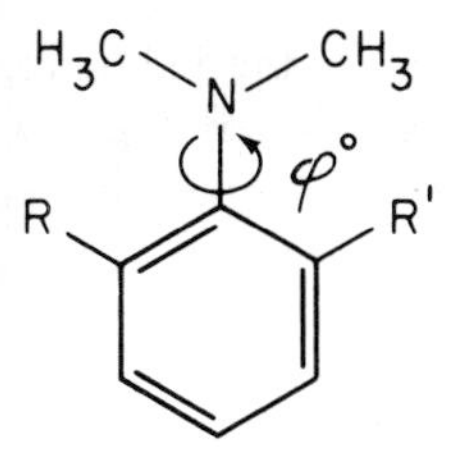

Figure 14.34 The twisting of a $N(CH_3)_2$ group out of the plane of the ring by bulky *ortho*-groups.

Figure 14.35 shows the change in energy of the Hückel orbital energies of aniline as a function of the angle between the planes of the benzene ring and the plane of the NH_2 group. Calculations were made with the parameters $\alpha_N = \alpha + 1.5\beta$, $\beta_{CN} = \beta \cos \varphi$, $\alpha'_C = \alpha + 0.15\beta$ where α and β are the standard Hückel parameters for carbon atoms, and α'_C is the Coulomb integral at the substituted carbon atom. For

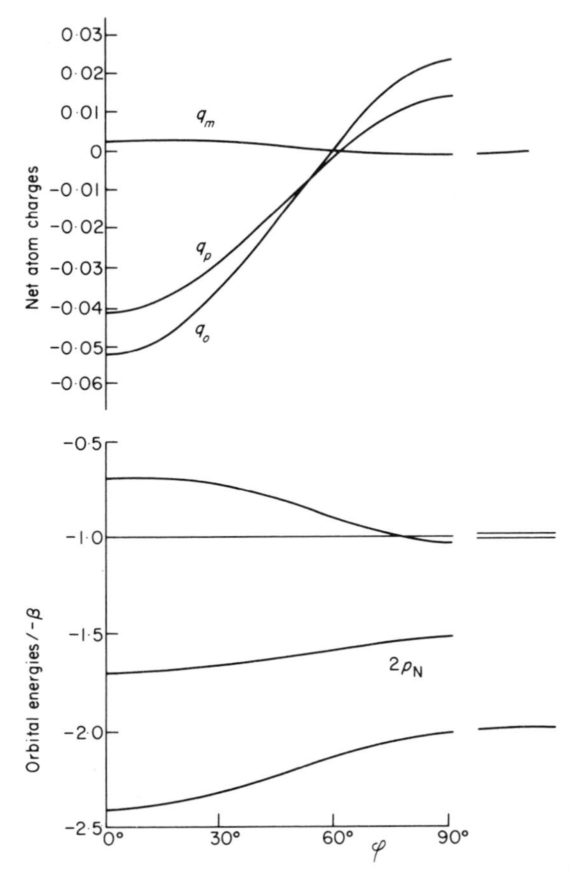

Figure 14.35 Hückel orbital energies and net atom charges for aniline as a function of the twist angle φ in Figure 14.34. The levels on the right are those of benzene.

$\varphi = \pi/2$ the only factor influencing the orbitals of the benzene is the inductive effect which has been introduced by making α'_C different from α: when $\beta_{CN} = 0$ there is no mixing of the nitrogen atomic orbital with the molecular orbitals of the ring. Figure 14.35 shows that the donor properties of the amino-group have a much larger influence on the orbital energies than the inductive effect, but both have an appreciable effect on the atom charges in the ring. The charge at the *meta* position is largely unaffected by the substituent.

It is difficult to make an unambiguous separation of inductive and donor/acceptor characteristics in *ab-initio* calculations because the SCF Hamiltonian contains contributions from electron repulsion. The electron redistribution in the molecule arising from the inductive effect will therefore interfere with the electron distribution arising from donor/acceptor properties. One cannot for example introduce a donor effect without at the same time introducing electrostatic fields which will themselves give rise to an inductive effect. We can however parallel the Hückel calculations to the extent of seeing how the energies and electron densities change on twisting the substituent and such calculations are shown for aniline in Figure 14.36.

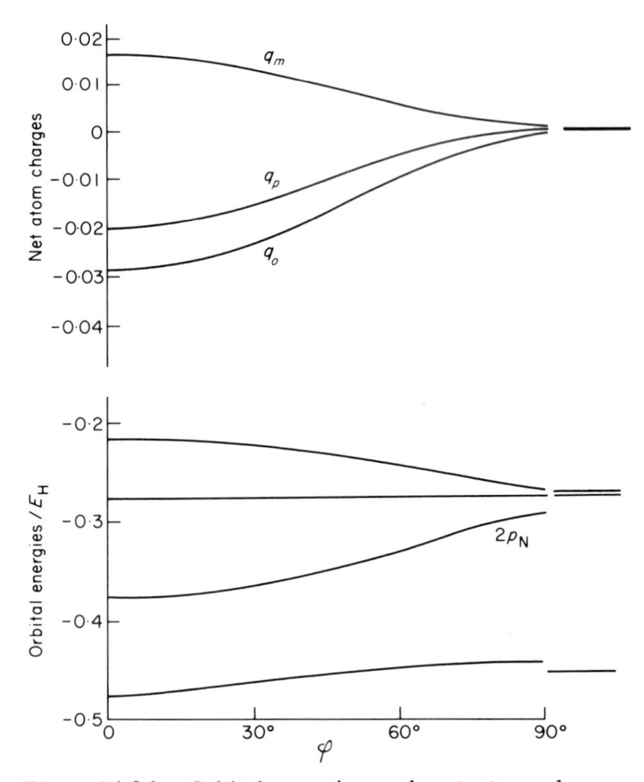

Figure 14.36 Orbital energies and net atom charges (relative to benzene, right) for aniline as a function of the twist angle, from *ab-initio* calculations.

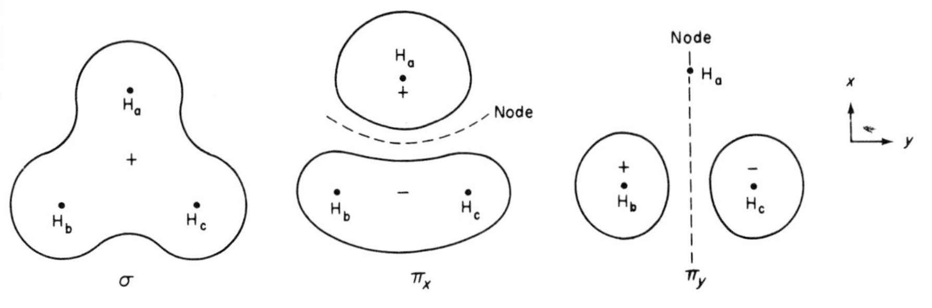

Figure 14.37 The group orbitals of the hydrogen atoms in a methyl substituent in, for example, toluene.

There is a satisfying, and perhaps surprizing agreement between the Hückel and the *ab-initio* calculations. The orbital energies show very similar trends. The changes in the carbon atom charges are not quite so similar. Most important is the fact that the *ab-initio* calculation does not show the inductive shifts for the 90° structure of the Hückel calculation. The Hückel calculation in turn does not show the depletion of charge at the *meta* position for the 0° structure obtained in the *ab-initio* calculation.

We have defined donor substituents as groups with filled π orbitals or non-bonding atomic orbitals whose symmetry is such that they can interact with the π orbitals of the rest of the molecule. There is evidence of a further type of weak donor and the simplest example is a methyl group, attached to a conjugated system (e.g. toluene). From the three hydrogen atoms of the methyl group we can form group orbitals of the type (c.f. equation 8.16)

$$\sigma = (h_a + h_b + h_c)$$

$$\pi(x) = (h_a - \tfrac{1}{2}h_b - \tfrac{1}{2}h_c)$$

$$\pi(y) = (h_b - h_c)$$

These group orbitals are depicted in Figure 14.37.

σ is symmetrical about the $C{-}C_6H_5$ axis and can only interact to form σ-type orbitals with the rest of the molecule. $\pi(x)$ is clearly similar to a normal π orbital, and the H_3 group can be thought of a pseudo-atom, with a π orbital which can conjugate with the benzene ring. $\pi(y)$ corresponds to a π type orbital at right angles to $\pi(x)$ and it would therefore be able to interact with the $p\pi(x)$ atomic orbitals of an acetylenic bond as in $H_3{\equiv}C{-}C{\equiv}C{-}H$.

Hyperconjugation, as this type of interaction is called, is not confined to the methyl group. Any group which has orbitals with components of π symmetry could in principle extend the conjugation of an unsaturated system. The carbon–carbon orbitals of the t-butyl group could be combined in symmetry types just as the hydrogen orbitals of a methyl group. Both chemical and spectroscopic evidence supports the conclusion that a methyl group is a weak donor as our treatment of hyperconjugation would suggest.

Chapter 15
Intermolecular Forces

15.1. Van der Waals' forces

In this chapter we discuss a number of topics under the heading 'weak chemical bonds'. There is no firm boundary between a weak bond and a strong bond, just as there is no firm boundary between ionic and covalent bonds. Thus whilst we can define precisely these concepts as the limits of bond type an actual bond may fall between these limits.

As a rough guide we take any bond whose energy is below 100 kJ mol^{-1} as weak and contrast these bonds with the typical ionic or covalent bond whose energy is >200 kJ mol^{-1}. Weak bonds are typically those formed between systems which separately have chemical stability.

All atoms and molecules under conditions of sufficiently low temperatures and/or high pressures will form liquid and then solid phases. Helium is the most difficult species to condense but does so at normal pressures at 4.12 K. There must be forces that hold even inert gas atoms or non-polar molecules together at low temperatures and these are referred to as Van der Waals' forces.

Before the advent of modern quantum mechanics the most successful theory of Van der Waals' forces was based on the electrostatic interaction of dipoles. Two dipoles separated by a distance R have an interaction energy which is proportional to R^{-3} (or a force proportional to R^{-4}). However, this energy depends upon the angles which the dipoles make with each other and with the vector distance, $\mathbf{R}$, between them. This is most compactly represented in vector notation by†

$$E_{DD} = \frac{\boldsymbol{\mu}_1 \cdot \boldsymbol{\mu}_2}{R^3} - \frac{3(\boldsymbol{\mu}_1 \cdot \mathbf{R})(\boldsymbol{\mu}_2 \cdot \mathbf{R})}{R^5} . \tag{15.1}$$

Figure 15.1 shows two attractive and two repulsive arrangements for a pair of dipoles. The average energy between rotating molecules with dipoles can be obtained by weighting each orientation by the Boltzmann probability factor $\exp(-E_{DD}/kT)$. This leads to the result that the average energy of attraction is proportioned to $R^{-6}T^{-1}$. From early studies there was some experimental support for this result. However, such a theory clearly fails to account for the forces between inert gas atoms or between non-dipolar molecules.

†The scalar product between two vectors, A and B, is written **A.B**, and is equal to $AB \cos \vartheta$ where A and B are the magnitudes of the two vectors and ϑ is the angle between them.

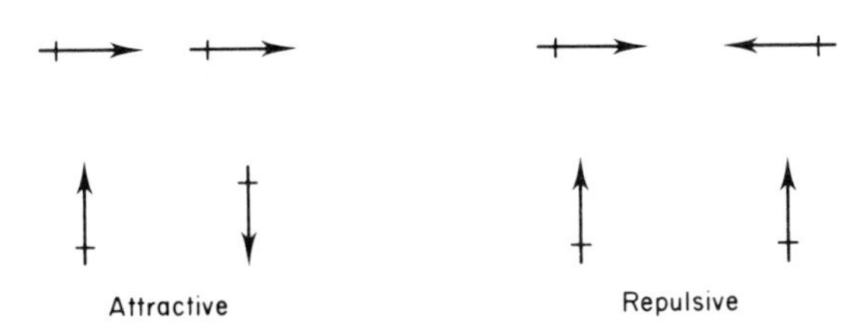

Figure 15.1 Attractive and repulsive arrangement of two dipoles.

Following closely upon the Heitler–London theory of the hydrogen molecule (Chapter 13), Wang calculated the interaction energy of two hydrogen atoms at large distances by including a term which allowed for the attraction between an *instantaneous* dipole of one atom and the *induced* dipole of the other atom, this induced dipole arising from the electric field of the instantaneous dipole. This energy we now call the *dispersion energy* and the force of attraction that it produces is the dispersion force.

A rigorous and comprehensive treatment of the dispersion energy was given by London in 1930 using the equations of second-order perturbation theory (11.27), but the picture suggested by the work of Wang provides a qualitatively acceptable model.

Although the time-average dipole moment of an inert gas atom (or a non-polar molecule) is zero, we can imagine that at any instant of time the centre of the electron density will not coincide with the nucleus so that the system will show an instantaneous dipole moment, $\mu_{ins}(t)$. The electrostatic field arising from this dipole is proportional to $\mu_{ins}(t)R^{-3}$. The electrons in a neighbouring atom will be displaced relative to the nucleus by such a field and a dipole moment will be induced. The magnitude of this induced dipole moment is proportional to the field and the proportionality constant is a quantity called the polarizability (α). We can therefore write for the induced dipole moment

$$\mu_{ind}(t) \propto \mu_{ins}(t)R^{-3}. \tag{15.2}$$

There will now be an interaction energy between μ_{ins} and μ_{ind} which, from (15.1), is

$$E_{DD}(t) = \frac{\mu_{ins}(t) \cdot \mu_{ind}(t)}{R^3} - \frac{3(\mu_{ins}(t) \cdot \mathbf{R})(\mu_{ind}(t) \cdot \mathbf{R})}{R^5}. \tag{15.3}$$

Because of the R^{-3} dependence of μ_{ind}, $E_{DD}(t)$ will be proportional to R^{-6}. An important fact is that the direction of $\mu_{ind}(t)$ is always such that $E_{DD}(t)$ is negative (attractive) for all orientations of $\mu_{ins}(t)$. This can be seen qualitatively in Figure 15.2. The time average of $E_{DD}(t)$, which is the dispersion energy, is therefore non-zero and proportional to R^{-6}. It must be emphasized that this energy is independent of temperature, unlike the rotational average of the permanent dipole interaction.

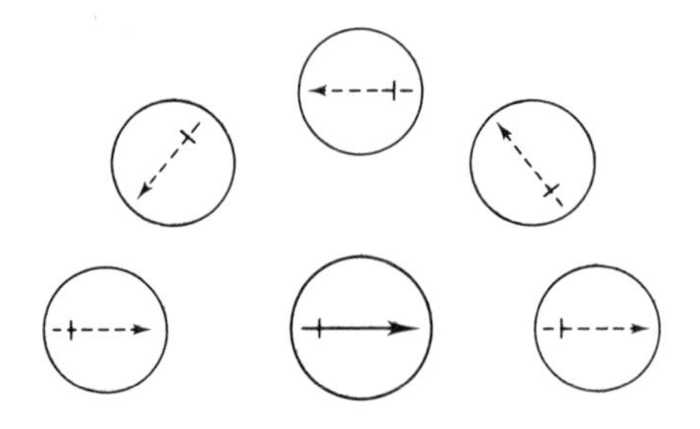

Figure 15.2 Interaction between an instantaneous dipole (solid arrow) and induced dipoles (broken arrows).

The magnitude of the dispersion energy depends on α and $\mu_{ins}(t)$. Because μ_{ins} is itself related to α (the more polarizable the atom the more likely is it that at any instant there will be a large value of μ_{ins}) the dispersion energy is proportional to the product of the polarizabilities of the two atoms.

An expression derived by London for the dispersion energy between *identical* atoms or molecules is

$$E_{disp} = -\frac{3}{4}\frac{\alpha^2 I}{R^6}, \tag{15.4}$$

where I is the ionization potential. However, so many drastic approximations have to be made to obtain this expression that it is now more common to quote the dispersion energy in the form

$$E_{disp} = -C_6 R^{-6}, \tag{15.5}$$

and values of C_6 which have been deduced empirically or by theoretical calculations are known for many systems. Some of the more important values are listed in Table 15.1.

In a perturbation treatment of the polarizability of an atom the electric field mixes together atomic orbitals which differ by one in their l quantum numbers. Thus s orbitals will mix with p, p will mix with s and d, etc. From expression

Table 15.1. Selected values of C_6^{AB} in atomic units [A. Dalgarno, *Adv. Chem. Phys.*, **12**, 143 (1967)]

A \ B	H	He	Ne	Li	H_2	N_2	CH_4
H	6.50	2.83	5.6	67	8.7	21	30
He	2.83	1.47	3.0	22	4.1	10	14
Ne	5.6	3.0	6.3	42	8.2	21	29
Li	67	22	42	1390	83	180	290
H_2	8.7	4.1	8.2	83	13	30	43
N_2	21	10	21	180	30	73	100
CH_4	30	14	29	290	43	100	150

(11.20) we see that two orbitals will mix strongly if the energy difference between them is small. The reason for the high value of C_6 for interactions involving Li is because Li is a very polarizable atom, and this in turn is due to the low $2s \rightarrow 2p$ promotion energy. In contrast, the inert gases have very low polarizabilities because their lowest $s \rightarrow p$ or $p \rightarrow s$ promotions involve an increase in the principal quantum number and hence require high energy.

A diatomic molecule, for example, is generally more polarizable along the internuclear axis than perpendicular to this axis. It follows that the coefficient C_6 between two molecules or between an atom and a molecule will depend on the mutual orientation of the two systems. At the present time there is little quantitative information about this orientational dependence, and the C_6 values quoted for molecules in Table 15.1 are meant to be rotational averages of the true C_6 parameters.

Although the dispersion energy decays quite rapidly with increasing internuclear distance (as R^{-6}) it is still referred to as a long-range interaction. This is to contrast it with valence energies which, dependent as they are on the overlap of orbitals, will decay exponentially (and therefore more rapidly than R^{-6}) as R is increased (see expressions 6.25 and 6.26 and Figure 6.2). The potential energy curve between two inert gas atoms will have a minimum where the attractive dispersion force just balances the repulsive force due to the overlap of the closed electron shells. This repulsive force is generally referred to as the exchange repulsion because it becomes important when there is sufficient overlap of the two sets of atomic orbitals that there is a significant probability of exchanging electrons between the two atoms.

The exchange repulsion can be calculated with equal ease in either simple molecular orbital or valence bond theory. For a pair of inert gas atoms the wavefunctions are in fact identical in the two descriptions as can be seen by the specific example of He_2.

For a molecular orbital description of He_2 we place the four electrons in the bonding $1\sigma_g$ and in the antibonding $1\sigma_u$ molecular orbitals whose wavefunctions are given by (6.18) and (6.19). Ignoring normalization factors their wavefunctions are

$$1\sigma_g = \phi_a + \phi_b \quad \text{and} \quad 1\sigma_u = \phi_a - \phi_b. \tag{15.6}$$

The ground state wavefunction of He_2 is therefore a Slater determinant (8.9) constructed from the four spin orbitals $1\sigma_g\alpha$, $1\sigma_g\beta$, $1\sigma_u\alpha$, $1\sigma_u\beta$.

We can make an equivalent orbital transformation of the orbitals (15.6) as described in Section 8.2 and illustrated in equation (8.11). These equivalent orbitals are clearly

$$\tfrac{1}{2}(1\sigma_g + 1\sigma_u) = \phi_a \quad \text{and} \quad \tfrac{1}{2}(1\sigma_g - 1\sigma_u) = \phi_b, \tag{15.7}$$

which are just the $1s$ orbitals of the two atoms. Thus an identical wavefunction can be formed as a Slater determinant with four electrons in spin orbitals $\phi_a\alpha$, $\phi_a\beta$, $\phi_b\alpha$, $\phi_b\beta$. However, such a function is just the valence-bond wavefunction for He_2 : an antisymmetrized product of the wavefunctions of two separate helium atoms.

Because these wavefunctions are identical, any quantity we calculate from them must be identical.

The exchange repulsion between closed shell systems is rather similar to the repulsion obtained for the triplet state of H_2 (expression 13.21 with the negative sign and Figure 13.1). The reason is that they both arise from the exchange of electrons between spin orbitals having the *same* spin.

In the limit of zero internuclear distance, usually called the united atom limit, the triplet wavefunction of H_2 (multiplying Ψ_- of 13.2 by one of the spin functions 13.6) would contravene the Pauli exclusion principle by having two electrons in what would be effectively the same spin orbital. Qualitatively we can associate exchange repulsion with this tendency to contravene the Pauli exclusion principle.

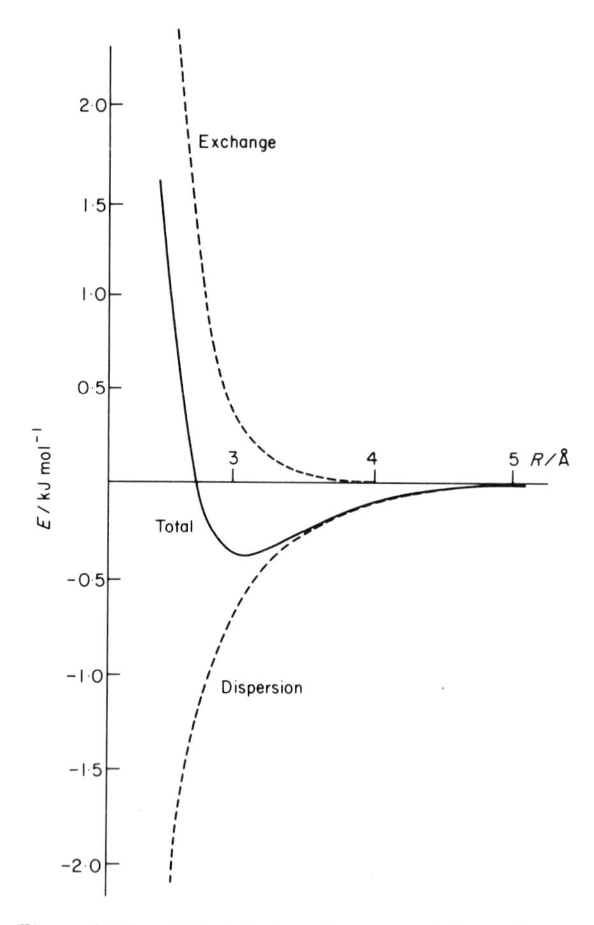

Figure 15.3 Calculated exchange and dispersion energies for Ne_2. A small Coulomb energy is included in the exchange term (J. N. Murrell and A. J. C. Varandas, *Molec. Phys.*, **30**, 223, 1975).

Figure 15.3 shows accurate calculations of the attractive dispersion energy and the repulsive exchange term for Ne_2. The minimum in the total potential occurs when the slopes of these two curves are equal and opposite and the value of R at this point is called the Van der Waals' distance. It can be seen that the depth of the minimum at the Van der Waals' distance is almost equal to the attractive part of the potential and this is because the repulsive part, once it starts to be important, rises very steeply.

The Van der Waals' distance can be calculated quite accurately for simple atoms, or can be determined experimentally in a number of different ways. For example, an X-ray study of a crystal structure will show the closest distances between atoms in *different* molecules and these can be taken as Van der Waals' distances for the atoms.

From accumulated data on Van der Waals' distances a table of Van der Waals' radii can be established such that the sum of the radii of two atoms is a close approximation to the Van der Waals' distance of the pair. Some of these radii are shown in Table 4.7. For those atoms for which both covalent and Van der Waals' radii can be defined it can be seen that the Van der Waals' radius is much the larger of the two. This is because for Van der Waals' interactions there are no attractive forces that depend on the overlap of orbitals on different atoms.

15.2. Steric forces

In some molecular configurations, atoms which are not joined together by a chemical bond approach to a distance closer than the sum of their Van der Waals' radii. At such distances the two will be in the repulsive region of the Van der Waals' interaction and chemists associate the concept *steric repulsion* with the higher than normal energies of such configurations.

As an example we consider the activation energy for the S_{N2} substitution

$$X + R^1R^2R^3CY \rightarrow Y + R^1R^2R^3CX \qquad (15.8)$$

This reaction goes via a transition state with a five-coordinated carbon atom in which the bonds are approximately those of a trigonal bipyramid, as shown in Figure 15.4. The rate of this reaction for different groups R appears to be mainly due to the size of the group. For example, the relative reactivities of bromides (Y = Br) to attack by I^- are shown in Table 15.2.

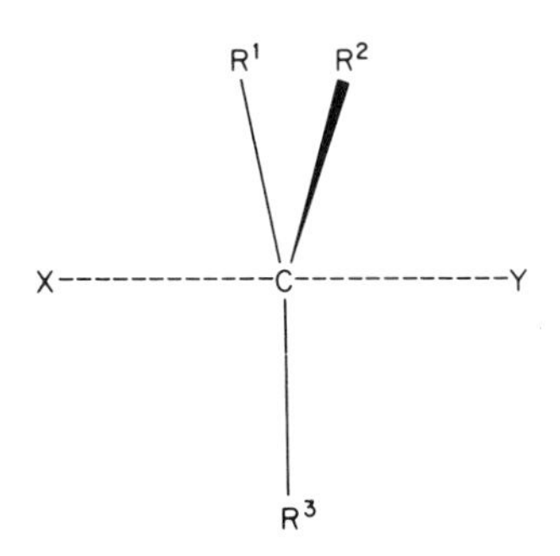

Figure 15.4 The transition state for an S_{N2} substitution reaction at a saturated carbon atom.

Table 15.2. Relative rates of the $S_{N}2$ reaction $I^- + R^1R^2R^3CBr \rightarrow R^1R^2R^3CI + Br^-$

R^1	R^2	R^3	Relative rate
H	H	H	150
H	H	CH_3	1
H	CH_3	CH_3	0.01
CH_3	CH_3	CH_3	0.001

From the data of Table 15.2 above one could put forward an alternative interpretation of the facts, namely, that the CH_3 group, through an inductive effect, is stabilizing the reactant more than the transition state. Indeed it is often difficult to isolate the steric effect from other effects of substituents. However, in this case one can rule out non-steric factors being important by the observation that the replacement of CH_3 by CF_3 has relatively little effect on the rate. These groups have similar Van der Waals' radii but the opposite sign for their inductive effects. Also replacing CH_3 by a much more bulky alkyl group such as t-butyl leads to a further large reduction in rate, whereas the inductive effects for these two groups should be very similar.

Steric repulsion can be calculated in a molecule by the use of empirical potential functions. One of the most commonly used is

$$V_{XY}(R) = -C_6R^{-6} + A\exp(-bR) \tag{15.9}$$

where the first term represents the attractive dispersion energy (15.5) and the second the exchange repulsion. For example, with such a function ($C_6 = 3.573$, $A = 10.51$, $b = 2.16$ atomic units), the effect of repulsion between the *ortho* hydrogens on the geometry of biphenyl has been calculated. This molecule is known to be planar in its crystalline form but the relief of steric repulsion occurs not only through the distortion of angle β (Figure 15.5), but also through small geometry changes in the rest of the molecule, as shown in Table 15.3. The final steric repulsion energy for the molecule after relaxation of the molecular geometry was found to be 15 kJ mol^{-1}. In solution the molecule relieves this strain by the

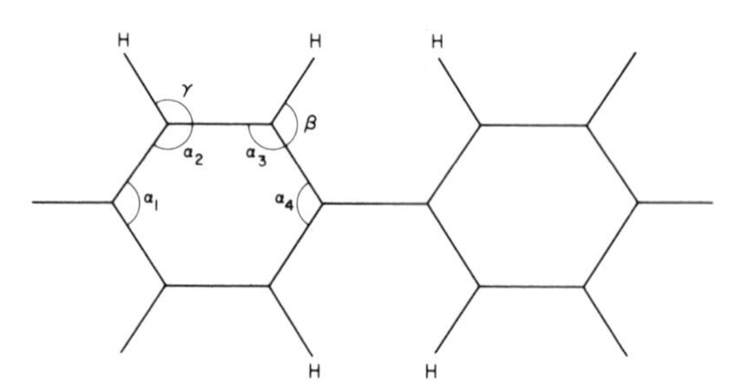

Figure 15.5 The geometry of biphenyl (see Table 15.3).

Table 15.3. Calculated changes in bond angles of planar biphenyl due to relief of steric strain between the *ortho* hydrogen atoms [K. Miller and J. N. Murrell, *Trans. Faraday Soc.*, **63**, 806 (1967)]

Angle (Figure 15.5)	
α_1	$-0.73°$
α_2	$0.37°$
α_3	$0.96°$
α_4	$-1.94°$
β	$2.33°$
γ	$-0.32°$

two rings twisting from planarity by an angle of 42°, but in the crystal this is prevented by packing forces that exist between one molecule and its neighbours.

15.3. Donor–acceptor complexes and the dative bond

A large family of molecular complexes exists whose binding energy is larger than that normally associated with Van der Waals' forces. These complexes are usually formed between one molecule which has the property that it readily gives up electrons (an electron donor or Lewis base) and another which readily accepts electrons (an electron acceptor or Lewis acid). The complexes are called donor–acceptor or charge-transfer complexes.

Donor–acceptor complexes are interesting because the bond holding the components together has properties that suggest it is intermediate between covalent and Van der Waals' bonds. Moreover, the complexes often have interesting physical properties such as being brightly coloured or of being semiconductors in the solid state.

In the Lewis theory of valence a special category was identified for covalent bonds in which the two electrons came from one atom. Such bonds were called coordinate bonds because of their relevance to the structures of transition metal coordination compounds, the theory of which had been developed by Werner.

In chemical formulae it is customary to represent a coordinate bond by an arrow whose direction is away from the donor atom. For example, the complex between BH_3 and NH_3 is represented

$$H_3B \leftarrow NH_3 \tag{15.20}$$

the donated electrons originating in the lone pair orbital of NH_3. An alternative structure that satisfies the valency of the atoms is

$$\begin{array}{ccccc} H & & & & H \\ H & — \overset{-}{B} — & \overset{+}{N} & — & H \\ H & & & & H \end{array} \qquad (15.21)$$

Both B^- and N^+ are isoelectronic with carbon and hence, by first transferring an electron from the nitrogen to the boron, we are able to draw a structure which is equivalent to that of ethane. Scheme (15.21) is often referred to as the dative structure for the molecule and the bond as a dative bond (dative comes from the Latin word for *give*).

A molecular orbital treatment of the coordinate bond does not require any new concepts. *Ab-initio* calculations have been made on $BH_3.NH_3$ a molecule that can be compared with the isoelectronic covalent species ethane.† The energy for dissociation to BH_3 and NH_3 has been calculated to be 123 kJ mol^{-1} which is one third of the dissociation energy for ethane into two methyl radicals. The B–N stretching force constant is one half that for the C–C bond. The calculated dipole moment for the complex is large, 5.7 D, and this is associated with the transfer of about half an electron from ammonia to borane.

Let us now compare this molecular orbital description with a qualitative valence bond description. There has unfortunately been no *ab-initio* valence bond calculation to allow us to make a fairer comparison. This qualitative valence bond picture forms the basis of the Mulliken treatment of donor–acceptor complexes which has been widely used since it was introduced in 1950.

Valence bond wavefunctions for the pyramidal structures of NH_3 and BH_3 would start with a hybridized basis for the nitrogen and boron atoms (approximately sp^3 hybrids) and in each case three of these would be combined with hydrogen $1s$ orbitals to give three electron-pair bonds. This would leave one sp^3 hybrid on each molecule which is not involved in bonding to the hydrogen atoms. In the case of NH_3 this contains two electrons (it is the lone pair orbital) and in the case of BH_3 it is empty.

Valence bond wavefunctions are formed by allocating electrons to atomic or hybrid orbitals and pairing their spins so as to obtain functions of the correct spin type. We will ignore the electrons making up the NH and BH bonds as the wavefunctions of these will be the same in all the structures we consider.

We are left with the allocation of electrons to the nitrogen lone pair hybrid and its boron equivalent which we give wavefunctions ϕ_d and ϕ_a respectively. At very large B–N distances the lowest energy structure will have both electrons in the nitrogen hybrid orbital as this corresponds to the ground states of the separate molecules. This has been called the no-bond state and it has the wavefunction

$$\Psi_1(D,A) = \phi_d(1)\phi_d(2)[\alpha(1)\beta(2) - \alpha(2)\beta(1)] \qquad (15.22)$$

where we symbolize NH_3 as the electron donor (D) and BH_3 as the electron acceptor (A).

†One such calculation is by S. D. Peyerimhoff and R. J. Buenker, *J. Chem. Phys.*, **49**, 312 (1968).

A wavefunction that corresponds to the structure (15.21) can be obtained by transferring one electron from ϕ_d to ϕ_a and forming a Heitler–London combination (equation 13.7)

$$\Psi_2(D^+-A^-) = [\phi_d(1)\phi_a(2) + \phi_a(1)\phi_d(2)]\,[\alpha(1)\beta(2) - \alpha(2)\beta(1)]\,. \qquad (15.23)$$

A further state can be formed by transferring two electrons from ϕ_d to ϕ_a to give a doubly ionic state but this will have very high energy and can be neglected.

As a better wavefunction for the molecule we can take a linear combination of the no-bond and dative structures and apply the variation principle to find the coefficients

$$\Psi = c_1\Psi_1(D,A) + c_2\Psi_2(D^+-A^-). \qquad (15.24)$$

Even without accurate calculations we can give a qualitative description of the way that the energies of the two components to the wavefunction (the diabatic curves) and the total energy vary with the distance between donor and acceptor. In the first place, at infinite separation the dative state will be greater in energy than the no-bond state by an amount equal to the energy required to remove an electron from ϕ_d, which is the ionization potential of ammonia, less the energy gained by adding an electron to ϕ_a, which is the electron affinity of the acceptor.† The ionization potential of NH_3 is 980 kJ mol^{-1} and the electron affinity of BH_3 must be ~200 kJ mol^{-1} hence the dative state will be greater in energy than the no-bond by ~800 kJ mol^{-1}. However, as the B–N distance is decreased the energy of the dative state will be decreased relative to that of the no-bond state by virtue of the Coulomb attraction between the charges: this amounts to $-1389/R$(Å) kJ mol^{-1}.

The situation we have so far described has a strong similarity to that for ionic molecules: see in particular the discussion of LiF, which was given in Section 6.7. The main difference is that the ionization potential minus the electron affinity is much greater than in the case of the alkali-metal halides, so that at the equilibrium separation the ionic state may still have a greater energy than the no-bond state. A second difference is that the non-ionic state does not have a covalent attraction because both electrons are in the same orbital, so that the potential curve for the no-bond structure will be repulsive or have at most a shallow minimum.

The interaction between the no-bond and dative structures can be analysed by first evaluating the overlap integral between them. We can ignore the spin parts of the wavefunctions, as they are the same for (15.22) and (15.23). The overlap integral will therefore be

$$\begin{aligned} S_{12} &= \int \phi_d(1)\phi_d(2)[\phi_d(1)\phi_a(2) + \phi_a(1)\phi_d(2)]\, dv_1\, dv_2 \\ &= \int \phi_d^2(1)\, dv_1 \int \phi_d(2)\phi_a(2)\, dv_2 + \int \phi_d^2(2)\, dv_2 \int \phi_d(1)\phi_a(1)\, dv_1 \\ &= 2S_{da}, \end{aligned} \qquad (15.25)$$

where S_{da} is the overlap between the donor and acceptor hybrid orbitals (both of

†BH_3 is almost certainly a planar molecule, although no electronic spectrum has been observed of this species which would confirm this. In the complex, BH_3 is pyramidal like NH_3 and it is the electron affinity of the pyramidal form that would be relevant for this calculation.

which have been assumed normalized). S_{da} will rise exponentially as the B—N distance is decreased.

Following the arguments given before (e.g. Chapter 6), and more specifically if we use the Mulliken formula (9.74), the Hamiltonian integral will be roughly proportional to the overlap integral. The result of the variational calculation (c.f. equations 6.64–6.66) will be to produce potential energy curves for the resulting states which are qualitatively of the form shown in Figure 15.6. Two cases are shown. In Figure 15.6A $I_D - A_A$ is large and the diabatic curves do not cross and in Figure 15.6B $I_D - A_A$ is smaller and they do cross.

The extent to which electron transfer has occurred in the complex is measured by the ratio of the probabilities of dative and no-bond structures in the ground state, that is by c_2^2/c_1^2. For the case of $BH_3.NH_3$ the quantitative molecular orbital calculation shows that this ratio is roughly unity.

Although the particular complex $BH_3 . NH_3$ that we have been discussing is perhaps too strongly bound to merit inclusion in this chapter it has served to contrast the *ab-initio* molecular orbital and qualitative valence bond approaches. Although it is difficult to make such valence bond calculations quantitative they have the great merit that they relate in a direct way the properties of the complex to some properties of the components, and they also show why such complexes are often brightly coloured.

The complex $BH_3 . NH_3$ is not coloured because the interaction between the no-bond and dative structures is so strong that the electronic transition between the two resulting states (indicated by arrows in Figure 15.6) is induced only by u.v. radiation. For weak complexes which occur between organic donors and acceptors

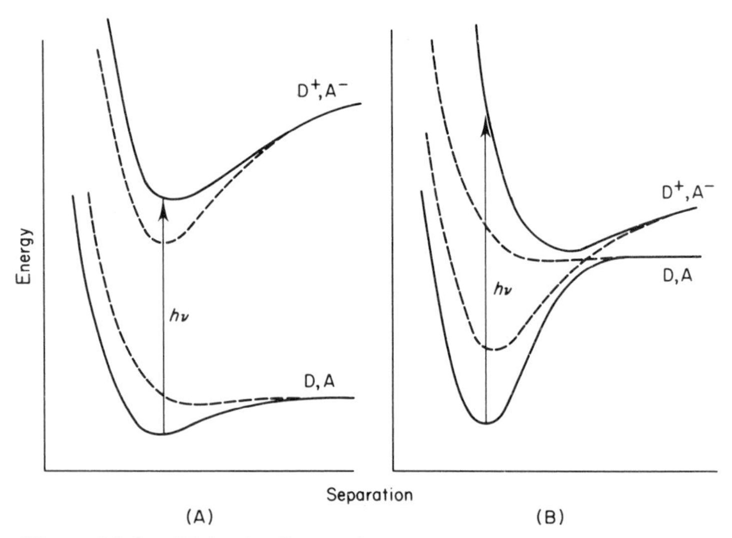

Figure 15.6 Diabatic (broken) and adiabatic (solid) potential energy curves for donor–acceptor complexes: (A) $I_D - A_A$ is large and (B) $I_D - A_A$ is small.

the energy difference between the two states is much lower. For example, hexamethylbenzene (HMB) as a donor and tetrachloroquinone (TCQ) as an accep'or form a sandwich complex whose geometry is roughly that shown in Figure 15.7, and this complex is red. In contrast the two components are colourless (HMB) and pale yellow (TCQ).

The Mulliken two-state resonance theory of donor acceptor complexes follows the treatment we have described for $BH_3 . NH_3$ except that the donor and acceptor orbitals are not associated with specific atomic or hybrid orbitals of any one atom but may be molecular orbitals of the two components. For example, both the highest occupied molecular orbital of HMB and the lowest unoccupied molecular orbital of TCQ are π molecular orbitals and their nodal characteristics are shown in Figure 15.7. These two orbitals will be the donor and acceptor orbitals and it is

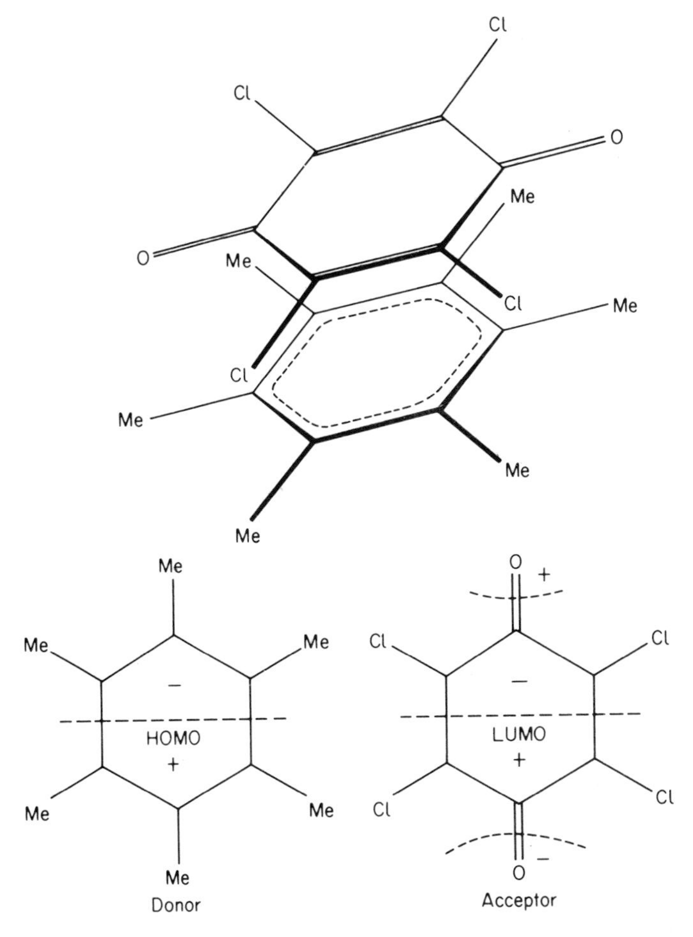

Figure 15.7 Structure of the hexamethylbenzene–tetrachloroquinone complex and the nodal properties of the donor and acceptor π orbitals.

clear that the overlap integral S_{da} is optimum for the sandwich structure of the complex.

For weak complexes the ratio c_2/c_1 in (15.24) will be small for the ground state and large for the excited state. That is, the ground state is almost completely described by the no-bond wavefunction, and the excited state almost completely by the dative wavefunction. The transition from the ground to the excited state is, therefore, accompanied by the transfer of almost one electron from the donor to the acceptor and the resulting spectroscopic absorption band is therefore called an electron-transfer or *charge-transfer* band. Such bands are also commonly assigned in the electronic spectra of transition metal complexes such as

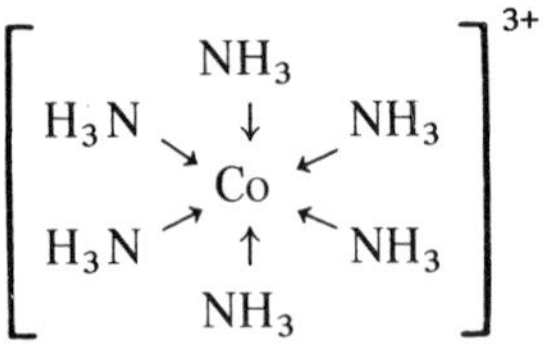

$$\left[\begin{array}{ccccc} & & NH_3 & & \\ H_3N & \searrow & \downarrow & \swarrow & NH_3 \\ & & Co & & \\ H_3N & \nearrow & \uparrow & \nwarrow & NH_3 \\ & & NH_3 & & \end{array}\right]^{3+}$$

The binding energies of some donor–acceptor complexes may be so small that they are no greater than the Van der Waals' energies between like molecules. For example, the complex between benzene (the donor) and iodine (the accceptor) which has been much studied for its spectroscopic interest has $\Delta H = 6$ kJ mol^{-1}. This complex has not been isolated in the solid state and is only inferred from the properties of benzene–iodine solutions. In such a case a large part of the binding energy must be due to Van der Waals' forces and the extra stability that arises from the resonance of the no-bond and dative states may not even be the dominant contribution. However, if we mix in solution a donor and an acceptor then on the basis of Van der Waals' forces alone we would also expect to see complexes DD and AA. The preponderance of DA pairs must be due to the extra energy that such complexes gain either though the resonance which we have discussed or through electrostatic interactions which may be present for DA pairs but not for DD or AA.

15.4. The hydrogen bond

In many cases where there is a strong association between molecules or between different groups within the same molecule this can be attributed to hydrogen atoms which show divalent character. Examples are the dimers of aliphatic acids, the bifluoride ion $(FHF)^-$ and the HF dimer whose structures are shown in Figure 15.8.

Hydrogen occurs as a bridging atom in another important class of compounds, the boron hydrides. The simplest member of the family is diborane (B_2H_6). However, these are not usually classed as hydrogen bonded systems because they cannot be broken into fragments which are themselves stable molecules. We include a brief discussion of them at this point in order to compare them with hydrogen bonded complexes.

Boron hydrides have been called *electron deficient* molecules because there are insufficient electrons to form the number of electron-pair bonds that seem to be

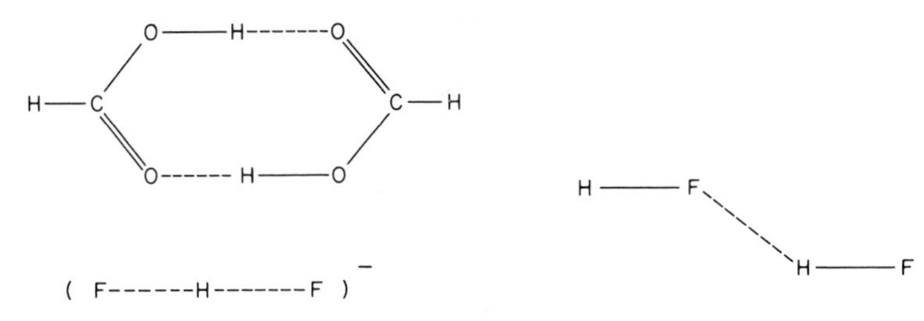

Figure 15.8 Some examples of hydrogen bonded structures.

implied by the molecular geometry. Diborane has eight B—H bonds but only fourteen valence electrons. Molecular orbital calculations show that the terminal bonds are normal electron pair bonds but the bridging bonds are to be described as 3-centre 2-electron bonds.

Figure 15.9 shows the localized orbitals that have been deduced from molecular orbital calculations, according to the method described in Section 8.2, and they confirm these conclusions.†

The term electron deficient as applied to the boron hydrides is perhaps anomalous because in all cases there are sufficient electrons to fill all the bonding molecular orbitals. Any molecule with empty bonding orbitals would be a poweful oxidizing agent (electron acceptor) and that is by no means a characteristic of these molecules.

The molecular orbital description of the bifluoride ion is quite different from that of diborane because there are *four* electrons to make up the bridging bonds. The equilibrium molecule is linear with the hydrogen at the mid-point of the F—F separation. The two highest occupied molecular are composed predominantly of the fluorine $2p\sigma$ and hydrogen $1s$ orbitals with some contribution from the fluorine $2s$. The lower of the two has σ_g symmetry and is bonding between the three atoms and the higher is σ_u (having a node through the hydrogen) and is antibonding between the fluorines. However, the fluorine atoms are sufficiently far apart for this antibonding effect to be small and the orbital has a negative energy (i.e. binds electrons) because of the large electronegativity of the fluorine atom.

It is possible to transform these σ_g and σ_u orbitals into equivalent orbitals $\theta_1 = \sigma_g + \sigma_u$, $\theta_2 = \sigma_g - \sigma_u$ which are localized in the two FH bonds and this shows the contrast with the boron hydride bridges whose orbitals cannot be bond localized.

The bifluoride ion is a rare example amongst hydrogen bonds in that the hydrogen atom is centred mid-way between the two heavy atoms. Typical hydrogen bonds have a much smaller binding energy than the bifluoride ion and the hydrogen is more closely associated with one atom than with the other, as in the formic acid dimer (Figure 15.8). In fact, the geometry of the two components making up the complex is little different from that in their uncomplexed states.

†E. Switkes, R. M. Stevens, W. N. Lipscomb, and M. D. Newton, *J. Chem. Phys.*, **51**, 2085 (1969).

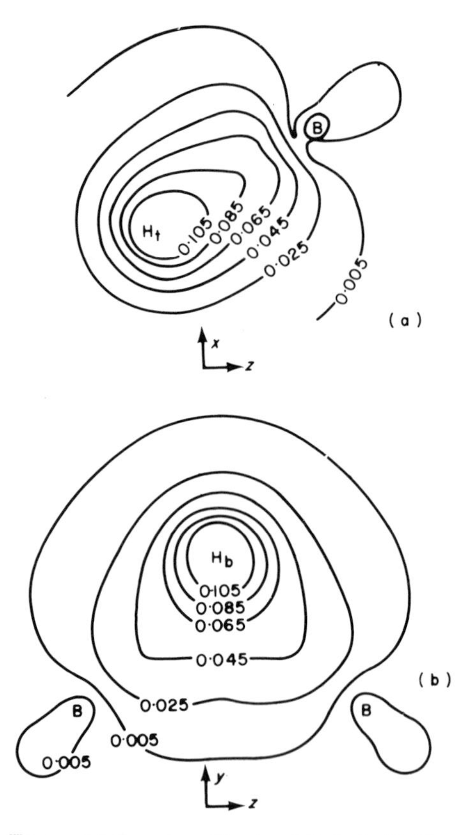

Figure 15.9 Localized orbitals in B_2H_6 (density contours for one electron $e/a_0{}^3$): (A) a terminal B–H bond; (B) a three-centre bond.

Most hydrogen bonded systems have been detected in solution or in crystals and factors such as solvation or long-range crystal forces may complicate the analysis of the binding. Fewer systems have been studied in the gas phase but their numbers include some of the simplest molecules and they are therefore important for study. Table 15.4 shows the dimerization energy for five gas-phase dimers.

The theory of weak hydrogen bonded complexes follows quite closely that of the donor–acceptor complexes given in the last section. Calculations show that there is a migration of electrons on complexing which is towards the hydrogen atom donor. However, the main difference is that there is a large electrostatic attraction in the no-bond state of the hydrogen bond which is usually absent for donor–acceptor complexes. There is now general agreement that this electrostatic energy is a large contribution to the binding energy and for weak complexes is probably dominant.

Figure 15.10 shows the correlation between the total binding energy from

Table 15.4. Dimerization energies for some gas-phase hydrogen bonded dimers [L. C. Allen, *J. Am. Chem. Soc.*, **97**, 6921 (1975)]

	Dim. energy/kJ mol^{-1}
$(HF)_2$	29 ± 4
$(H_2O)_2$	22 ± 6
$(NH_3)_2$	19 ± 2
$(HCl)_2$	9 ± 1
$(H_2S)_2$	7 ± 1

ab-initio molecular orbital calculations and the electrostatic potentials of the two partners in the complex.† In the top half of the graph are data for a series of complexes with ammonia as a common hydrogen bond acceptor ($A—H \cdots NH_3$) and the electrostatic potential has been calculated at 2 Å from the proton along the A—H bond. The strongest complex is $F—H \cdots NH_3$ and the weakest is $H_3C—H \cdots NH_3$. In the bottom half of the graph are data for complexes with HF as a common hydrogen bond donor ($F—H \cdots B$). The potential was calculated at 2.12 Å from the acceptor heavy atom when this is a first-row element and 2.65 Å when this is a second-row element: this allows for the fact that the $H \cdots B$ distances are longer for second row elements. The strongest complex is again $F—H \cdots NH_3$ and the weakest is $H—F \cdots P{\equiv}C—H$.

Although these potentials have been calculated at fixed distances from the atoms involved and not at the distances actually pertaining in the complex, the correlation shown in Figure 15.10 is impressive. There may, of course, be other important factors which themselves show a correlation with these potentials, so that Figure 15.10 alone does not indicate that the electrostatic contributions alone are important.

The charge-transfer contribution to the bond energy will depend on the ionization potential of B and the electron affinity of AH according to the two-state resonance model given in the last section. There are conflicting views about the amount of charge transfer energy in any one complex but it appears to increase in proportion to the Coulomb energy as the strength of the complex increases. Kollman and Allen‡ have analysed molecular orbital calculations on the water dimer and deduce that the sum of the Coulomb attraction and exchange repulsion energy is -19 kJ mol^{-1}, the sum of the charge-transfer energy and the polarization energy (the energy arising from the polarization of one component in the field of the other) is -13 kJ mol^{-1}, and the dispersion energy is -6 kJ mol^{-1}.

†P. Kollman, J. McKelvey, A. Johansson, and S. Rothenberg, *J. Am. Chem. Soc.*, 97, 955 (1975).
‡P. Kollman and L. C. Allen, *Chem. Rev.*, 72, 283 (1972).

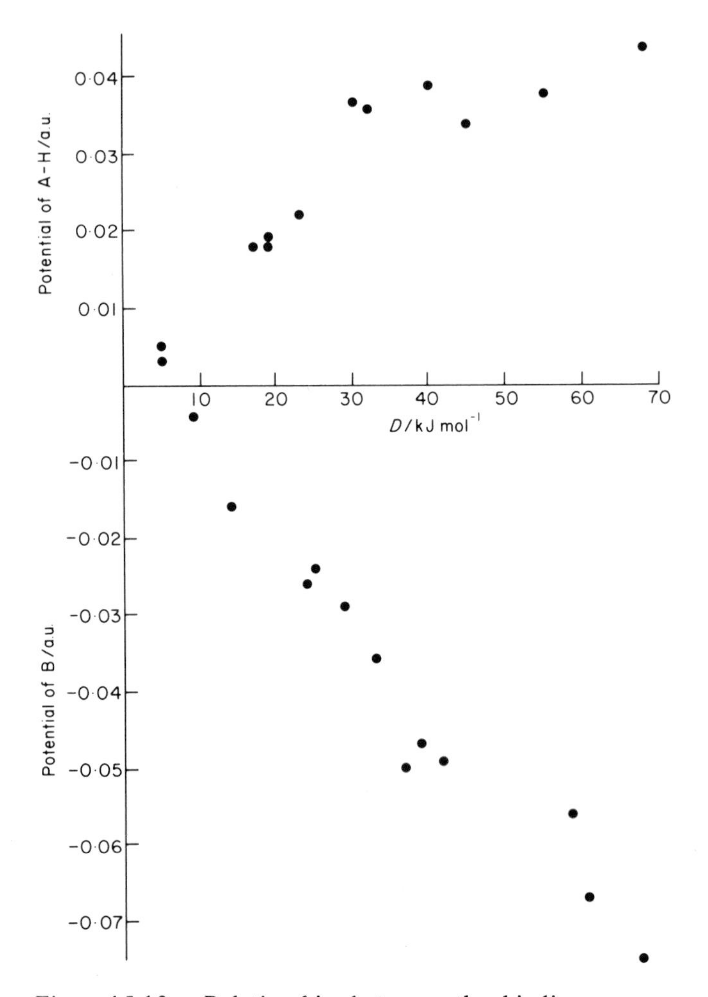

Figure 15.10 Relationship between the binding energy (D) of a complex AH–B and the electrostatic potentials of AH and of B. The top half of the plot is for various AH with the same B (NH_3), and the bottom half is for various B with HF.

The geometry of hydrogen bonded complexes is a matter of considerable interest. In almost all cases the hydrogen atom lies along the line of centres of the two heavy atoms although the energy required to displace it off this line must be very small. Calculations suggest that in the gas-phase the molecules are very floppy.

Both the Coulomb attraction and the charge-transfer energy are angle dependent and there must be a balance between the two. To optimize the charge-transfer energy the hydrogen atom should approach along the line of maximum electron density of the lone pair orbital of B, so maximizing the overlap integral S_{da} (15.25). However, this factor alone has been shown not to account for geometries because such a line is not necessarily optimum for the Coulomb energy.

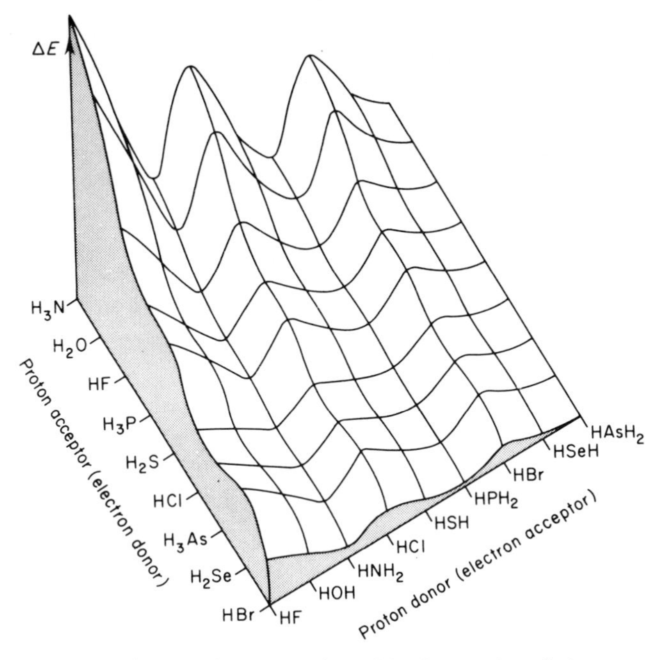

Figure 15.11 Binding energies of hydrogen bonded complexes, deduced from expression (15.27).

Because of the wide range of strength of hydrogen bond complexes it is difficult to find simple rules that apply to all. Allen has proposed an empirical formula for the binding energy D, based upon analysis of molecular orbital calculations and experiment,† which is

$$D = K\mu_{AH} \times \Delta I_B/R. \tag{15.27}$$

ΔI_B is the ionization potential of the hydrogen atom acceptor measured relative to that of the isoelectronic inert gas atom (for example H_2O relative to Ne). Calculations show that systems like FH $\cdots$ Ne have binding energies that can be explained solely by dispersion forces. μ_{AH} is the dipole moment of the AH bond. There is some uncertainty in this quantity as only the total dipole moment of the molecule is unambiguously defined. It can, however, be estimated from the localized orbitals of the molecule (Section 8.2). R is the internuclear distance between atoms A and B, and K is a constant (units charge^{-1}).

Figure 15.11 shows the 'binding energy surface' for eighty one complexes calculated from expression (15.27) which gives a visual relationship between the complexes. A value $K = 1.87e^{-1}$ reproduces with reasonable accuracy the energies of the gas-phase dimers quoted in Table 15.4.

†L. C. Allen, *J. Am. Chem. Soc.*, 97, 6921 (1975).

Subject Index

Chemical Formula and Substance Index

Molecules with six or less atoms are listed by formula using the commonly accepted order, larger molecules are listed by name.